Biology: Exploring Life

SECOND EDITION

Biology: Exploring Life

Gil Brum
California State Polytechnic University, Pomona

Larry McKane
California State Polytechnic University, Pomona

Gerry Karp
Formerly of the University of Florida, Gainesville

JOHN WILEY & SONS, INC.
New York / Chichester / Brisbane / Toronto / Singapore

Acquisitions Editor *Sally Cheney*
Developmental Editor *Rachel Nelson*
Marketing Manager *Catherine Faduska*
Associate Marketing Manager *Deb Benson*
Senior Production Editor *Katharine Rubin*
Text Designer/"Steps" Illustration Art Direction *Karin Gerdes Kincheloe*
Manufacturing Manager *Andrea Price*
Photo Department Director *Stella Kupferberg*
Senior Illustration Coordinator *Edward Starr*
"Steps to Discovery" Art Illustrator *Carlyn Iverson*
Cover Design *Meryl Levavi*
Text Illustrations *Network Graphics/Blaize Zito Associates, Inc.*
Photo Editor/Cover Photography Direction *Charles Hamilton*
Photo Researchers *Hilary Newman, Pat Cadley, Lana Berkovitz*
Cover Photo *James H. Carmichael, Jr./The Image Bank*

This book was set in New Caledonia by Progressive Typographers and printed and bound by Von Hoffmann Press. The cover was printed by Lehigh. Color separations by Progressive Typographers and Color Associates, Inc.

Recognizing the importance of preserving what has been written, it is a policy of John Wiley & Sons, Inc. to have books of enduring value published in the United States printed on acid-free paper, and we exert our best efforts to that end.

The paper in this book was manufactured by a mill whose forest management programs include sustained yield harvesting of its timberlands. Sustained yield harvesting principles ensure that the number of trees cut each year does not exceed the amount of new growth.

Library of Congress Cataloging in Publication Data:
Brum, Gilbert D.
Biology : exploring life/Gil Brum, Larry McKane, Gerry Karp.—2nd ed.
p. cm.
Includes bibliographical references and index.
ISBN 0-471-54408-6 (cloth)
1. Biology. I. McKane, Larry. II. Karp, Gerald. III. Title.
QH308.2.B78 1993
574—dc20 93-23383
CIP

Unit I ISBN 0-471-01827-9 (pbk)
Unit II ISBN 0-471-01831-7 (pbk)
Unit III ISBN 0-471-01830-9 (pbk)
Unit IV ISBN 0-471-01829-5 (pbk)
Unit V ISBN 0-471-01828-7 (pbk)
Unit VI ISBN 0-471-01832-5 (pbk)

Printed in the United States of America

10 9 8 7 6 5 4 3

For the Student, we hope this book helps you discover the thrill of exploring life and helps you recognize the important role biology plays in your everyday life.

To Margaret, Jan, and Patsy, who kept loving us even when we were at our most unlovable.

To our children, Jennifer, Julia, Christopher, and Matthew, whose fascination with exploring life inspires us all. And especially to Jenny—we all wish you were here to share the excitement of this special time in life.

Preface to the Instructor

Biology: Exploring Life, Second Edition is devoted to the process of investigation and discovery. The challenge and thrill of understanding how nature works ignites biologists' quests for knowledge and instills a desire to share their insights and discoveries. The satisfactions of knowing that the principles of nature can be understood and sharing this knowledge are why we teach. These are also the reasons why we created this book.

Capturing and holding student interest challenges even the best of teachers. To help meet this challenge, we have endeavored to create a book that makes biology relevant and appealing, that reveals biology as a dynamic process of exploration and discovery, and that emphasizes the widening influence of biologists in shaping and protecting our world and in helping secure our futures. We direct the reader's attention toward principles and concepts to dispel the misconception of many undergraduates that biology is nothing more than a very long list of facts and jargon. Facts and principles form the core of the course, but we have attempted to show the *significance* of each fact and principle and to reveal the important role biology plays in modern society.

From our own experiences in the introductory biology classroom, we have discovered that

- emphasizing principles, applications, and scientific exploration invigorates the teaching and learning process of biology and helps students make the significant connections needed for full understanding and appreciation of the importance of biology; and
- students learn more if a book is devoted to telling the story of biology rather than a recitation of facts and details.

Guided by these insights, we have tried to create a process-oriented book that still retains the facts, structures, and terminology needed for a fundamental understanding of biology. With these goals in mind, we have interwoven into the text

1. an emphasis on the ways that science works,
2. the underlying adventure of exploration,
3. five fundamental biological themes, and
4. balanced attention to the human perspective.

This book should challenge your students to think critically, to formulate their own hypotheses as possible explanations to unanswered questions, and to apply the approaches learned in the study of biology to understanding (and perhaps helping to solve) the serious problems that affect every person, indeed every organism, on this planet.

THE DEVELOPMENT STORY

The second edition of *Biology: Exploring Life* builds effectively on the strengths of the First Edition by Gil Brum and Larry McKane. For this edition, we added a third author, Gerry Karp, a cell and molecular biologist. Our complementary areas of expertise (genetics, zoology, botany, ecology, microbiology, and cell and molecular biology) as well as awards for teaching and writing have helped us form a balanced team. Together, we exhaustively revised and refined each chapter until all three of us, each with our different likes and dislikes, sincerely believed in the result. What evolved from this process was a satisfying synergism and a close friendship.

THE APPROACH

The elements of this new approach are described in the upcoming section "To the Student: A User's Guide." These pedagogical features are embedded in a book that is written in an informal, accessible style that invites the reader to explore the process of biology. In addition, we have tried to keep the narrative focused on *processes*, rather than on static facts, while creating an underlying foundation that helps students make the connections needed to tie together the information into a greater understanding than that which comes from memorizing facts alone. One way to help students make these connections is to relate the fundamentals of biology to humans, revealing the human perspective in each biological principle, from biochemicals to ecosystems. With each such insight, students take a substantial step toward becoming the informed citizens that make up responsible voting public.

We hope that, through this textbook, we can become partners with the instructor and the student. The biology teacher's greatest asset is the basic desire of students to understand themselves and the world around them. Unfortunately, many students have grown detached from this natural curiosity. Our overriding objective in creating this book was to arouse the students' fascination with exploring life, building knowledge and insight that will enable them to make real-life judgments as modern biology takes on greater significance in everyday life.

THE ART PROGRAM

The diligence and refinement that went into creating the text of *Biology: Exploring Life,* Second Edition characterizes the art program as well. Each photo was picked specifically for its relevance to the topic at hand and for its aesthetic and instructive value in illustrating the narrative concepts. The illustrations were carefully crafted under the guidance of the authors for accuracy and utility as well as aesthetics. The value of illustrations cannot be overlooked in a discipline as filled with images and processes as biology. Through the use of cell icons, labeled illustrations of pathways and processes, and detailed legends, the student is taken through the world of biology, from its microscopic chemical components to the macroscopic organisms and the environments that they inhabit.

SUPPLEMENTARY MATERIALS

In our continuing effort to meet all of your individual needs, Wiley is pleased to offer the various topics covered in this text in customized paperback "splits." For more details, please contact your local Wiley sales representative. We have also developed an integrated supplements package that helps the instructor bring the study of biology to life in the classroom and that will maximize the students' use and understanding of the text.

The *Instructor's Manual,* developed by Michael Leboffe and Gary Wisehart of San Diego City College, contains lecture outlines, transparency references, suggested lecture activities, sample concept maps, section concept map masters (to be used as overhead transparencies), and answers to study guide questions.

Gary Wisehart and Mark Mandell developed the test bank, which consists of four types of questions: fill-in questions, matching questions, multiple-choice questions, and critical thinking questions. A computerized test bank is also available.

A comprehensive visual ancillary package includes four-color transparencies (200 figures from the text), *Process of Science* transparency overlays that break down various biological processes into progressive steps, a video library consisting of tapes from Coronet MTI, and the *Bio Sci* videodisk series from Videodiscovery, covering topics in biochemistry, botany, vertebrate biology, reproduction, ecology, animal behavior, and genetics. Suggestions for integrating the videodisk material in your classroom discussions are available in the instructor's manual.

A comprehensive study guide and lab manual are also available and are described in more detail in the User's Guide section of the preface.

Acknowledgments

It was a delight to work with so many creative individuals whose inspiration, artistry, and vital steam guided this complex project to completion. We wish we were able to acknowledge each of them here, for not only did they meet nearly impossible deadlines, but each willingly poured their heart and soul into this text. The book you now hold in your hands is in large part a tribute to their talent and dedication.

There is one individual whose unique talent, quick intellect, charm, and knowledge not only helped to make this book a reality, but who herself made an enormous contribution to the content and pedagogical strength of this book. We are proud to call Sally Cheney, our biology editor, a colleague. Her powerful belief in this textbook's new approaches to teaching biology helped instill enthusiasm and confidence in everyone who worked on it. Indeed, Sally is truly a force of positive change in college textbook publishing–she has an uncommon ability to think both like a biologist and an editor; she knows what biologists want and need in their classes and is dedicated to delivering it; she recognizes that the future of biology education is more than just publishing another look-alike text; and she is knowledgeable and persuasive enough to convince publishers to stick their necks out a little further for the good of educational advancement. Without Sally, this text would have fallen short of our goal. With Sally, it became even more than we envisioned.

Another individual also helped make this a truly special book, as well as made the many long hours of work so delightful. Stella Kupferberg, we treasure your friendship, applaud your exceptional talent, and salute your high standards. Stella also provided us with two other important assets, Charles Hamilton and Hilary Newman. Stella and Charles tirelessly applied their skill, and artistry to get us images of incomparable effectiveness and beauty, and Hilary's diligent handling helped to insure there were no oversights.

Our thanks to Rachel Nelson for her meticulous editing, for maintaining consistency between sometimes dissimilar writing styles of three authors, and for keeping track of an incalculable number of publishing and biological details; to Katharine Rubin for expertly and gently guiding this project through the myriad levels of production, and for putting up with three such demanding authors; to Karin Kincheloe for a stunningly beautiful design; to Ishaya Monokoff and Ed Starr for orchestrating a brilliant art program; to Network Graphics, especially John Smith and John Hargraves, who executed our illustrations with beauty and style without diluting their conceptual strength or pedagogy, and to Carlyn Iverson, whose artistic talent helped us visually distill our "Steps to Discovery" episodes into images that bring the process of science to life.

We would also like to thank Cathy Faduska and Alida Setford, their creative flair helped us to tell the story behind this book, as well as helped us convey what we tried to accomplish. And to Herb Brown, thank you for your initial confidence and continued support. A very special thank you to Deb Benson, our marketing manager. What a joy to work with you, Deb, your energy, enthusiasm, confidence, and pleasant personality bolstered even our spirits.

We wish to acknowledge Diana Lipscomb of George Washington University for her invaluable contributions to the evolution chapters, Judy Goodenough of the University of Massachusetts, Amherst, for contributing an outstanding chapter on Animal Behavior, and Dorothy Rosenthal for contributing the end–of–chapter "Critical Thinking Questions."

To the reviewers and instructors who used the First Edition, your insightful feedback helped us forge the foundation for this new edition. To the reviewers, and workshop and conference participants for the Second Edition, thank you for your careful guidance and for caring so much about your students.

Dennis Anderson, *Oklahoma City Community College*
Sarah Barlow, *Middle Tennessee State University*
Robert Beckman, *North Carolina State University*
Timothy Bell, *Chicago State University*
David F. Blaydes, *West Virginia University*
Richard Bliss, *Yuba College*
Richard Boohar, *University of Nebraska, Lincoln*
Clyde Bottrell, *Tarrant County Junior College*
J. D. Brammer, *North Dakota State University*
Peggy Branstrator, *Indiana University, East*
Allyn Bregman, *SUNY, New Paltz*
Daniel Brooks, *University of Toronto*
Gary Brusca, *Humboldt State University*
Jack Bruk, *California State University, Fullerton*
Marvin Cantor, *California State University, Northridge*
Richard Cheney, *Christopher Newport College*
Larry Cohen, *California State University, San Marcos*
David Cotter, *Georgia College*
Robert Creek, *Eastern Kentucky University*
Ken Curry, *University of Southern Mississippi*
Judy Davis, *Eastern Michigan University*
Loren Denny, *southwest Missouri State University*
Captain Donald Diesel, *U. S. Air Force Academy*
Tom Dickinson, *University College of the Cariboo*

Mike Donovan, *Southern Utah State College*
Robert Ebert, *Palomar College*
Thomas Emmel, *University of Florida*
Joseph Faryniarz, *Mattatuck Community College*
Alan Feduccia, *University of North Carolina, Chapel Hill*
Eugene Ferri, *Bucks County Community College*
Victor Fet, *Loyola University, New Orleans*
David Fox, *Loyola University, New Orleans*
Mary Forrest, *Okanagan University College*
Michael Gains, *University of Kansas*
S. K. Gangwere, *Wayne State University*
Dennis George, *Johnson County Community College*
Bill Glider, *University of Nebraska*
Paul Goldstein, *University of North Carolina, Charlotte*
Judy Goodenough, *University of Massachusetts, Amherst*
Nels Granholm, *South Dakota State University*
Nathaniel Grant, *Southern Carolina State College*
Mel Green, *University of California, San Diego*
Dana Griffin, *Florida State University*
Barbara L. Haas, *Loyola University of Chicago*
Richard Haas, *California State University, Fresno*
Fredrick Hagerman, *Ohio State University*
Tom Haresign, *Long Island University, Southampton*
W. R. Hawkins, *Mt. San Antonio College*
Vernon Hendricks, *Brevard Community College*
Paul Hertz, *Barnard College*
Howard Hetzle, *Illinois State University*
Ronald K. Hodgson, *Central Michigan University*
W. G. Hopkins, *University of Western Ontario*
Thomas Hutto, *West Virginia State College*
Duane Jeffrey, *Brigham Young University*
John Jenkins, *Swarthmore College*
Claudia Jones, *University of Pittsburgh*
R. David Jones, *Adelphi University*
J. Michael Jones, *Culver Stockton College*
Gene Kalland, *California State University, Dominiquez Hills*
Arnold Karpoff, *University of Louisville*
Judith Kelly, *Henry Ford Community College*
Richard Kelly, *SUNY, Albany*
Richard Kelly, *University of Western Florida*
Dale Kennedy, *Kansas State University*
Miriam Kittrell, *Kingsborough Community College*
John Kmeltz, *Kean College New Jersey*
Robert Krasner, *Providence College*
Susan Landesman, *Evergreen State College*
Anton Lawson, *Arizona State University*
Lawrence Levine, *Wayne State University*
Jerri Lindsey, *Tarrant County Junior College*
Diana Lipscomb, *George Washington University*
James Luken, *Northern Kentucky University*
Ted Maguder, *University of Hartford*
Jon Maki, *Eastern Kentucky University*
Charles Mallery, *University of Miami*
William McEowen, *Mesa Community College*
Roger Milkman, *University of Iowa*
Helen Miller, *Oklahoma State University*
Elizabeth Moore, *Glassboro State College*
Janice Moore, *Colorado State University*
Eston Morrison, *Tarleton State University*
John Mutchmor, *Iowa State University*
Jane Noble-Harvey, *University of Delaware*
Douglas W. Ogle, *Virginia Highlands Community College*
Joel Ostroff, *Brevard Community College*
James Lewis Payne, *Virginia Commonwealth University*
Gary Peterson, *South Dakota State University*
MaryAnn Phillippi, *Southern Illinois University, Carbondale*
R. Douglas Powers, *Boston College*
Robert Raikow, *University of Pittsburgh*
Charles Ralph, *Colorado State University*
Aryan Roest, *California State Polytechnic Univ., San Luis Obispo*
Robert Romans, *Bowling Green State University*
Raymond Rose, *Beaver College*
Richard G. Rose, *West Valley College*
Donald G. Ruch, *Transylvania University*
A. G. Scarbrough, *Towson State University*
Gail Schiffer, *Kennesaw State University*
John Schmidt, *Ohio State University*
John R. Schrock, *Emporia State University*
Marilyn Shopper, *Johnson County Community College*
John Smarrelli, *Loyola University of Chicago*
Deborah Smith, *Meredith College*
Guy Steucek, *Millersville University*
Ralph Sulerud, *Augsburg College*
Tom Terry, *University of Connecticut*
James Thorp, *Cornell University*
W. M. Thwaites, *San Diego State University*
Michael Torelli, *University of California, Davis*
Michael Treshow, *University of Utah*
Terry Trobec, *Oakton Community College*
Len Troncale, *California State Polytechnic University, Pomona*
Richard Van Norman, *University of Utah*
David Vanicek, *California State University, Sacramento*
Terry F. Werner, *Harris-Stowe State College*
David Whitenberg, *Southwest Texas State University*
P. Kelly Williams, *University of Dayton*
Robert Winget, *Brigham Young University*
Steven Wolf, *University of Missouri, Kansas City*
Harry Womack, *Salisbury State University*
William Yurkiewicz, *Millersville University*

Gil Brum
Larry McKane
Gerry Karp

Brief Table of Contents

PART 1 / *Biology: The Study of Life* 1

Chapter 1 Biology: Exploring Life 3
Chapter 2 The Process of Science 29

PART 2 / *Chemical and Cellular Foundations of Life* 45

Chapter 3 The Atomic Basis of Life 47
Chapter 4 Biochemicals: The Molecules of Life 65
Chapter 5 Cell Structure and Function 87
Chapter 6 Energy, Enzymes, and Metabolic Pathways 115
Chapter 7 Movement of Materials Across Membranes 135
Chapter 8 Processing Energy: Photosynthesis and Chemosynthesis 151
Chapter 9 Processing Energy: Fermentation and Respiration 171
Chapter 10 Cell Division: Mitosis 193
Chapter 11 Cell Division: Meiosis 211

PART 3 / *The Genetic Basis of Life* 227

Chapter 12 On the Trail of Heredity 229
Chapter 13 Genes and Chromosomes 245
Chapter 14 The Molecular Basis of Genetics 265
Chapter 15 Orchestrating Gene Expression 291
Chapter 16 DNA Technology: Developments and Applications 311
Chapter 17 Human Genetics: Past, Present, and Future 333

PART 4 / *Form and Function of Plant Life* 355

Chapter 18 Plant Tissues and Organs 357
Chapter 19 The Living Plant: Circulation and Transport 383
Chapter 20 Sexual Reproduction of Flowering Plants 399
Chapter 21 How Plants Grow and Develop 423

PART 5 / *Form and Function of Animal Life* 445

Chapter 22 An Introduction to Animal Form and Function 447
Chapter 23 Coordinating the Organism: The Role of the Nervous System 465
Chapter 24 Sensory Perception: Gathering Information about the Environment 497
Chapter 25 Coordinating the Organism: The Role of the Endocrine System 517
Chapter 26 Protection, Support, and Movement: The Integumentary, Skeletal, and Muscular Systems 539
Chapter 27 Processing Food and Providing Nutrition: The Digestive System 565
Chapter 28 Maintaining the Constancy of the Internal Environment: The Circulatory and Excretory Systems 585
Chapter 29 Gas Exchange: The Respiratory System 619
Chapter 30 Internal Defense: The Immune System 641
Chapter 31 Generating Offspring: The Reproductive System 661
Chapter 32 Animal Growth and Development: Acquiring Form and Function 685

PART 6 / *Evolution* 711

Chapter 33 Mechanisms of Evolution 713
Chapter 34 Evidence for Evolution 741
Chapter 35 The Origin and History of Life 759

PART 7 / *The Kingdoms of Life: Diversity and Classification* *781*

Chapter 36 The Monera Kingdom and Viruses 783
Chapter 37 The Protist and Fungus Kingdoms 807
Chapter 38 The Plant Kingdom 829
Chapter 39 The Animal Kingdom 857

PART 8 / *Ecology and Animal Behavior* *899*

Chapter 40 The Biosphere 901
Chapter 41 Ecosystems and Communities 931
Chapter 42 Community Ecology: Interactions between Organisms 959
Chapter 43 Population Ecology 983
Chapter 44 Animal Behavior 1005

Appendix A Metric and Temperature Conversion Charts A-1
Appendix B Microscopes: Exploring the Details of Life B-1
Appendix C The Hardy-Weinberg Principle C-1
Appendix D Careers in Biology D-1

Contents

PART 1
Biology: The Study of Life 1

Chapter 1 / Biology: Exploring Life 3

Steps to Discovery:
Exploring Life—The First Step 4

Distinguishing the Living from the Inanimate 6
Levels of Biological Organization 11
What's in a Name 11
Underlying Themes of Biology 15
Biology and Modern Ethics 25

Reexamining the Themes 25
Synopsis 26

Chapter 2 / The Process of Science 29

Steps to Discovery:
What Is Science? 30

The Scientific Approach 32
Applications of the Scientific Process 35
Caveats Regarding "The" Scientific Method 40

Synopsis 41

BIOETHICS
Science, Truth, and Certainty:
Is a Theory "Just a Theory"? 40

PART 2
Chemical and Cellular Foundations of Life 45

Chapter 3 / The Atomic Basis of Life 47

Steps to Discovery:
The Atom Reveals Some of Its Secrets 48

The Nature of Matter 50
The Structure of the Atom 51
Types of Chemical Bonds 55
The Life-Supporting Properties of Water 59
Acids, Bases, and Buffers 61

Reexamining the Themes 62
Synopsis 63

BIOLINE
Putting Radioisotopes to Work 54

THE HUMAN PERSPECTIVE
Aging and Free Radicals 56

Chapter 4 / Biochemicals: The Molecules of Life 65

Steps to Discovery:
Determining the Structure of Proteins 66

The Importance of Carbon in Biological Molecules 68
Giant Molecules Built From Smaller Subunits 69
Four Biochemical Families 69
Macromolecular Evolution and Adaptation 83

Reexamining the Themes 84
Synopsis 85

THE HUMAN PERSPECTIVE
Obesity and the Hungry Fat Cell 76

Chapter 5 / Cell Structure and Function 87

Steps to Discovery: The Nature of the Plasma Membrane 88

Discovering the Cell 90
Basic Characteristics of Cells 91
Two Fundamentally Different Types of Cells 91
The Plasma Membrane: Specialized for Interaction with the Environment 93
The Eukaryotic Cell: Organelle Structure and Function 95
Just Outside the Cell 108
Evolution of Eukaryotic Cells: Gradual Change or an Evolutionary Leap? 110

Reexamining the Themes 111
Synopsis 112

THE HUMAN PERSPECTIVE

Lysosome-Related Diseases 101

Chapter 6 / Energy, Enzymes, and Metabolic Pathways 115

Steps to Discovery: The Chemical Nature of Enzymes 116

Acquiring and Using Energy 118
Enzymes 123
Metabolic Pathways 127
Macromolecular Evolution and Adaptation of Enzymes 131

Reexamining the Themes 132
Synopsis 132

BIOLINE

Saving Lives by Inhibiting Enzymes 126

THE HUMAN PERSPECTIVE

The Manipulation of Human Enzymes 131

Chapter 7 / Movement of Materials Across Membranes 135

Steps to Discovery: Getting Large Molecules Across Membrane Barriers 136

Membrane Permeability 138
Diffusion: Depending on the Random Movement of Molecules 138
Active Transport 142
Endocytosis 144

Reexamining the Themes 148
Synopsis 148

THE HUMAN PERSPECTIVE

LDL Cholesterol, Endocytosis, and Heart Disease 146

Chapter 8 / Processing Energy: Photosynthesis and Chemosynthesis 151

Steps to Discovery: Turning Inorganic Molecules into Complex Sugars 152

Autotrophs and Heterotrophs: Producers and Consumers 154
An Overview of Photosynthesis 155
The Light-Dependent Reactions: Converting Light Energy into Chemical Energy 156
The Light-Independent Reactions 163
Chemosynthesis: An Alternate Form of Autotrophy 167

Reexamining the Themes 168
Synopsis 169

BIOLINE

Living on the Fringe of the Biosphere 167

THE HUMAN PERSPECTIVE

Producing Crop Plants Better Suited to Environmental Conditions 165

Chapter 9 / Processing Energy: Fermentation and Respiration 171

Steps to Discovery: The Machinery Responsible for ATP Synthesis 172

Fermentation and Aerobic Respiration: A Preview of the Strategies 174
Glycolysis 175
Fermentation 178
Aerobic Respiration 180

Balancing the Metabolic Books: Generating and Yielding Energy 185
Coupling Glucose Oxidation to Other Pathways 188

Reexamining the Themes 190
Synopsis 190

BIOLINE

The Fruits of Fermentation 179

THE HUMAN PERSPECTIVE

The Role of Anaerobic and Aerobic Metabolism in Exercise 186

Chapter 10 / Cell Division: Mitosis 193

Steps to Discovery: Controlling Cell Division 194

Types of Cell Division 196
The Cell Cycle 199
The Phases of Mitosis 202
Cytokinesis: Dividing the Cell's Cytoplasm and Organelles 206

Reexamining the Themes 207
Synopsis 208

THE HUMAN PERSPECTIVE

Cancer: The Cell Cycle Out of Control 202

Chapter 11 / Cell Division: Meiosis 211

Steps to Discovery: Counting Human Chromosomes 212

Meiosis and Sexual Reproduction: An Overview 214
The Stages of Meiosis 216
Mitosis Versus Meiosis Revisited 222

Reexamining the Themes 223
Synopsis 223

THE HUMAN PERSPECTIVE

Dangers That Lurk in Meiosis 221

PART 3
The Genetic Basis of Life 227

Chapter 12 / On the Trail of Heredity 229

Steps to Discovery: The Genetic Basis of Schizophrenia 230

Gregor Mendel: The Father of Modern Genetics 232
Mendel Amended 239
Mutation: A Change in the Genetic Status Quo 241

Reexamining the Themes 242
Synopsis 242

Chapter 13 / Genes and Chromosomes 245

Steps to Discovery: The Relationship Between Genes and Chromosomes 246

The Concept of Linkage Groups 248
Lessons from Fruit Flies 249
Sex and Inheritance 253
Aberrant Chromosomes 257
Polyploidy 260

Reexamining the Themes 261
Synopsis 262

BIOLINE

The Fish That Changes Sex 254

THE HUMAN PERSPECTIVE

Chromosome Aberrations and Cancer 260

Chapter 14 / The Molecular Basis of Genetics 265

Steps to Discovery: The Chemical Nature of the Gene 266

Confirmation of DNA as the Genetic Material 268
The Structure of DNA 268
DNA: Life's Molecular Supervisor 272
The Molecular Basis of Gene Mutations 286

Reexamining the Themes 287
Synopsis 288

THE HUMAN PERSPECTIVE

The Dark Side of the Sun 280

Chapter 15 / Orchestrating Gene Expression 291

Steps to Discovery: Jumping Genes: Leaping into the Spotlight 292

Why Regulate Gene Expression? 294
Gene Regulation in Prokaryotes 298
Gene Regulation in Eukaryotes 302
Levels of Control of Eukaryotic Gene Expression 303

Reexamining the Themes ***308***
Synopsis ***309***

BIOLINE
RNA as an Evolutionary Relic 306

THE HUMAN PERSPECTIVE
Clones: Is There Cause for Fear? 296

Chapter 16 / DNA Technology: Developments and Applications 311

Steps to Discovery: DNA Technology and Turtle Migration 312

Genetic Engineering 314
DNA Technology I: The Formation and Use of Recombinant DNA Molecules 321
DNA Technology II: Techniques That Do Not Require Recombinant DNA Molecules 325
Use of DNA Technology in Determining Evolutionary Relationships 329

Reexamining the Themes ***329***
Synopsis ***330***

BIOLINE
DNA Fingerprints and Criminal Law 328

THE HUMAN PERSPECTIVE
Animals That Develop Human Diseases 320

BIOETHICS
Patenting a Genetic Sequence 315

Chapter 17 / Human Genetics: Past, Present, and Future 333

Steps to Discovery: Developing a Treatment for an Inherited Disorder 334

The Consequences of an Abnormal Number of Chromosomes 336
Disorders that Result from a Defect in a Single Gene 338
Patterns of Transmission of Genetic Disorders 343
Screening Humans for Genetic Defects 346

Reexamining the Themes ***350***
Synopsis ***351***

BIOLINE
Mapping the Human Genome 340

THE HUMAN PERSPECTIVE
Correcting Genetic Disorders by Gene Therapy 348

PART 4
Form and Function of Plant Life ***355***

Chapter 18 / Plant Tissues and Organs 357

Steps to Discovery: Fruit of the Vine and the French Economy 358

The Basic Plant Design 360
Plant Tissues 362
Plant Tissue Systems 365
Plant Organs 372

Reexamining the Themes ***379***
Synopsis ***379***

BIOLINE
Nature's Oldest Living Organisms 366

THE HUMAN PERSPECTIVE
Agriculture, Genetic Engineering, and Plant Fracture Properties 375

Chapter 19 / The Living Plant: Circulation and Transport 383

Steps to Discovery: Exploring the Plant's Circulatory System 384

Xylem: Water and Mineral Transport 386
Phloem: Food Transport 393

Reexamining the Themes ***396***
Synopsis ***397***

BIOLINE
Mycorrhizae and Our Fragile Deserts 388

THE HUMAN PERSPECTIVE
Leaf Nodules and World Hunger: An Unforeseen Connection 396

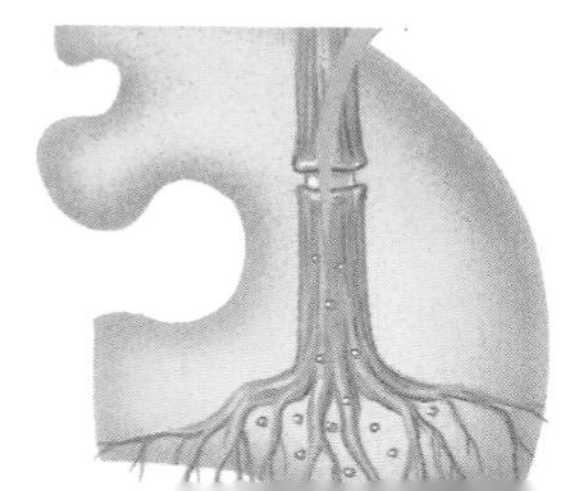

Chapter 20 / Sexual Reproduction of Flowering Plants 399

Steps to Discovery: What Triggers Flowering? 400

Flower Structure and Pollination 402
Formation of Gametes 407
Fertilization and Development 408
Fruit and Seed Dispersal 412
Germination and Seedling Development 416
Asexual Reproduction 417
Agricultural Applications 418

Reexamining the Themes 419
Synopsis 420

BIOLINE
The Odd Couples: Bizarre Flowers and Their Pollinators 404

THE HUMAN PERSPECTIVE
The Fruits of Civilization 414

Chapter 21 / How Plants Grow and Develop 423

Steps to Discovery: Plants Have Hormones 424

Levels of Control of Growth and Development 426
Plant Hormones: Coordinating Growth and Development 427
Timing Growth and Development 434
Plant Tropisms 438

Reexamining the Themes 440
Synopsis 440

THE HUMAN PERSPECTIVE
Saving Tropical Rain Forests 428

PART 5
Form and Function of Animal Life 445

Chapter 22 / An Introduction to Animal Form and Function 447

Steps to Discovery: The Concept of Homeostasis 448

Homeostatic Mechanisms 450
Unity and Diversity Among Animal Tissues 451
Unity and Diversity Among Organ Systems 455
The Evolution of Organ Systems 457
Body Size, Surface Area, and Volume 460

Reexamining the Themes 462
Synopsis 463

Chapter 23 / Coordinating the Organism: The Role of the Nervous System 465

Steps to Discovery: A Factor Promoting the Growth of Nerves 466

Neurons and Their Targets 468
Generating and Conducting Neural Impulses 470
Neurotransmission: Jumping the Synaptic Cleft 474
The Nervous System 479
Architecture of the Human Central Nervous System 481
Architecture of the Peripheral Nervous System 489
The Evolution of Complex Nervous Systems from Single-Celled Roots 491

Reexamining the Themes 493
Synopsis 494

BIOLINE
Deadly Meddling at the Synapse 478

THE HUMAN PERSPECTIVE
Alzheimer's Disease: A Status Report 484

BIOETHICS
Blurring the Line Between Life and Death 489

Chapter 24 / Sensory Perception: Gathering Information About the Environment 497

Steps to Discovery: Echolocation: Seeing in the Dark 498

The Response of a Sensory Receptor to a Stimulus 500
Somatic Sensation 500
Vision 502
Hearing and Balance 506

Chemoreception: Taste and Smell 509
The Brain: Interpreting Impulses into Sensations 510
Evolution and Adaptation of Sense Organs 511

Reexamining the Themes 514
Synopsis 514

BIOLINE
Sensory Adaptations to an Aquatic Environment 512

THE HUMAN PERSPECTIVE
Two Brains in One 504

Chapter 25 / Coordinating the Organism: The Role of the Endocrine System 517

Steps to Discovery: The Discovery of Insulin 518

The Role of the Endocrine System 520
The Nature of the Endocrine System 521
A Closer Look at the Endocrine System 523
Action at the Target 533
Evolution of the Endocrine System 534

Reexamining the Themes 536
Synopsis 537

BIOLINE
Chemically Enhanced Athletes 528

THE HUMAN PERSPECTIVE
The Mysterious Pineal Gland 533

Chapter 26 / Protection, Support, and Movement: The Integumentary, Skeletal, and Muscular Systems 539

Steps to Discovery: Vitamin C's Role in Holding the Body Together 540

The Integument: Covering and Protecting the Outer Body Surface 542
The Skeletal System: Providing Support and Movement 544
The Muscles: Powering the Motion of Animals 551
Body Mechanics and Locomotion 558
Evolution of the Skeletomuscular System 560

Reexamining the Themes 561
Synopsis 562

THE HUMAN PERSPECTIVE
Building Better Bones 550

Chapter 27 / Processing Food and Providing Nutrition: The Digestive System 565

Steps to Discovery: The Battle Against Beri-Beri 566

The Digestive System: Converting Food into a Form Available to the Body 568
The Human Digestive System: A Model Food-Processing Plant 568
Evolution of Digestive Systems 576

Reexamining the Themes 582
Synopsis 583

BIOLINE
Teeth: A Matter of Life and Death 570

THE HUMAN PERSPECTIVE
Human Nutrition 578

Chapter 28 / Maintaining the Constancy of the Internal Environment: the Circulatory and Excretory Systems 585

Steps to Discovery: Tracing the Flow of Blood 586

The Human Circulatory System: Form and Function 588
Evolution of Circulatory Systems 603
The Lymphatic System: Supplementing the Functions of the Circulatory System 606
Excretion and Osmoregulation: Removing Wastes and Maintaining the Composition of the Body's Fluids 606
Thermoregulation: Maintaining a Constant Body Temperature 614

Reexamining the Themes 615
Synopsis 615

BIOLINE
The Artificial Kidney 611

THE HUMAN PERSPECTIVE
The Causes of High Blood Pressure 591

Chapter 29 / Gas Exchange: The Respiratory System 619

Steps to Discovery:
Physiological Adaptations in Diving Mammals 620

Properties of Gas Exchange Surfaces 622
The Human Respiratory System: Form and Function 623
The Exchange of Respiratory Gases and the Role of Hemoglobin 627
Regulating the Rate and Depth of Breathing to Meet the Body's Needs 629
Adaptations for Extracting Oxygen from Water Versus Air 632
Evolution of the Vertebrate Respiratory System 636

Reexamining the Themes 637
Synopsis 637

THE HUMAN PERSPECTIVE
Dying for a Cigarette 630

Chapter 30 / Internal Defense: The Immune System 641

Steps to Discovery:
On the Trail of a Killer: Tracking the AIDS Virus 642

Nonspecific Mechanisms: A First Line of Defense 644
The Immune System: Mediator of Specific Mechanisms of Defense 646
Antibody Molecules: Structure and Formation 650
Immunization 654
Evolution of the Immune System 655

Reexamining the Themes 658
Synopsis 658

BIOLINE
Treatment of Cancer with Immunotherapy 648

THE HUMAN PERSPECTIVE
Disorders of the Human Immune System 656

Chapter 31 / Generating Offspring: The Reproductive System 661

Steps to Discovery:
The Basis of Human Sexuality 662

Reproduction: Asexual Versus Sexual 664
Human Reproductive Systems 667
Controlling Pregnancy: Contraceptive Methods 677
Sexually Transmitted Diseases 681

Reexamining the Themes 682
Synopsis 682

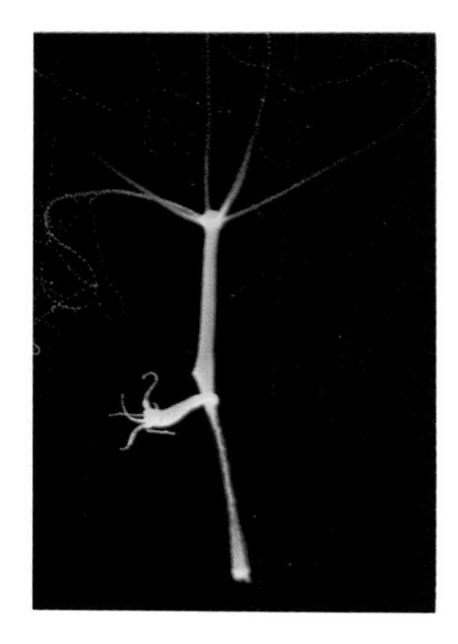

BIOLINE
Sexual Rituals 668

THE HUMAN PERSPECTIVE
Overcoming Infertility 679

BIOETHICS
Frozen Embryos and Compulsive Parenthood 680

Chapter 32 / Animal Growth and Development: Acquiring Form and Function 685

Steps to Discovery:
Genes That Control Development 686

The Course of Embryonic Development 688
Beyond Embryonic Development: Continuing Changes in Body Form 699
Human Development 700
Embryonic Development and Evolution 705

Reexamining the Themes 706
Synopsis 707

THE HUMAN PERSPECTIVE
The Dangerous World of a Fetus 704

PART 6
Evolution 711

Chapter 33 / Mechanisms of Evolution 713

Steps to Discovery:
Silent Spring Revisited 714

The Basis of Evolutionary Change 717
Speciation: The Origin of Species 731
Patterns of Evolution 734
Extinction: The Loss of Species 736
The Pace of Evolution 737

Reexamining the Themes 738
Synopsis 738

BIOLINE

A Gallery of Remarkable Adaptations 728

Chapter 34 / Evidence for Evolution 741

Steps to Discovery:
An Early Portrait of the Human Family 742

Determining Evolutionary Relationships 744
Evidence for Evolution 746
The Evidence of Human Evolution: The Story Continues 752

Reexamining the Themes 756
Synopsis 757

Chapter 35 / The Origin and History of Life 759

Steps to Discovery:
Evolution of the Cell 760

Formation of the Earth: The First Step 762
The Origin of Life and Its Formative Stages 763
The Geologic Time Scale: Spanning Earth's History 764

Reexamining the Themes 775
Synopsis 778

BIOLINE

The Rise and Fall of the Dinosaurs 772

PART 7
The Kingdoms of Life: Diversity and Classification 781

Chapter 36 / The Monera Kingdom and Viruses 783

Steps to Discovery:
The Burden of Proof: Germs and Infectious Diseases 784

Kingdom Monera: The Prokaryotes 787
Viruses 800

Reexamining the Themes 804
Synopsis 805

BIOLINE

Living with Bacteria 799

THE HUMAN PERSPECTIVE

Sexually Transmitted Diseases 794

Chapter 37 / The Protist and Fungus Kingdoms 807

Steps to Discovery:
Creating Order from Chaos 808

The Protist Kingdom 810
The Fungus Kingdom 820

Reexamining the Themes 826
Synopsis 827

Chapter 38 / The Plant Kingdom 829

Steps to Discovery:
Distinguishing Plant Species: Where Should the Dividing Lines Be Drawn? 830

Major Evolutionary Trends 833
Overview of the Plant Kingdom 834
Algae: The First Plants 835
Bryophytes: Plants Without Vessels 842
Tracheophytes: Plants with Fluid-Conducting Vessels 843

Reexamining the Themes 854
Synopsis 854

BIOLINE

The Exceptions: Plants That Don't Photosynthesize 836

THE HUMAN PERSPECTIVE

Brown Algae: Natural Underwater Factories 840

Chapter 39 / The Animal Kingdom 857

Steps to Discovery:
The World's Most Famous Fish 858

Evolutionary Origins and Body Plans 861
Invertebrates 863
Vertebrates 886

Reexamining the Themes 895
Synopsis 895

BIOLINE

Adaptations to Parasitism among Flatworms 872

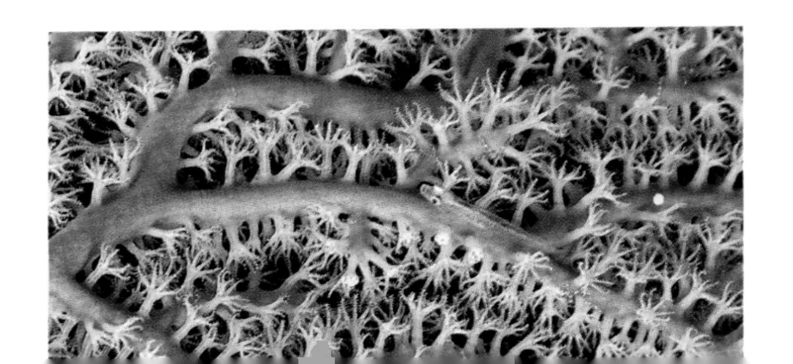

PART 8
Ecology and Animal Behavior 899

Chapter 40 / The Biosphere 901

Steps to Discovery: The Antarctic Ozone Hole 902

Boundaries of the Biosphere 904
Ecology Defined 905
The Earth's Climates 905
The Arenas of Life 908

Reexamining the Themes 927
Synopsis 928

THE HUMAN PERSPECTIVE
Acid Rain and Acid Snow: Global Consequences of Industrial Pollution 918

BIOETHICS
Values in Ecology and Environmental Science: Neutrality or Advocacy? 909

Chapter 41 / Ecosystems and Communities 931

Steps to Discovery: The Nature of Communities 932

From Biomes to Microcosms 934
The Structure of Ecosystems 934
Ecological Niches and Guilds 939
Energy Flow Through Ecosystems 942
Biogeochemical Cycles: Recycling Nutrients in Ecosystems 947
Succession: Ecosystem Change and Stability 952

Reexamining the Themes 955
Synopsis 956

BIOLINE
Reverberations Felt Throughout an Ecosystem 936

THE HUMAN PERSPECTIVE
The Greenhouse Effect: Global Warming 950

Chapter 42 / Community Ecology: Interactions Between Organisms 959

Steps to Discovery: Species Coexistence: The Unpeaceable Kingdom 960

Symbiosis 962
Competition: Interactions That Harm Both Organisms 963
Interactions that Harm One Organism and Benefit the Other 965
Commensalism: Interactions that Benefit One Organism and Have No Affect on the Other 976
Interactions That Benefit Both Organisms 977

Reexamining the Themes 979
Synopsis 979

Chapter 43 / Population Ecology 983

Steps to Discovery: Threatening a Giant 984

Population Structure 986
Factors Affecting Population Growth 988
Factors Controlling Population Growth 996
Human Population Growth 997

Reexamining the Themes 1002
Synopsis 1002

BIOLINE
Accelerating Species Extinction 997

THE HUMAN PERSPECTIVE
Impacts of Poisoned Air, Land, and Water 993

Chapter 44 / Animal Behavior 1005

Steps to Discovery: Mechanisms and Functions of Territorial Behavior 1006

Mechanisms of Behavior 1008
Learning 1011
Development of Behavior 1014
Evolution and Function of Behavior 1015
Social Behavior 1018
Altruism 1022

Reexamining the Themes 1026
Synopsis 1027

BIOLINE
Animal Cognition 1024

Appendix A Metric and Temperature Conversion Charts A-1
Appendix B Microscopes: Exploring the Details of Life B-1
Appendix C The Hardy-Weinberg Principle C-1
Appendix D Careers in Biology D-1
Glossary G-1
Photo Credits P-1
Index I-1

To The Student: A User's Guide

Biology is a journey of exploration and discovery, of struggle and breakthrough. It is enlivened by the thrill of understanding not only what living things do but also how they work. We have tried to create such an experience for you.

Excellence in writing, visual images, and broad biological coverage form the core of a modern biology textbook. But as important as these three factors are in making difficult concepts and facts clear and meaningful, none of them reveals the excitement of biology—the adventure that unearths what we know about life. To help relate the true nature of this adventure, we have developed several distinctive features for this book, features that strengthen its biological core, that will engage and hold your attention, that reveal the human side of biology, that enable every reader to understand how science works, that stimulate critical thinking, and that will create the informed citizenship we all hope will make a positive difference in the future of our planet.

Steps to Discovery

The process of science enriches all parts of this book. We believe that students, like biologists, themselves, are intrigued by scientific puzzles. Every chapter is introduced by a "Steps to Discovery" narrative, the story of an investigation that led to a scientific breakthrough in an area of biology which relates to that chapter's topic. The "Steps to Discovery" narratives portray biologists as they really are: human beings, with motivations, misfortunes, and mishaps, much like everyone experiences. We hope these narratives help you better appreciate biological investigation, realizing that it is understandable and within your grasp.

Throughout the narrative of these pieces, the writing is enlivened with scientific work that has provided knowledge and understanding of life. This approach is meant not just to pay tribute to scientific giants and Nobel prize winners, but once again to help you realize that science does not grow by itself. Facts do not magically materialize. They are the products of rational ideas, insight, determination, and, sometimes, a little luck. Each of the "Steps to Discovery" narratives includes a painting that is meant primarily as an aesthetic accompaniment to the adventure described in the essay and to help you form a mental picture of the subject.

STEPS TO DISCOVERY

A Factor Promoting the Growth of Nerves

Rita Levi-Montalcini received her medical degree from the University of Turin in Italy in 1936, the same year that Benito Mussolini began his anti-Semitic campaign. By 1939, as a Jew, Levi-Montalcini had been barred from carrying out research and practicing medicine, yet she continued to do both secretly. As a student, Levi-Montalcini had been fascinated with the structure and function of the nervous system. Unable to return to the university, she set up a simple laboratory in her small bedroom in her family's home. As World War II raged throughout Europe, and the Allies systematically bombed Italy, Levi-Montalcini studied chick embryos in her bedroom, discovering new information about the growth of nerve cells from the spinal cord into the nearby limbs. In her autobiography *In Praise of Imperfection*, she writes: "Every time the alarm sounded, I would carry down to the precarious safety of the cellars the Zeiss binocular microscope and my most precious silver stained embryonic sections." In September 1943, Germa troops arrived in Turin to support the Italian Fascists. Le Montalcini and her family fled southward to Floren where they remained in hiding for the remainder of the w

After the war ended, Levi-Montalcini continued research at the University of Turin. In 1946, she acce an invitation from Viktor Hamburger, a leading expe the development of the chick nervous system, to co Washington University in St. Louis to work with him f semester; she remained at Washington University years.

A chick embryo and one of its nerve cells helped scientists discover nerve growth factor (NGF).

One of Levi-Montalcini's first projects was the reexamination of a previous experiment of Elmer Bueker, a former student of Hamburger's. Bueker had removed a limb from a chick embryo, replaced it with a fragment of a mouse connective tissue tumor, and found that nerve fibers grew into this mass of implanted tumor cells. When Levi-Montalcini repeated the experiment she made an unexpected discovery: One part of the nervous system of these experimental chick embryos—the sympathetic nervous system—had grown five to six times larger than had its counterpart in a normal chick embryo. (The sympathetic nervous system helps control the activity of internal organs, such as the heart and digestive tract.) Close examination revealed that the small piece of tumor tissue that had been grafted onto the embryo had caused sympathetic nerve fibers to grow "wildly" into all of the chick's internal organs, even causing some of the blood vessels to become obstructed by the invasive fibers. Levi-Montalcini hypothesized that the tumor was releasing some soluble substance that induced the remarkable growth of this part of the nervous system. Her hypothesis was soon confirmed by further experiments. She called the active substance **nerve growth factor (NGF).**

The next step was to determine the chemical nature of NGF, a task that was more readily performed by growing the tumor cells in a culture dish rather than an embryo. But Hamburger's laboratory at Washington University did not have the facilities for such work. To continue the project, Levi-Montalcini boarded a plane, with a pair of tumor-bearing mice in the pocket of her overcoat, and flew to Brazil, where she had a friend who operated a tissue culture laboratory. When she placed sympathetic nervous tissue in the proximity of the tumor cells in a culture dish, the nervous tissue sprouted a halo of nerve fibers that grew toward the tumor cells. When the tissue was cultured in the absence of NGF, no such growth occurred.

For the next 2 years, Levi-Montalcini's lab was devoted to characterizing the substance in the tumor cells that possessed the ability to cause nerve outgrowth. The work was carried out primarily by a young biochemist, Stanley Cohen, who had joined the lab. One of the favored approaches to studying the nature of a biological molecule is to determine its sensitivity to enzymes. In order to determine if nerve growth factor was a protein or a nucleic acid, Cohen treated the active material with a small amount of snake venom, which contains a highly active enzyme that degrades nucleic acid. It was then that chance stepped in.

Cohen expected that treatment with the venom w ther destroy the activity of the tumor cell fraction was a nucleic acid) or leave it unaffected (if NG protein). To Cohen's surprise, treatment with the *increased* the nerve-growth promoting activity of the rial. In fact, treatment of sympathetic nerve tissue w venom alone (in the absence of the tumor extract) in the growth of a halo of nerve fibers! Cohen soon disco why: The snake venom possessed the same nerve g factor as did the tumor cells, but at much higher conce tion. Cohen soon demonstrated that NGF was a prote

Levi-Montalcini and Cohen reasoned that since s venom was derived from a *modified* salivary gland, other salivary glands might prove to be even better sou of the protein. This hypothesis proved to be correct. W Levi-Montalcini and Cohen tested the salivary glands fr male mice, they discovered the richest source of NGF ye source 10,000 times more active than the tumor cells a ten times more active than snake venom.

A crucial question remained: Did NGF play a role the normal development of the embryo, or was its ability stimulate nerve growth just an accidental property of the molecule? To answer this question, Levi-Montalcini an Cohen injected embryos with an antibody against NGF which they hoped would inactivate NGF molecules wherever they were present in the embryonic tissues. The embryos developed normally, with one major exception: They virtually lacked a sympathetic nervous system. The researchers concluded that NGF must be important during normal development of the nervous system; otherwise, inactivation of NGF could not have had such a dramatic effect.

By the early 1970s, the amino acid sequence of NGF had been determined, and the protein is now being synthesized by recombinant DNA technology. During the past decade, Fred Gage, of the University of California, has found that NGF is able to revitalize aged or damaged nerve cells in rats. Based on these studies, NGF is currently being tested as a possible treatment of Alzheimer's disease. For their pioneering work, Rita Levi-Montalcini and Stanley Cohen shared the 1987 Nobel Prize in Physiology and Medicine.

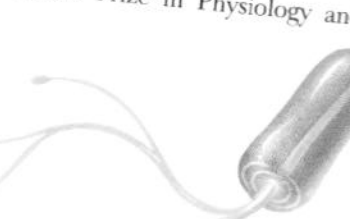

A Thematic Approach

Many students are overwhelmed by the diversity of living organisms and the multitude of seemingly unrelated facts that they are forced to learn in an introductory biology course. Most aspects of biology, however, can be thought of as examples of a small number of recurrent themes. Using the thematic approach, the details and principles of biology can be assembled into a body of knowledge that makes sense, and is not just a collection of disconnected facts. Facts become ideas, and details become parts of concepts as you make connections between seemingly unrelated areas of biology, forging a deeper understanding.

All areas of biology are bound together by evolution, the central theme in the study of life. Every organism is the product of evolution, which has generated the diversity of biological features that distinguish organisms from one another and the similarities that all organisms share. From this basic evolutionary theme emerge several other themes that recur throughout the book:

- **Relationship between Form and Function**
- **Biological Order, Regulation, and Homeostatis**
- **Acquiring and Using Energy**
- **Unity Within Diversity**
- **Evolution and Adaptation**

We have highlighted the prevalent recurrence of each theme throughout the text with an icon, shown above. The icons can be used to activate higher thought processes by inviting you to explore how the fact or concept being discussed fits the indicated theme.

Reexamining the Themes

Each chapter concludes with a "Reexamining the Themes" section, which revisits the themes and how they emerge within the context of the chapter's concepts and principles. This section will help you realize that the same themes are evident at all levels of biological organization, whether you are studying the molecular and cellular aspects of biology or the global characteristics of biology.

When two organisms have the same protein, the difference in amino acid sequence of that protein can be correlated with the evolutionary relatedness of the organisms. The amino acid sequence of hemoglobin, for example, is much more similar between humans and monkeys—organisms that are closely related—than between humans and turtles, who are only distantly related. In fact, the evolutionary tree that emerges when comparing the structure of specific proteins from various animals very closely matches that previously constructed from fossil evidence.

The fact that the amino acid sequences of proteins change as organisms diverge from one another reflects an underlying change in their genetic information. Even though a DNA molecule from a mushroom, a redwood tree, and a cow may appear superficially identical, the sequences of nucleotides that make up the various DNA molecules are very different. These differences reflect evolutionary changes resulting from natural selection (Chapter 34).

Virtually all differences among living organisms can be traced to evolutionary changes in the structure of their various macromolecules, originating from changes in the nucleotide sequences of their DNA. (See CTQ #7.)

REEXAMINING THE THEMES

Relationship between Form and Function

The structure of a macromolecule correlates with a particular function. The unbranched, extended nature of the cellulose molecule endows it with resistance to pulling forces, an important property of plant cell walls. The hydrophobic character of lipids underlies many of their biological roles, explaining, for example, how waxes are able to provide plants with a waterproof covering. Protein function is correlated with protein shape. Just as a key is shaped to open a specific lock, a protein is shaped for a particular molecular interaction. For example, the shape of each polypeptide chain of hemoglobin enables a molecule of oxygen to fit perfectly into its binding site. A single alteration in the amino acid sequence of a hemoglobin chain can drastically reduce the molecule's oxygen-carrying capacity.

Biological Order, Regulation, and Homeostasis

Both blood sugar levels and body weight in humans are controlled by complex homeostatic mechanisms. The level of glucose in your blood is regulated by factors acting on the liver, which stimulate either glycogen breakdown (which increases blood sugar) or glycogen formation (which decreases blood sugar). Your body weight is, at least partly, determined by factors emanating from fat cells which either increase metabolic rate (which tends to decrease body weight) or slow down metabolic rate (which tends to increase body weight).

Acquiring and Utilizing Energy

The chemical energy that fuels biological activities is stored primarily in two types of macromolecules: polysaccharides and fats. Polysaccharides, including starch in plants and glycogen in animals, function primarily in the short-term storage of chemical energy. These polysaccharides can be rapidly broken down to sugars, such as glucose, which are readily metabolized to release energy. Gram-for-gram, fats contain even more energy than polysaccharides and function primarily as a long-term storage of chemical energy.

Unity within Diversity

All organisms, from bacteria to humans, are composed of the same four families of macromolecules, illustrating the unity of life—even at the biochemical level. The precise nature of these macromolecules and the ways they are or ganized into higher structures differ from organism to orga nism, thereby building diversity. Plants, for example, p lymerize glucose into starch and cellulose, while anim polymerize glucose into glycogen. Similarly, many prote (such as hemoglobin) are present in a variety of organis but the precise amino acid sequence of the protein va from one species to the next.

Evolution and Adaptation

Evolution becomes very apparent at the mol level when we compare the structure of macromol among diverse organisms. Analysis of the amino a quences of proteins and the nucleotide sequences cleic acids reveals a gradual change over time in th ture of macromolecules. Organisms that are closely have proteins and nucleic acids whose sequences similar than are those of distantly related organis large degree, the differences observed among d ganisms derives from the evolutionary difference nucleic acid and protein sequences.

The segregation of alleles and their independent assortment during meiosis increase genotype diversity by promoting new combinations of genes. But the shuffling of existing genes alone does not explain the presence of such a vast diversity of life. If all organisms descended from a common ancestor, with its relatively small complement of genes, where did all the genes present in today's millions of species come from? The answer is mutation.

Most mutant alleles are detrimental; that is, they are more likely to disrupt a well-ordered, smoothly functioning organism than to increase the organism's fitness. For example, a mutation might change a gene so that it produces an inactive enzyme needed for a critical life function. Occasionally, however, one of these stable genetic changes creates an advantageous characteristic that fitness of the offspring. In this way, mutation raw material for evolution and the diversificati earth.

One of the requirements for genes is stability; remain basically the same from generation to ge the fitness of organisms would rapidly deterior same time, there must be some capacity for change; otherwise, there would be no potential tion. Alterations in genes do occur, albeit rarely, changes (mutations) represent the raw material tion. (See CTQ #7.)

REEXAMINING THE THEMES

Biological Order, Regulation, and Homeostasis

Mendel discovered that the transmission of genetic factors followed a predictable pattern, indicating that the processes responsible for the formation of gametes, including the segregation of alleles, must occur in a highly ordered manner. This orderly pattern can be traced to the process of meiosis and the precision with which homologous chromosomes are separated during the first meiotic division. Mendel's discovery of independent assortment can also be connected with the first meiotic division, when each pair of homologous chromosomes becomes aligned at the metaphase plate in a manner that is independent of other pairs of homologues.

Unity within Diversity

All eukaryotic, sexually reproducing organisms follow the same "rules" for transmitting inherited traits. Although Mendel chose to work with peas, he could have come to the same conclusions had he studied fruit flies or mice or had he scrutinized a family's medical records on the transmission of certain genetic diseases, such as cystic fibrosis. Alt the mechanism by which genes are transmitted is univ the genes themselves are highly diverse from one orga to the next. It is this genetic difference among specie forms the very basis of biological diversity.

Evolution and Adaptation

Mendel's findings provided a critical link in our kno edge of the mechanism of evolution. A key tenet in theory of evolution is that favorable genetic variations crease the likelihood that an individual will survive to productive age and that its offspring will exhibit these sam favorable characteristics. Mendel's demonstration th units of inheritance pass from parents to offspring withou being blended revealed the means by which advantageous traits could be preserved in a species over many generations. The subsequent discovery of genetic change by mutation revealed how new genes appeared in a population, thus providing the raw material for evolution.

SYNOPSIS

Gregor Mendel discovered the pattern by which inherited traits are transmitted from parents to offspring. Mendel discovered that inherited traits were controlled by pairs of factors (genes). The two factors for a given trait in an individual could be identical (homozygous) or different (heterozygous). In heterozygotes, one of the gene variants (alleles) may be dominant over the other, recessive allele. Because of dominance, the appearance (phenotype) of the heterozygote (genotype of *Aa*) is identical to that of the homozygote with two dominant alleles

The Human Perspective

Students will naturally find many ways in which the material presented in any biology course relates to them. But it is not always obvious how you can use biological information for better living or how it might influence your life. Your ability to see yourself in the course boosts interest and heightens the usefulness of the information. This translates into greater retention and understanding.

To accomplish this desirable outcome, the entire book has been constructed with you—the student—in mind. Perhaps the most notable feature of this approach is a series of boxed essays called "The Human Perspective" that directly reveals the human relevance of the biological topic being discussed at that point in the text. You will soon realize that human life, including your own, is an integral part of biology.

• PART 2 / Chemical and Cellular Foundations of Life

◁ THE HUMAN PERSPECTIVE ▷

Obesity and the Hungry Fat Cell

FIGURE 1
Actor Robert DeNiro in (*left*) a scene from the movie *Raging Bull* and (*right*) a recent photograph.

It has become increasingly clear in recent years that people who are exceedingly overweight—that is, obese—are at increased risk of serious health problems, including heart disease and cancer. By most definitions, a person is obese if he or she is about 20 percent above "normal" or desirable body weight. Approximately 35 percent of adults in the United States are considered obese by this definition, twice as many as at the turn of the century. Among young adults, high blood pressure is five times more prevalent and diabetes three times more prevalent in a group of obese people than in a group of people who are at normal weight. Given these statistics, together with the social stigma facing the obese, there would seem to be strong motivation for maintaining a "normal" body weight. Why, then, are so many of us so overweight? And, why is it so hard to lose unwanted pounds and yet so easy to gain them back? The answers go beyond our fondness for high-calorie foods.

Excess body fat is stored in fat cells (*adipocytes*) located largely beneath the skin. These cells can change their volume more than a hundredfold, depending on the amount of fat they contain. As a person gains body fat, his or her fat cells become larger and larger, accounting for the bulging, sagging body shape. If the person becomes sufficiently overweight, and their fat cells approach their maximum fat-carrying capacity, chemical messages are sent through the blood, causing formation of new fat cells that are "hungry" to begin accumulating their own fat. Once a fat cell is formed, it may expand or contract in volume, but it appears to remain in the body for the rest of the person's life.

Although the subject remains controversial, current research findings suggest that body weight is one of the properties subject to physiologic regulation in humans. Apparently, each person has a particular weight that his or her body's regulatory machinery acts to maintain. This particular value—whether 40 kilograms (80 pounds) or 200 kilograms (400 pounds)—is referred to as the person's **set-point.**

People maintain their body weight at a relatively constant value by balancing energy intake (in the form of food calories) with energy expenditure (in the form of calories burned by metabolic activities or excreted). Obese individuals are thought to have a higher set-point than do persons of normal weight. In many cases, the set-point value appears to have a strong genetic component. For instance, studies reveal there is no correlation between the body mass of adoptees and their adoptive parents, but there is a clear relationship between adoptees and their biological parents, with whom they have not lived.

The existence of a body-weight set-point is most evident when the body weight of a person is "forced" to deviate from the regulated value. Individuals of normal body weight who are fed large amounts of high-calorie foods under experimental conditions tend to gain increasing amounts of weight. If these people cease their energy-rich diets, however, they return quite rapidly to their previous levels, at which point further weight loss stops. This is illustrated by actor Robert DeNiro, who reportedly gained about 50 pounds for the filming of the movie "Raging Bull" (Figure 1), and then lost the weight prior to his next acting role. Conversely, a person who is put on a strict, low-calorie diet will begin to lose weight. The drop in body weight soon triggers a decrease in the person's resting metabolic rate; that is, the amount of calories burned when the person is not engaged in physical activity. The drop in metabolic rate is the body's compensatory measure for the decreased food intake. In other words, it is the body's attempt to halt further weight loss. This effect is particularly pronounced among obese people who diet and lose large amounts of weight: Their pulse rate and blood pressure drop markedly, their fat cells shrink to "ghosts" of their former selves, and they tend to be continually hungry. If these obese individuals go back to eating a *normal* diet, they tend to regain the lost weight rapidly. The drive of these formerly obese persons to increase their food intake is probably a response to chemical signals emanating from the fat cells as they shrink below their previous size.

630 • PART 5 / Form and Function of Animal Life

◁ THE HUMAN PERSPECTIVE ▷

Dying for a Cigarette?

...e average, smoking cigarettes will cut ...imately 6 to 8 years off your life, ...han 5 minutes for every cigarette ...! Cigarette smoking is the greatest ...preventable death in the United ...ccording to a 1991 report by the ...for Disease Control (CDC), ...000 Americans die each year ...ng-related causes. Smoking ac... 87 percent of all lung-cancer ...smokers are more susceptible ...he esophagus, larynx, mouth, ...l bladder than are nonsmok...reased incidence of lung ...mong smokers compared to ...hown in Figure 1*a*, and the ...d by quitting is shown in ...ffects of smoking on lung ...Figure 2. Atherosclero... and peptic ulcers also ...greater frequency than ...rs. For example, long-...5 times more likely to ...terial disease than are ...ysema (a condition ...ction of lung tissue, ...culty in breathing) ...mmation of the air-...e prevalent among

...ger other people. ...sponsible for the ...nocent bystand-...re the same air ...passive (invol-...own; second-... seriously ill ...rs have dou-...ry infections ...posed to to-...ng married ...us; 20 per-...ong non-...ributable to inhaling other people's tobacco smoke. Another "innocent bystander" is a fetus developing in the uterus of a woman who smokes. Smoking increases the incidence of miscarriage and stillbirth and decreases the birthweight of the infant. Once born, these babies suffer twice as many respiratory infections as do babies of nonsmoking mothers.

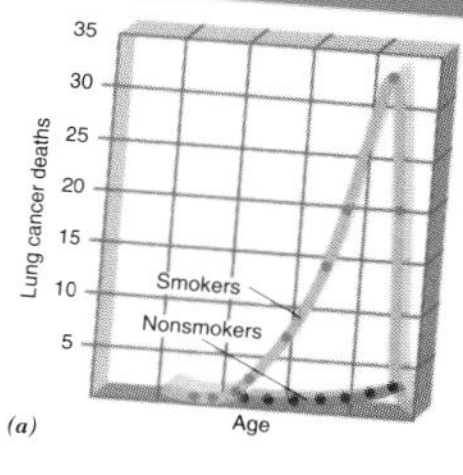

(*a*)

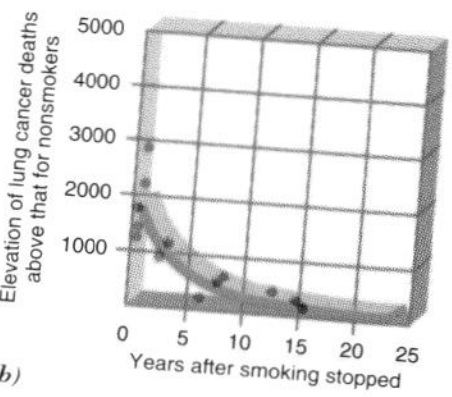

(*b*)

Why is smoking so bad for your health? The smoke emitted from a burning cigarette contains more than 2,000 identifiable substances, many of which are either irritants or carcinogens. These compounds include carbon monoxide, sulfur dioxide, formaldehyde, nitrosamines, toluene, ammonia, and radioactive isotopes. Autopsies of respiratory tissues from smokers (and from nonsmokers who have lived for long periods with smokers) show widespread cellular changes, including the presence of precancerous cells (cells that may become malignant, given time) and a marked reduction in the number of cilia that play a vital role in the removal of bacteria and debris from the airways.

Of all the compounds found in tobacco (including smokeless varieties), the most important is nicotine, not because it is carcinogenic, but because it is so addictive. Nicotine is addictive because it acts like a neurotransmitter by binding to certain acetylcholine receptors (page 477), stimulating postsynaptic neurons. The physiological effects of this stimulation include the release of epinephrine, an increase in blood sugar, an elevated heart rate, and the constriction of blood vessels, causing elevated blood pressure. A smoker's nervous system becomes "accustomed" to the presence of nicotine and decreases the output of the natural neurotransmitter. As a result, when a person tries to stop smoking, the sudden absence of nicotine, together with the decreased level of the natural transmitter, decreases stimulation of postsynaptic neurons, which creates a craving for a cigarette—a "nicotine fit." Ex-smokers may be so conditioned to the act of smoking that the craving for cigarettes can continue long after the physiological addiction disappears.

The "Biolines" are boxed essays that highlight fascinating facts, applications, and real-life lessons, enlivening the mainstream of biological information. Many are remarkable stories that reveal nature to be as surprising and interesting as any novelist could imagine.

◁ BIOLINE ▷
DNA Fingerprints and Criminal Law

On February 5, 1987, a woman and her 2-year-old daughter were found stabbed to death in their apartment in the New York City borough of the Bronx. Following a tip, the police questioned a resident of a neighboring building. A small bloodstain was found on the suspect's watch, which was sent to a laboratory for DNA fingerprint analysis. The DNA from the white blood cells in the stain was amplified using the PCR technique and was digested with a restriction enzyme. The restriction fragments were then separated by electrophoresis, and a pattern of labeled fragments was identified with a radioactive probe. The banding pattern produced by the DNA from the suspect's watch was found to be a perfect match to the pattern produced by DNA taken from one of the victims. The results were provided to the opposing attorneys, and a pretrial hearing was called in 1989 to discuss the validity of the DNA evidence.

During the hearing, a number of expert witnesses for the prosecution explained the basis of the DNA analysis. According to these experts, no two individuals, with the exception of identical twins, ave the same nucleotide sequence in their NA. Moreover, differences in DNA se- ence can be detected by comparing the gths of the fragments produced by re- tion-enzyme digestion of different A samples. The patterns produce a A fingerprint" (Figure 1) that is as e to an individual as is a set of con- nal fingerprints lifted from a glass. In NA fingerprints had already been more than 200 criminal cases in the States and had been hailed as the portant development in forensic (the application of medical facts to legal problems) in decades. The widespread use of DNA fingerprinting evidence in court had been based on its general acceptability in the scientific community. According to a report from the company performing the DNA analysis, the likelihood that the same banding patterns could be obtained by chance from two *different* individuals in the community was only one in 100 million.

FIGURE 1
Alec Jeffreys of the University of Leicester, England, examining a DNA fingerprint. Jeffreys was primarily responsible for developing the DNA fingerprint technique and was the scientist who confirmed the death of Josef Mengele.

What made this case (known as the Castro case, after the defendant) memorable and distinct from its predecessors was that the defense also called on expert witnesses to scrutinize the data and to present their opinions. While these experts con firmed the capability of DNA fingerprint ing to identify an individual out of a hug population, they found serious technica flaws in the analysis of the DNA sample used by the prosecution. In an unprece dented occurrence, the experts who had earlier testified *for the prosecution* agreed that the DNA analysis in this case was unreliable and should not be used as evidence! The problem was not with the technique itself but in the way it had been carried out in this particular case. Consequently, the judge threw out the evidence.

In the wake of the Castro case, the use of DNA fingerprinting to decide guilt or innocence has been seriously questioned. Several panels and agencies are working to formulate guidelines for the licensing of forensic DNA laboratories and the certification of their employees. In 1992, a panel of the National Academy of Sciences released a report endorsing the general reliability of the technique but called for the institution of strict standards *to be set by scientists.*

Meanwhile, another issue regarding DNA fingerprinting has been raised and hotly debated. Two geneticists, Richard Lewontin of Harvard University and Daniel Hartl of Washington University, coauthored a paper published in December 1991, suggesting that scientists do not have enough data on genetic variation within different racial or ethnic groups to calculate the odds that two individuals—a suspect and a perpetrator of the crime—are one and the same on the basis of an identical DNA fingerprint. The matter remains an issue of great concern in both the scientific and legal communities and has yet to be resolved.

◁ BIOLINE ▷
The Fish That Changes Sex

In vertebrates, gender is generally a biologically inflexible commitment: An indi- vidual develops into either a male or a fe- male as dictated by the sex chromosomes acquired from one's parents. Yet, even among vertebrates, there are organisms that can reverse their sexual commitment. The Australian cleaner fish (Figure 1), a small animal that sets up "cleaning stations" to which larger fishes come for parasite removal, can change its gender in response to environmental demands. Most male cleaner fish travel alone rather than with a school. Except for a single male, schools of cleaner fish are comprised entirely of females. Although it might seem logical to conclude that maleness engenders solo travel, it is actually the other way around: Being alone fosters maleness. A cleaner fish that develops away from a school *becomes* a male, whereas the same fish developing in a school would have become a female.

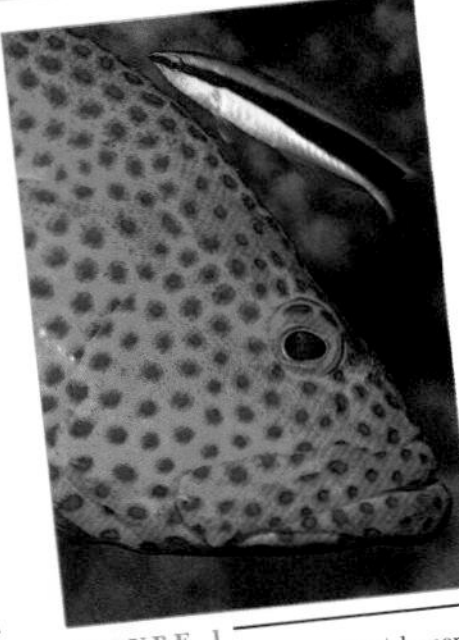

FIGURE 1
The small Australian wrasse (cleaner fish) is seen on a much larger grouper.

But what of the one male in the school—the one with the harem? He may have developed as a solo fish and then found a school in need of his spermatogenic services. But there is another way a school may acquire a male. If the male in a school dies (or is removed experimentally), one of the females, the one at the top of a behavioral hierarchy that exists in each school, becomes uncharacteristically aggressive and takes over the behavioral role of the missing male. She begins to develop male gonads, and within a few weeks, the female becomes a reproductively competent male, indistinguishable from other males. Furthermore, the sex change is reversible. If a fully developed male enters the school during the sexual transition, the almost-male fish developmentally backpedals, once again assuming the biological and behavioral role of a female.

Not all organisms follow the mammalian pattern of sex determination. In some animals, most notably birds, the opposite pattern is found: The female's cells have an X and a Y chromosome, while the male's cells have two Xs. An exception to this rule of a strict relation between sex and chromosomes is discussed in the Bioline: The Fish That Changes Sex. Although some plants possess sex chromosomes and gender distinctions between individuals, most have only autosomes; consequently, each individual produces both male and female parts.

SEX LINKAGE

For fruit flies and humans alike, there are hundreds of genes on the X chromosome that have no counterpart on the smaller Y chromosome. Most of these genes have nothing to do with determining gender, but their effect on phenotype usually *depends on* gender. For example, in females, a recessive allele on one X chromosome will be masked (and not expressed) if a dominant counterpart resides on the other X chromosome. In males, it only takes one recessive allele on the single X chromosome to determine the individual's phenotype since there is no corresponding allele on the Y chromosome. Inherited characteristics determined by genes that reside on the X chromosome are called **X-linked characteristics.**

So far, some 200 human X-linked characteristics have been described, many of which produce disorders that are found almost exclusively in men. These include a type of heart-valve defect (*mitral stenosis*), a particular form of mental retardation, several optical and hearing impairments, muscular dystrophy, and red-green colorblindness (Figure 13-8).

One X-linked recessive disorder has altered the course of history. The disease is **hemophilia,** or "bleeder's disease," a genetic disorder characterized by the inability to produce a clotting factor needed to halt blood flow quickly following an injury. Nearly all hemophiliacs are males. Although females can inherit two recessive alleles for hemophilia, this occurrence is extremely rare. In general, women who have acquired the rare defective allele are heterozygous **carriers** for the disease. The phenotype of a carrier

Bioethics Essays

Several ethical issues are discussed in the Bioethics essays which add provocative pauses throughout the text. Biological Science does not operate in a vacuum but has profound consequences on the general community. Because biologists study life, the science is peppered with ethical considerations. The moral issues discussed in these essays are neither simple nor easy to resolve, and we do not claim to have any certain answers. Our goal is to encourage you to consider the bioethical issues that you will face now and in the future.

◁ BIOETHICS ▷

Blurring the Line between Life and Death

By **ARTHUR CAPLAN**
Division of the Center for Biomedical Ethics at the University of Minnesota

Theresa Ann Campo Pearson didn't have a very long life. When she died in 1992, she was only 10 days old. Despite her short life, she became the center of a very strange, sad, and wrenching ethical controversy. Theresa died because her brain had failed to form. She had anencephaly, a condition in which only the brainstem, located at the top of the spinal cord, is present. Her parents wanted to donate Theresa's organs; the courts said no. Some people found it strange that Theresa's parents, Laura Campo and Justin Pearson, did not get their way. Why not allow donation, when every day in North America a baby dies because there is no heart, lung, or liver available for transplantation?

Anencephaly is best described as completely "unabling," not disabling. Children born with anencephaly cannot think, feel, sense, or be aware of the world. Many are stillborn; the majority of the rest die within days of birth. A mere handful live for a few weeks. Theresa's parents knew all this. But rather than abort the pregnancy, they chose to have their baby. In fact, the baby was born by Caesarean section, at least partly in the hope that it would be born alive, thereby making organ donation possible. When Theresa died at Broward General Medical Center in Fort Lauderdale, Florida, however, no organs were taken. Two Florida courts ruled that the baby could not be used as a source of organs unless she was brain-dead, and Theresa Ann Campo was never pronounced brain-dead.

Brain death refers to a situation in which the brain has irreversibly lost all function and activity. Babies born with anencephaly have some brain function in their brainstem so, while they cannot think or feel, they are alive. According to Florida law—and the law in more than 40 other states—only those individuals declared brain-dead can donate organs. The courts of Florida had no other option but to deny the request for organ donation.

One obvious solution is to change the law so that states could decide that organs can be removed upon parental consent from either those who are born brain-dead or from babies who are born with anencephaly. Another solution is to rewrite the definition of death to say that death occurs either when the brain has totally ceased to function or if a baby is born anencephalic. Do you feel that either of these changes should be made? Some may argue that medicine will fudge the line between life and death in order to get organs for transplant. Do you agree with this concern? How do you think redefining death will affect a person's decision to check off the donation box on the back of a driver's license? Do you think people may worry that if they are known to be potential donors they won't be aggressively treated at the hospital? In your opinion, would changing the definition of death to include anencephaly be beneficial or deleterious?

Like the brain, the spinal cord is composed of white matter (myelinated axons) and gray matter (dendrites and cell bodies). However, the arrangement of these types of matter is reversed in the spinal cord, compared to their arrangement in the brain: The spinal cord's white matter surrounds the gray matter (Figure 23-16).

The human central nervous system is the most complex and highly evolved assembly of matter. Among its functions are the processing of sensory information collected from both the external and internal environment; the regulation of internal physiological activities; the coordination of complex motor activities; and the endowment of such intangible "mental" qualities as emotions, creativity, language, and the ability to think, learn, and remember. (See CTQ #6.)

ARCHITECTURE OF THE PERIPHERAL NERVOUS SYSTEM

The peripheral nervous system provides the neurological bridge between the central nervous system and the various parts of the body. The peripheral nervous system is made up of paired nerves that extend into the periphery from the CNS at various levels along the body. Each nerve is composed of a large bundle of myelinated axons surrounded by a connective tissue sheath. Twelve pairs of **cranial nerves** emerge from the central stalk of the human brain, and 31 pairs of **spinal nerves** extend from the spinal cord out between the vertebrae of humans (Figure 23-16). For the most part, the cranial nerves *innervate* (supply nerves to) tissues and organs of the head and neck, whereas the spinal nerves innervate the chest, abdomen, and limbs.

Additional Pedagogical Features

We have worked to assure that each chapter in this book is an effective teaching and learning instrument. In addition to the pedagogical features discussed above, we have included some additional tried-and-proven-effective tools.

KEY POINTS

Key points follow each major section and offer a condensation of the relevant facts and details as well as the concepts discussed. You can use these key points to reaffirm your understanding of the previous reading or to alert you to misunderstood material before moving on to the next topic. Each key point is tied to a Critical Thinking Question found at the end of the chapter; together, they encourage you to analyze the information, taking it beyond mere memorization.

the corresponding polypeptide. The cumulative effect of gradual changes in polypeptides over evolutionary time has been the generation of life's diversity.

Evolution and Adaptation

Evolutionary change from generation to generation depends on genetic variability. Much of this variability arises from reshuffling maternal and paternal genes during meiosis, but somewhere along the way *new* genetic information must be introduced into the population. Ne netic information arises from mutations in existing g Some of these mutations arise during replication; o occur as the result of unrepaired damage as the DNA i "sitting" in a cell. Mutations that occur in an individ germ cells can be considered the raw material on w natural selection operates; whereas harmful mutations duce offspring with a reduced fitness, beneficial mutat produce offspring with an increased fitness.

SYNOPSIS

Experiments in the 1940s and 1950s established conclusively that DNA is the genetic material. These experiments included the demonstration that DNA was capable of transforming bacteria from one genetic strain to another; that bacteriophages injected their DNA into a host cell during infection; and that the injected DNA was transmitted to the bacteriophage progeny.

DNA is a double helix. DNA is a helical molecule consisting of two chains of nucleotides running in opposite directions, with their backbones on the outside, and the nitrogenous bases facing inward like rungs on a ladder. Adenine-containing nucleotides on one strand always pair with thymine-containing nucleotides on the other strand, likewise for guanine- and cytosine-containing nucleotides. As a result, the two strands of a DNA molecule are complementary to one another. Genetic information is encoded in the specific linear sequence of nucleotides that make up the strands.

DNA replication is semiconservative. During replication, the double helix separates, and each strand serves as a template for the formation of a new, complementary strand. Nucleotide assembly is carried out by the enzyme DNA polymerase, which moves along the two strands in opposite directions. As a result, one of the strands is synthesized continuously, while the other is synthesized in segments that are covalently joined. Accuracy is maintained by a proofreading mechanism present within the polymerase.

Information flows in a cell from DNA to RNA to protein. Each gene consists of a linear sequence of nucleotides that determines the linear sequence of amino acids in a polypeptide. This is accomplished in two maj steps: transcription and translation.

During transcription, the information spelled out b the gene's nucleotide sequence is encoded in a mole cule of messenger RNA (mRNA). The mRNA contain a series of codons. Each codon consists of three nucleotides Of the 64 possible codons, 61 specify an amino acid, and the other 3 stop the process of protein synthesis.

During translation, the sequence of codons in the mRNA is used as the basis for the assembly of a chain of specific amino acids. Translating mRNA messages occurs on ribosomes and requires tRNAs, which serve as decoders. Each tRNA is folded into a cloverleaf structure with an anticodon at one end—which binds to a complementary codon in the mRNA—and a specific amino acid at the other end—which becomes incorporated into the growing polypeptide chain. Amino acids are added to their appropriate tRNAs by a set of enzymes. The sequential interaction of charged tRNAs with the mRNA results in the assembly of a chain of amino acids in the precise order dictated by the DNA.

Mutation is a change in the genetic message. Gene mutations may occur as a single nucleotide substitution, which leads to the insertion of an amino acid different from that originally encoded. In contrast, the addition of one or two nucleotides throws off the reading frame of the ribosome as it moves along the mRNA, leading to the incorporation of incorrect amino acids "downstream" from the point of mutation. Exposure to mutagens increases the rate of mutation.

Many plants replenish old and dying cells with vigorous new cells. But since each plant cell has a surrounding cell wall (Chapter 7) old plant cells do not just wither and disappear when they die. Instead, dead plant cells leave cellular "skeletons" where they once lived. As a result, the longer a plant lives, the more complex its anatomy becomes. **Annuals** are plants that live for 1 year or less, such as corn and marigolds. Because they live for such a brief period, these plants do not completely replace old cells. As a result, annuals are anatomically less complex than are **biennials**—plants that live for 2 years—and **perennials**—herbs, shrubs, and trees that live longer than 2 years. Biennials (carrots, Queen Anne's lace) and perennials (rosebushes, apple trees) are able to live longer than annuals because they produce new cells to replace those that cease functioning or die, providing a continual supply of young, vigorous cells.

In this chapter, we will focus on the body construction of flowering plants, the most familiar, most evolutionarily advanced, and structurally complex of any group in the plant kingdom. All flowering plants are **vascular plants;** that is, they contain specialized cells that circulate water, minerals, and food (organic molecules) throughout the plant. Botanists divide flowering plants into two main groups: **dicotyledons,** or dicots (*di* = two, *cotyledon* = embryonic seed leaf), and **monocotyledons,** or monocots (*mono* = one). Table 18-1 illustrates the many differences that distinguish dicots from monocots and will be used as a reference throughout the chapter.

TABLE 18-1

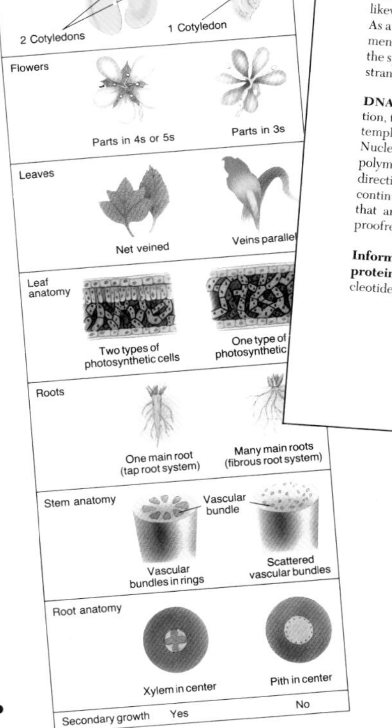

Dicot and Monocot Comparison	DICOT	MONOCOT
Embryo	2 Cotyledons	1 Cotyledon
Flowers	Parts in 4s or 5s	Parts in 3s
Leaves	Net veined	Veins paralle
Leaf anatomy	Two types of photosynthetic cells	One type of photosynthetic
Roots	One main root (tap root system)	Many main roots (fibrous root system)
Stem anatomy	Vascular bundles in rings	Scattered vascular bundles
Root anatomy	Xylem in center	Pith in center
Secondary growth	Yes	No

SHOOTS AND ROOTS

The flowering plant body is a study in contradictions. A typical plant grows through the soil and the air simultaneously, two very different habitats with very different conditions. As a result, the two main parts of the plant differ dramatically in form (anatomy) and function (physiology): The underground **root system** anchors the plant in the soil and absorbs water and nutrients, while the aerial **shoot system** absorbs sunlight and gathers carbon dioxide for photosynthesis (Figure 18-2). The shoot system also produces stems, leaves, flowers, and fruits. Interconnected vascular tissues transport materials between the aerial shoot system and the underground root system. These connections allow water and minerals absorbed by the root to be conducted to shoot tissues, and for food produced by the shoot to be transported to root tissues. We will discuss the various components of these two systems in more detail later in the chapter.

Over 90 percent of all plant species are flowering plants. Flowering plants are the most recently evolved plant group, having undergone rapid evolution during the past 1 million to 2 million years as environmental conditions on land became more variable. (See CTQ #2.)

SYNOPSIS

The synopsis section offers a convenient summary of the chapter material in a readable narrative form. The material is summarized in concise paragraphs that detail the main points of the material, offering a useful review tool to help reinforce recall and understanding of the chapter's information.

REVIEW QUESTIONS

Along with the synopsis, the Review Questions provide a convenient study tool for testing your knowledge of the facts and processes presented in the chapter.

STIMULATING CRITICAL THINKING

Each chapter contains as part of its end material a diverse mix of Critical Thinking Questions. These questions ask you to apply your knowledge and understanding of the facts and concepts to hypothetical situations in order to solve problems, form hypotheses, and hammer out alternative points of view. Such exercises provide you with more effective thinking skills for competing and living in today's complex world.

224 • PART 2 / *Chemical and Cellular Foundations of Life*

Key Terms

zygote (p. 214)
meiosis (p. 214)
life cycle (p. 214)
germ cell (p. 214)
somatic cell (p. 214)
meiosis I (p. 216)
reduction division (p. 216)
synapsis (p. 216)
tetrad (p. 216)
crossing over (p. 216)
genetic recombination (p. 216)
synaptonemal complex (p. 218)
maternal chromosome (p. 219)
paternal chromosome (p. 219)
independent assortment (p. 219)
meiosis II (p. 219)

Review Questions

1. Match the activity with the phase of meiosis in which it occurs.

a. synapsis	1. prophase I
b. crossing over	2. metaphase I
c. kinetochores split	3. anaphase I
d. independent assortment	4. telophase I
e. homologous chromosomes separate	5. prophase II
f. cytokinesis	6. anaphase II
	7. telophase II

2. How do crossing over and independent assortment increase the genetic variability of a species?
3. Why is meiosis I (and not meiosis II) referred to as the reduction division?
4. Suppose that one human sperm contains x amount of DNA. How much DNA would a cell just entering meiosis contain? A cell entering meiosis II? A cell just completing meiosis II? Which of these three cells would have a haploid number of chromosomes? A diploid number of chromosomes?

Critical Thinking Questions

1. Why are disorders, such as Down syndrome, that arise from abnormal chromosome numbers, characterized by a number of seemingly unrelated abnormalities?
2. A gardener's favorite plant had white flowers and long seed pods. To add some variety to her garden, she transplants some plants of the same type, but with pink flowers and short seed pods from her neighbor's garden. To her surprise, in a few generations, she grows plants with white flowers and short seed pods and plants with pink flowers and long seed pods, as well as the original combinations. What are two ways in which these new combinations could have arisen?
3. Set up the meiosis template in the diagram below on a large sheet of paper. Then use pieces of colored yarn or pipe cleaners to simulate chromosomes and make a model of the phases of meiosis. *(See template on opposite page)*
4. Would you expect two genes on the same chromosome, such as yellow flowers and short stems, always to be exchanged during crossing over? How might they remain together *in spite of* crossing over?
5. Suppose paternal chromosomes always lined up on the same side of the metaphase plate of cells in meiosis I. How would this affect genetic variability of offspring? Would they all be identical? Why or why not?

Additional Readings

Chandley, A. C. 1988. Meiosis in man. *Trends in Gen.* 4:79–83. (Intermediate)

Hsu, T. C. 1979. *Human and mammalian cytogenetics.* New York: Springer-Verlag. (Intermediate)

John, B. 1990. *Meiosis.* New York: Cambridge University Press. (Advanced)

Moens, P. B. 1987. *Meiosis.* Orlando: Academic. (Advanced)

Patterson, D. 1987. The causes of Down syndrome. *Sci. Amer.* Feb:52–60. (Intermediate-Advanced)

White, M. J. D. 1973. *The chromosomes.* Halsted. (Advanced)

ADDITIONAL READINGS

Supplementary readings relevant to the Chapter's topics are provided at the end of every chapter. These readings are ranked by level of difficulty (introductory, intermediate, or advanced) so that you can tailor your supplemental readings to your level of interest and experience.

Careers in Biology

The appendices of this edition include "Careers in Biology," a frequently overlooked aspect of our discipline. Although many of you may be taking biology as a requirement for another major (or may have yet to declare a major), some of you are already biology majors and may become interested enough to investigate the career opportunities in life sciences. This appendix helps students discover how an interest in biology can grow into a livelihood. It also helps the instructor advise students who are considering biology as a life endeavor.

APPENDIX D

Careers in Biology

Although many of you are enrolled in biology as a requirement for another major, some of you will become interested enough to investigate the career opportunities in life sciences. This interest in biology can grow into a satisfying livelihood. Here are some facts to consider:

- Biology is a field that offers a very wide range of possible science careers
- Biology offers high job security since many aspects of it deal with the most vital human needs: health and food
- Each year in the United States, nearly 40,000 people obtain bachelor's degrees in biology. But the number of newly created and vacated positions for biologists is increasing at a rate that exceeds the number of new graduates. Many of these jobs will be in the newer areas of biotechnology and bioservices.

Biologists not only enjoy job satisfaction, their work often changes the future for the better. Careers in medical biology help combat diseases and promote health. Biologists have been instrumental in preserving the earth's life-supporting capacity. Biotechnologists are engineering organisms that promise dramatic breakthroughs in medicine, food production, pest management, and environmental protection. Even the economic vitality of modern society will be increasingly linked to biology.

Biology also combines well with other fields of expertise. There is an increasing demand for people with backgrounds or majors in biology complexed with such areas as business, art, law, or engineering. Such a distinct blend of expertise gives a person a special advantage.

The average starting salary for all biologists with a Bachelor's degree is $22,000. A recent survey of California State University graduates in biology revealed that most were earning salaries between $20,000 and $50,000. But as important as salary is, most biologists stress job satisfaction, job security, work with sophisticated tools and scientific equipment, travel opportunities (either to the field or to scientific conferences), and opportunities to be creative in their job as the reasons they are happy in their career.

Here is a list of just a few of the careers for people with degrees in biology. For more resources, such as lists of current openings, career guides, and job banks, write to Biology Career Information, John Wiley and Sons, 605 Third Avenue, New York, NY 10158.

A SAMPLER OF JOBS THAT GRADUATES HAVE SECURED IN THE FIELD OF BIOLOGY*

Agricultural Biologist
Agricultural Economist
Agricultural Extension Officer
Agronomist
Amino-acid Analyst
Analytical Biochemist
Anatomist
Animal Behavior Specialist
Anticancer Drug Research Technician
Antiviral Therapist
Arid Soils Technician
Audio-neurobiologist
Author, Magazines & Books
Behavioral Biologist
Bioanalyst
Bioanalytical Chemist
Biochemical/Endocrine Toxicologist
Biochemical Engineer
Pharmacology Distributor
Pharmacology Technician
Biochemist
Biogeochemist
Biogeographer
Biological Engineer
Biologist
Biomedical Communication Biologist
Biometerologist
Biophysicist
Biotechnologist
Blood Analyst
Botanist
Brain Function Researcher
Cancer Biologist
Cardiovascular Biologist
Cardiovascular/Computer Specialist
Chemical Ecologist
Chromatographer
Clinical Pharmacologist
Coagulation Biochemist
Cognitive Neuroscientist
Computer Scientist
Dental Assistant
Ecological Biochemist
Electrophysiology/Cardiovascular Technician
Energy Regulation Officer
Environmental Biochemist
Environmental Center Director
Environmental Engineer
Environmental Geographer
Environmental Law Specialist
Farmer
Fetal Physiologist
Flavorist
Food Processing Technologist
Food Production Manager
Food Quality Control Inspector
Flower Grower
Forest Ecologist
Forest Economist
Forest Engineer
Forest Geneticist
Forest Manager

Study Guide

Written by Gary Wisehart and Michael Leboffe of San Diego City College, the *Study Guide* has been designed with innovative pedagogical features to maximize your understanding and retention of the facts and concepts presented in the text. Each chapter in the *Study Guide* contains the following elements.

Concepts Maps

In Chapter 1 of the *Study Guide*, the beginning of a concept map stating the five themes is introduced. In each subsequent chapter, the concept map is expanded to incorporate topics covered in each chapter as well as the interconnections between chapters and the five themes. "Connector" phrases are used to link the concepts and themes, and the text icons representing the themes are incorporated into the concept maps.

Go Figure!

In each chapter, questions are posed regarding the figures in the text. Students can explore their understanding of the figures and are asked to think critically about the figures based on their understanding of the surrounding text and their own experiences.

Self-Tests

Each chapter includes a set of matching and multiple-choice questions. Answers to the Study Guide questions are provided.

Concept Map Construction

The student is asked to create concept maps for a group of terms, using appropriate connector phrases and adding terms as necessary.

Laboratory Manual

Biology: Exploring Life, Second Edition is supplemented by a comprehensive *Laboratory Manual* containing approximately 60 lab exercises chosen by the text authors from the National Association of Biology Teachers. These labs have been thoroughly class-tested and have been assembled from various scientific publications. They include such topics as

- Chaparral and Fire Ecology: Role of Fire in Seed Germination *(The American Biology Teacher)*
- A Model for Teaching Mitosis and Meiosis *(American Biology Teacher)*
- Laboratory Study of Climbing Behavior in the Salt Marsh Snail *(Oceanography for Landlocked Classrooms)*
- Down and Dirty DNA Extraction *(A Sourcebook of Biotechnology Activities)*
- Bioethics: The Ice-Minus Case *(A Sourcebook of Biotechnology Activities)*
- Using Dandelion Flower Stalks for Gravitropic Studies *(The American Biology Teacher)*
- pH and Rate of Enzymatic Reactions *(The American Biology Teacher)*

CHAPTER

40

The Biosphere

STEPS TO DISCOVERY
The Antarctic Ozone Hole

BOUNDARIES OF THE BIOSPHERE

ECOLOGY DEFINED

Ecology and Evolution

THE EARTH'S CLIMATES

THE ARENAS OF LIFE

Aquatic Ecosystems

Biomes: Patterns of Life on Land

THE HUMAN PERSPECTIVE

Acid Rain and Acid Snow: Global Consequences of Industrial Pollution

BIOETHICS

Values in Ecology and Environmental Science: Neutrality or Advocacy?

STEPS TO DISCOVERY
The Antarctic Ozone Hole

Virtually all the earth's organisms owe their lives to the filtering effects of the ozone layer 20 to 50 kilometers (12 to 30 miles) above. About 99 percent of the sun's lethal ultraviolet rays are absorbed by this invisible layer; as a result, organisms are spared overexposure to these killer rays that destroy many biological molecules, including DNA. Recent studies have revealed a very alarming trend, however: The ozone layer is rapidly being destroyed. As this protective layer becomes depleted, we are bound to see more cases of skin cancer, cataracts, and immune deficiencies, as well as reduced crop yields and other serious consequences.

The prospect that the ozone layer may be at risk was forwarded in 1974 by F. Sherwood Rowland and Mario Molina, two atmospheric chemists at the University of California, Irvine. Rowland and Molina created laboratory conditions resembling those found in the mid- to outer stratosphere, the region of the earth's atmosphere where the protective ozone layer is located. The chemists examined the effects of a group of manmade compounds called chorofluorocarbons (CFCs) on ozone molecules (O_3). They discovered that CFCs destroyed ozone molecules with alarming efficiency. CFCs had been widely used as propellents in aerosol products, such as deodorants and air fresheners, as refrigerants, and to inflate the bubbles in styrofoam.

When products containing CFCs are used, CFCs are released into the air as a gas. In their research, Rowland and Molina discovered that when CFC molecules were irradiated with ultraviolet light equivalent to that found in sun-

Photographs taken from space by weather satellites revealed the earth's protective ozone layer was deteriorating

light, the CFC molecule broke apart, releasing a free chlorine atom (Cl). The free chlorine reacted with an ozone molecule (O_3), splitting it into $ClO + O_2$. The ClO molecule then combined with free oxygen (O) to form Cl and O_2, thereby rereleasing the chlorine atom. The free chlorine then split yet another ozone molecule, and so on. Based on their observations, Rowland and Molina projected that a single chlorine atom could destroy between 10,000 and 100,000 ozone molecules in the stratosphere before being washed out of the atmosphere by rain. At that rate, Rowland and Molina hypothesized that CFCs could eventually damage the protective ozone layer. In fact, they projected that CFCs could destroy between 20 and 30 percent of the ozone layer, threatening all life on earth.

Many scientists were skeptical of Rowland and Molina's projections, most believing that a decline of only 2 to 4 percent would be more likely, sometime in the next century. In addition, although Rowland and Molina had clearly demonstrated the ozone-destroying capabilities of CFCs under laboratory conditions, they had not examined whether CFCs actually reached the stratosphere, where they would pose a threat to the ozone layer.

In the early 1980s, satellite studies were initiated to assess stratospheric ozone and CFC levels. These studies verified that CFCs indeed reached the stratosphere. Atmospheric balloons, each carrying flasks that sampled the gases in the air of the stratosphere, were also released. When recovered, the flasks contained both CFCs and chlorine gas, supporting Rowland's and Molina's hypothesis. These samples yielded some very disturbing data: The CFC concentration in the stratosphere had doubled in only 10 years. This led EPA (Environmental Protection Agency) scientists to project a 60 percent decline in ozone levels by 2050 if CFC use and production continued to grow at the current annual rate of 4.5 percent.

Once studies established that significant levels of CFCs were reaching the stratosphere, scientists began to measure changes in the thickness of the ozone layer. The British Antarctic Survey collected monthly samples of ozone in the stratosphere. They found that each spring, ozone concentrations fell sharply and that ozone depletion was growing worse each year. The scientists were also able to correlate rapid ozone depletion to periods of increased usage of CFCs.

In 1987, an international panel of more than 100 scientists was convened to study all of the data and make a judgment about the threat to the ozone layer. In their report, the scientists confirmed that ozone had declined by 1.7 percent to 3.0 percent over the Northern Hemisphere since 1969, and from 5 percent to 10 percent over Antarctica during the same period. In September 1987, the United Nations sponsored negotiations to reduce CFC production worldwide; 24 nations signed what became known as the Montreal Protocol, agreeing to cut CFC production in half by 1999. Since CFC-containing products continue to be used and since CFCs persist in the atmosphere for long periods, however, CFCs are expected to remain in the stratosphere for decades to come.

Satellite photographs clearly depict the trend of ozone depletion. In the late 1980s, ozone had become depleted by as much as 60 percent at certain altitudes over Antarctica. By 1990, ozone levels had dropped by as much as 95 percent. In 1992, the World Meteorological Organization reported that there were regions over Antarctica where no ozone could be detected at all. The report summarized the fact that the ozone hole over Antarctica had enlarged to a record size of over 9 million square miles, about three times the size of the continental United States. This was about 25 percent larger than reported in previous years.

These findings are very important because they reveal that the rate of ozone depletion is even more rapid than originally forecasted by scientists, including Rowland and Molina. In an October 1992 interview, Rowland was asked for this response to these new findings. He said: "What we are looking at is ozone depletion caused by CFCs that were released back in 1987 and 1988. The expectation is that it will probably continue to get worse in the stratosphere for another decade or so." Ultimately, life hangs in the balance.

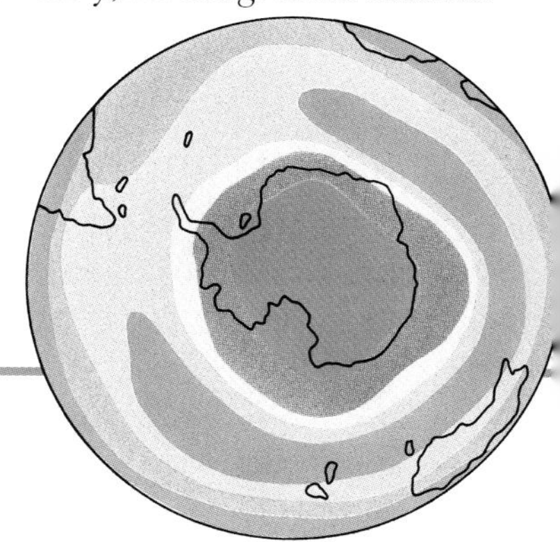

at an alarming rate.

The Apollo astronauts watched the earth recede in the distance as they sped toward the moon (Figure 40-1). The artificial life-support systems that sustained the astronauts' lives while on board the spacecraft temporarily replaced their natural life-support system, the earth.

Without artificial support systems, life as we know it could not exist beyond the earth's atmosphere or deep beneath its solid surface. All life is restricted to a relatively narrow zone of air, water, and land, called the **biosphere,** the thin envelope in which all living organisms are found (see Chapter 1). The biosphere is only 22.4 kilometers (14 miles) thick, from the upper limits of life in the atmosphere to the depths of the dark ocean trenches. In relation to the size of the earth, the biosphere is only about as thick as the skin on an apple. All life as we know it exists within this thin layer that envelops the earth.

▼ ▼ ▼

BOUNDARIES OF THE BIOSPHERE

The biosphere includes portions of the earth's hydrosphere (waters), lithosphere (crust), and atmosphere (gases). Of the three spheres, the **hydrosphere** of oceans, seas, lakes, ponds, rivers, and streams harbors the greatest quantity of life. Most aquatic organisms are found in shallow waters along the shorelines, where sunlight penetrates, providing the energy for photosynthesis. Similarly, most land organisms live near the illuminated surfaces of the **lithosphere,** the rocky crust that forms the earth's rigid plates and terrestrial habitats. In the gaseous **atmosphere,** organisms are found living at altitudes below 7 kilometers (4.3 miles).

There are a few exceptions to these general boundaries. For example, pollen, dormant bacteria, and fungal spores have been found in the atmosphere at altitudes above 62 kilometers (100 miles). In addition, gravity pulls dead organisms and their wastes downward in the hydrosphere, providing food for life in the darkened depths of the earth's waters. Thus, the biosphere is not uniform in width; rather, its thickness varies from one location to another.

Life as we know it is supported within a relatively thin envelope of the earth's air, water, and soil. This life-sustaining envelope is called the biosphere. (See CTQ #2.)

FIGURE 40-1
The earth from space. Portions of Africa, Madagascar, and Antarctica can be seen beneath the swirling cloud layer. Seemingly calm at this distance, the earth teems with millions of species of organisms.

ECOLOGY DEFINED

Ecology is the study of the biosphere and its components. Although the term is a familiar one, the word itself is somewhat new. It was coined a little more than 125 years ago by the German zoologist Ernst Haeckel to refer to the total relations between an animal and its organic and inorganic environment. Haeckel derived the term "oecology" from the Greek, *oikos,* meaning home, and *logos,* meaning the study of. At a national conference held in 1893, American scientists adopted a simpler spelling by dropping the letter "o."

Since the late 1800s, more restrictive definitions have been proposed. For example, in his 1927 book on *Animal Ecology,* Charles Elton defines ecology as "scientific natural history." Nearly 4 decades later, in 1963, in his book *Ecology,* Eugene Odum defined ecology as "the study of the structure and function of nature," a reminder that the unifying theme of form and function applies to ecology as well as to all other biological topics. Many biologists simply define ecology as the "study of ecosystems" because an ecosystem includes both living organisms and the inanimate physical environment.

These three definitions are all somewhat vague. A more exact definition is offered by Charles J. Krebs in his 1985 text, *Ecology.* Krebs wrote, "Ecology is the scientific study of the interactions that determine the distribution and abundance of organisms." Based on Krebs's definition, ecology studies

- where organisms are found,
- how many organisms occur there, and
- why organisms occur where they do.

The distribution and abundance of organisms are affected by the ways in which organisms interact with one another in the biotic community (e.g., competition) and with the surrounding physical abiotic environment (e.g., availability of light or nutrients). These are the topics we will explore in the final five chapters of this text.

ECOLOGY AND EVOLUTION

To study ecology is to study evolution because ecology and evolution are interlinked. Evolution takes place in the ecological arena. Organisms occur where they do because, through evolution, they have acquired adaptations that enable them to survive and reproduce in particular habitats. These adaptations are the result of natural selection, the process that drives evolution. Recall that natural selection is the result of a number of factors, including limited resources, competition for space or mates, adverse changes in the physical environment, the introduction of new organisms into an ecosystem, and so on. These "natural selection factors" are also ecological factors. In other words, natural selection is ecology in action.

No organism can ever be separated from its environment. As a result, organisms are continually affecting and, in turn, are continually being affected by their environment. This constant ecological interplay is the foundation for evolution. (See CTQ #3.)

THE EARTH'S CLIMATES

Climate—the prevailing weather in an area—is the chief physical factor in determining where organisms are distributed within the biosphere. Not only does climate differ greatly over the surface of the earth today, but global climates have changed dramatically over the almost 4 billion years since life originated on earth. These climatic changes have resulted in the formation of a great diversity of biomes as well as the evolution of millions of kinds of organisms with adaptations suited for these unique environments.

The earth's climates are shaped by major circulation patterns that develop in the earth's atmosphere and oceans. Air and ocean currents result from three primary factors: (1) the fact that different parts of the earth receive different amounts of incoming solar radiation; (2) the daily rotation of the earth and its annual orbit around the sun; and (3) the distribution and elevation of the earth's land masses.

Because the earth is round, different parts of the earth receive different amounts of solar radiation (Figure 40-2*a*). These differences in incoming radiation heat the earth unevenly, producing warm tropics near the equator, where sunlight hits the earth more directly, and progressively cooler regions in higher latitudes, as the intensity of sunlight is reduced. The north and south poles are the two coldest regions in the biosphere because they receive the least amount of sunlight, about five times less than the amount that reaches the equator. This variation in incoming sunlight not only helps produce different climates and biomes at different latitudes, it also affects the activity of organisms that are found there. This explains why tropical areas have much higher productivity than do biomes found in cooler, higher latitudes.

In addition to its round shape, the earth's annual orbit around the sun and its constant 23.5 degree tilt cause annual changes in incoming solar radiation to those parts of the earth farther away from the equator (Figure 40-2*b*). These annual differences trigger a progression of seasons away from the equator. During the summer months, for example, the Northern Hemisphere of the earth is tipped toward the sun (and therefore receives greater amounts of radiation), while during winter, it is tipped away from the sun. The differences in incoming radiation help explain why summer months are warmer and why organisms are more active during the summer than during the winter months. Because of the earth's constant tilt, the seasons are reversed in the Southern Hemisphere.

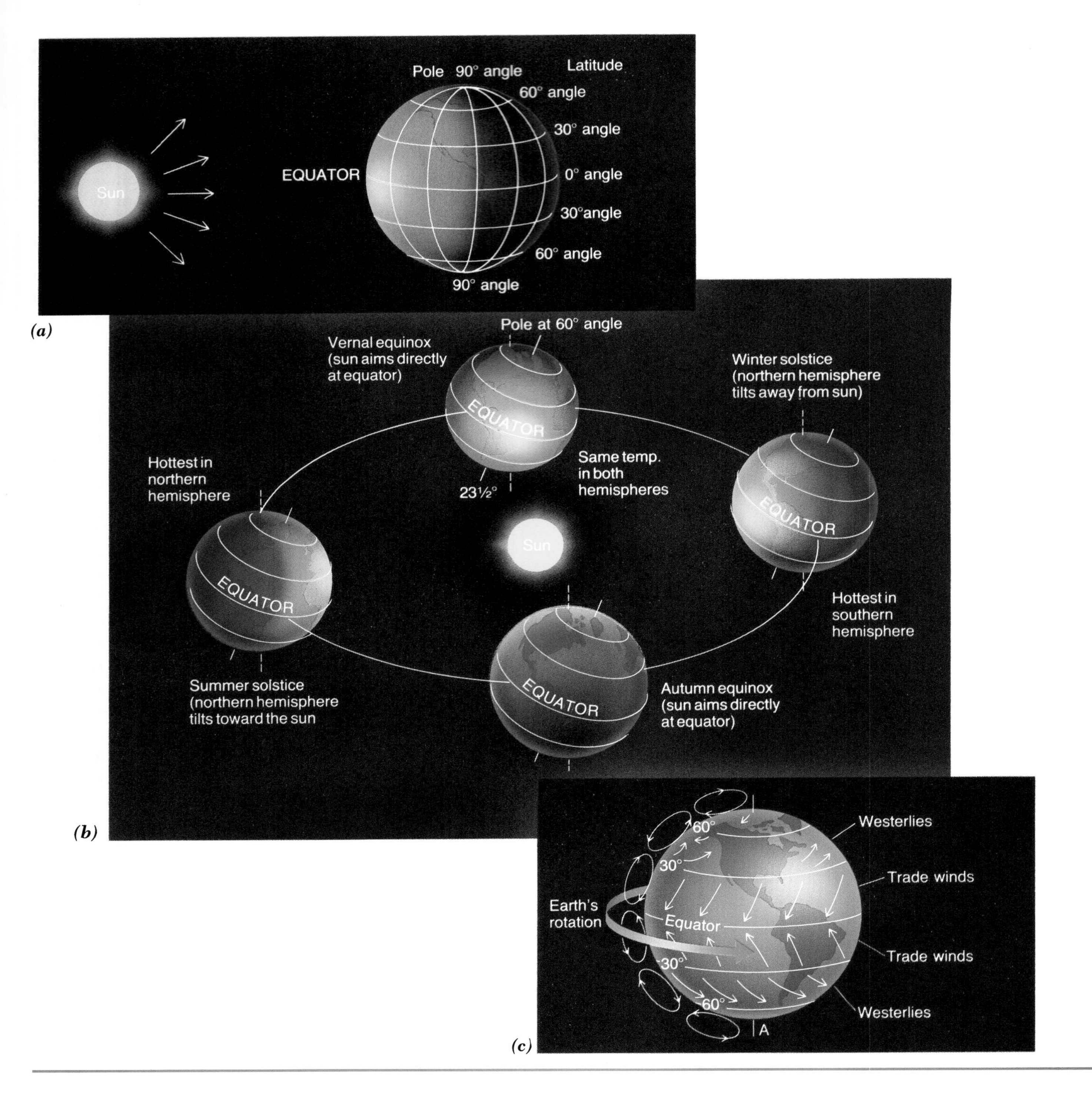

The variation in the amount of sunlight that strikes different parts of the earth at different times of the day and year heats the earth's air and oceans unevenly. Warm air near the equator rises and flows toward the poles, where cooler air sinks. Because of the earth's rotation, however, the moving air mass breaks into six circulating coils—three in the Northern Hemisphere, and three in the Southern Hemisphere (Figure 40-2*c*).

Intense sunlight at the equator causes the air to rise. As air rises, it cools, releasing its moisture and producing abundant rainfall. The cool poleward-moving air masses sink and become reheated at about 30° north and south latitude, creating zones of decreased rainfall. This is where the earth's largest desert regions are found. As the air in the coils moves over the surface of the rotating earth, prevailing winds are formed. *Trade winds* forms as air moves back toward the equator, and the *westerlies* form as air moves northward from 30° toward 60° latitude. At 60° north and south latitude, the air rises and cools, and, as at the equator, it releases its moisture as rainfall. This region of cool tem-

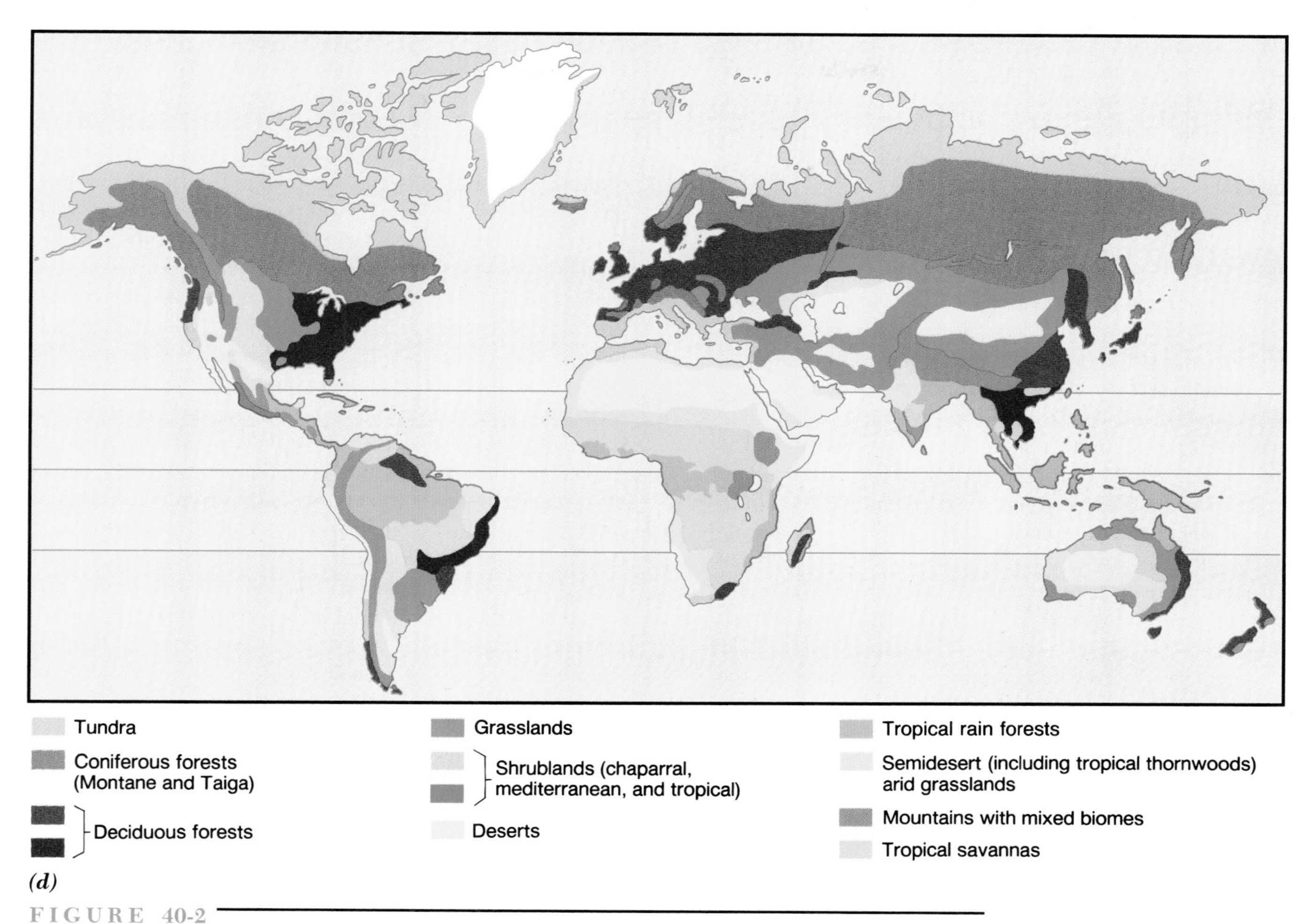

FIGURE 40-2

Climate and the earth's biomes. ***(a)*** The round shape of the earth results in differences in the amount of incoming sunlight in different regions. ***(b)*** The earth's yearly orbit around the sun and its constant tilt result in differences in the amount of incoming sunlight at different times of the year, generating the earth's seasons. ***(c)*** Six coils of air circulation create the earth's prevailing winds, as well as zones of high and low rainfall. ***(d)*** Climate is the principal factor in determining where plants grow, which, in turn, is the principal factor in determining where animals live. Not surprisingly, the locations of the earth's biomes closely follow it's 11 main climate types.

peratures and relatively high rainfall produces expansive temperate forests of North and South America, Europe, and Asia.

Prevailing winds blowing over the ocean surface create currents in the sea's upper layers. Continents deflect these water movements, creating slow, circular ocean currents. Circulation mixes the oceans' waters, bringing warmer waters that originated in tropical areas to higher latitudes, and the cooler waters from higher latitudes to tropical areas. These exchanges redistribute heat around the earth, contributing to the formation of the earth's climates (Figure 40-2*d*).

The earth's many climates are the result of the combined effects of air and water currents; the rotation, tilt, and orbit of the earth; and the distribution and elevation of the continents. Climate is the principal factor that affects where organisms are found within the biosphere. (See CTQ #4).

ARENAS OF LIFE

In Chapter 1, we mentioned that the biosphere is the highest level of biological organization (see Figure 1-4). The biosphere consists of many thousands of *ecosystems,* from conifer forests to estuaries to marshes to alpine meadows, each one a functional unit of biological and physical organization. The earth's major ecosystems are either aquatic or terrestrial.

AQUATIC ECOSYSTEMS

Nearly 75 percent of the earth's surface is covered with water. Most of this water (71 percent) is salty, forming the earth's marine habitats of seas and oceans. Other aquatic habitats include freshwater lakes, ponds, rivers, streams, and estuaries of mixed fresh and salt water. Aquatic organisms are affected by the chemistry of the water. In the open oceans, salinity (salt concentration) is relatively even and, as a result, does not limit the distribution and abundance of organisms that naturally occur there. In marine habitats, sunlight and mineral nutrients vary more than does salinity, creating environmental complexity that influences the community of marine organisms. In contrast, salinity levels vary greatly near the shorelines and in those areas where rivers flow into oceans. In these habitats, the salt concentration of the environment has the greatest effect on the organisms that live there.

Marine Habitats

Marine ecologists disagree as to whether distinct ecosystems form in open oceans or whether the ocean is simply one giant ecosystem that is constantly homogenized by fluctuating water currents. The former point of view is supported by studies that clearly reveal that there are specific locations in the oceans where nutrients circulate upward, producing a localized community that has more kinds of organisms, in greater numbers, than are found in adjacent areas. Each of these localized areas displays a unique biological order and regulation. Because of their unique physical and biological composition, some ecologists contend that these areas are distinct ocean ecosystems.

Other studies have established that distinct communities and features also exist in other ocean regions. In a paper written in 1988, however, Richard Barber argues that since these boundaries may easily disappear, perhaps marine communities should not be considered separate ecosystems. For instance, Barber points out that in 1983, at the peak of the El Niño, large-scale changes in temperature and ocean circulation obliterated boundaries. Since an El Niño recurs every 5 years or so, Barber suggests that the concept of ecosystems with well-defined, constant boundaries is more applicable to land than to the oceans. Of course, when the effects of an El Niño subside, marine ecosystem boundaries reappear.

Open Oceans Marine ecosystems are found throughout the earth's oceans, in shallow waters around continents, islands and reefs; in intertidal zones (the area between high and low tides); and in the **pelagic zone,** or open oceans (Figure 40-3). Biologists subdivide the vast pelagic zone into three, vertical layers: (1) the upper, sunlit **epipelagic** *(photic)* **zone;** (2) the dimly lit, intermediate **mesopelagic zone;** and (3) the continually dark, bottom **bathypelagic** *(aphotic)* **zone.** The sea floor is called the **benthic zone.**

In the open ocean two important physical factors—sunlight and nutrients—vary from location to location. Sunlight does not usually penetrate the ocean below 100 meters (320 feet). Considering that the average ocean depth is 3.9 kilometers (12,500 feet), the sunlit zone is indeed very thin, permitting photosynthesis in only the upper 2 percent of the ocean's volume. In the sunlit epipelagic zone, where nutrients are abundant, large populations of *phytoplankton* flourish. Phytoplankton are microscopic photosynthetic bacteria and algae that drift with the ocean currents and provide the energy and nutrients for the animal species that dwell in the open ocean. Most phytoplankton are eaten by *zooplankton,* tiny crustaceans (mostly copepods and shrimplike krill), larvae of invertebrates, and fish that are small enough to be swept along by ocean currents. Larger fishes and other animals feed on the tiny zooplankton or on both phytoplankton and zooplankton (or simply, plankton). An adult blue whale, for example, guzzles an average of 3 tons of plankton in a single day.

Since the dim midwaters of the mesopelagic zone do not receive enough light to power photosynthesis, phytoplankton do not reside there. Inhabitants of this zone must therefore make daily migrations either up to the epipelagic zone or down to the bathypelagic zone to feed. Large fishes, whales, and squid are the principal animal predators of the mesopelagic zone. They eat smaller fishes that feed on the wastes or carcasses of organisms from the sunlit zone.

The benthic zone contains no energy-capturing plants or bacteria, except for a few unusual deep-sea communities of organisms that inhabit the warm waters surrounding fissures and cracks in the ocean floor (see Bioline: Living on the Fringe of the Biosphere, page 167). This pitch-black zone is populated primarily by heterotrophic bacteria and scavengers that feed on a constant rain of organic debris, wastes, and corpses that settle to the bottom as well as predators that eat the scavengers and one another. These bottom dwellers include sponges, sea anemones, sea cucumbers, worms, sea stars, and crustaceans, as well as a collection of odd-looking fish, some with dangling lanterns that light up to attract a meal or a potential mate (Figure 40-4).

Coastal Waters The greatest concentration of marine life inhabits the shallow **coastal waters,** or **neritic** ("near shore") **zone,** along the edges of continents and reefs (Figure 40-5). In addition to abundant light, coastal waters are

◃ BIOETHICS ▹

Values in Ecology and Environmental Science: Neutrality or Advocacy?

By **ANN S. CAUSEY**
Prescott College

Ecology, along with all other branches of science, is usually characterized as a neutral endeavor. Ecologists gather and interpret data and then attempt to describe and explain relationships between organisms and their environments. In recent years, however, some have sought to challenge this picture of ecology as a descriptive, neutral science. Instead, ecology is portrayed as a subversive activity that seeks to impose a conservative (anti-growth) social doctrine on an unsuspecting public. These critics contend that ecology is not merely *descriptive*, but *prescriptive* as well. Ecologists, they say, are not merely gathering facts but are drawing value judgements from their interpretations of those facts and prescribing courses of action that often obstruct technological and economic progress.

Which portrayal is accurate? What is the proper role, if any, of values in ecology? And, how does environmental science enter into this dispute? To understand the source of this controversy, we need to take a closer look at science and values.

The strict separation of facts from values is a key component of the success of science in objectively analyzing reality. The objectivity of science is not a function of the objectivity of scientists however; it is a function of the rules of the game, of the scientific method itself. Granted, the findings of ecology may be particularly relevant to judgements of value made in many arenas of society. However, these findings are obtained by a method designed to separate objective analysis of nature from subjective value judgements; they are, in principle, value-neutral.

Ecologists, then, *in their science*, do not properly become advocates; that is, ecology is not necessarily a thinly disguised form of environmental advocacy, or *environmentalism*, as some have charged. Nevertheless, certain scientific findings, if highly corroborated, may indeed call for advocacy on particular issues. Thus began a movement, born in the 1960s and maturing through the past 2 decades, to apply ecological principles and theories not only to wild plants and animals in natural environments but to humans, as well.

This application of ecology to human society has become known as *environmental science*. Environmental scientists seek to understand the interactions of humans with other species and with the nonliving environment. Thus, these scientists must integrate knowledge from the natural sciences (ecology, physics, chemistry, biology, geology, etc.) with that from technology and the social sciences (demography, economics, politics, etc.). This interdisciplinary field seeks not only to understand how the various levels of life operate and interact but also to identify, integrate, and apply principles of sustainability to human society.

The blending of ecology and environmentalism to create environmental science has resulted in the charge that science is now value-laden. This characterization may be incorrect. While environmental science rests on ecology as its main scientific foundation, the values reflected by environmental science do not come to it by way of ecology. In other words, ecology can only tell us how nature works, not what humans should to. It may also help us accurately predict the consequences of certain human activities, though it cannot tell us whether or not we should curtail those activities since such decisions always involve weighing competing values and interests. For instance, we know from ecological studies that no population can reproduce uncontrollably. Ultimately, natural resistances, such as disease and starvation, bring a rapidly growing population back into balance with available resources. Environmental scientists assure us that this mechanism will ultimately check the growth of human populations. No science or scientists can determine whether we should now take steps to control human population growth, however, or whether we should simply wait and let nature take its course. That is a value judgement—an ethical decision.

Environmental scientists use information from the natural and social sciences to tell us how we can best minimize our impact on the natural systems that support humans and other species and how we can develop ecologically and economically sustainable societies. They work under the legitimate *assumption* that such goals are desirable. Their work may even make these goals more appealing and widespread, graphically showing us the undesirable consequences of current unsustainable practices such as deforestation or continued production of ozone-destroying chemicals. Nevertheless, science cannot tell us what objectives to strive for or what our goals should be. These are decisions that reflect larger societal values and standards, standards that are only partly shaped by our scientific understanding.

Do you consider ecology and environmental science advocacy? Some say no. They see ecology as appropriately responsive to environmental crises too urgent to ignore. Modern ecologists and environmental scientists can make a great contribution toward solving many environmental problems by helping us base our policies and judgements on reality, rather than on wish, conjecture, or the political agenda of any particular interest group. According to this view, values do not threaten the objectivity of scientists or the validity of their work. Instead, they help guide modern science in the service of human needs and wants, an increasingly important goal in these times of environmental ignorance and deterioration.

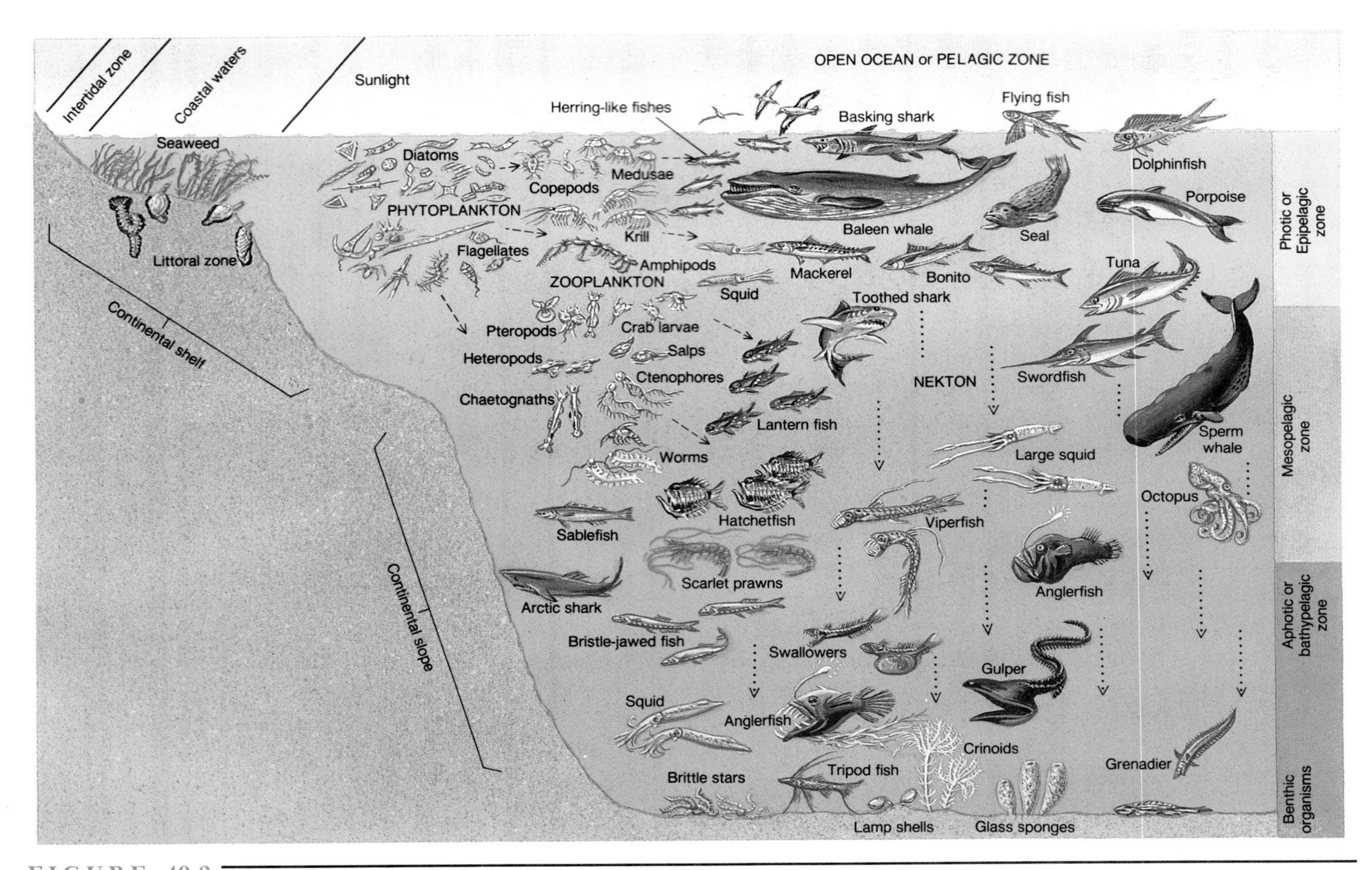

FIGURE 40-3
Marine habitats and representative marine organisms. (Sizes of organisms are not drawn to same scale.)

generally rich in nutrients that are available as a continuous drain from the surrounding land. Waves, winds, and tides constantly stir coastal waters, distributing the nutrients. Bathed in light and nutrients, photosynthetic organisms grow at fantastic rates and in great profusion, providing ample food and habitats for a multitude of fishes, arthropods, mollusks, worms, and mammals.

The richest coastal waters occur in regions of *upwelling*, where nutrient-laden water from below circulates to the surface. Upwellings form along the coasts of Peru, Portugal, Africa, and California and create conditions where light and nutrients are abundant. Under these conditions, it is not surprising that regions of upwellings form the earth's most fertile fishing waters.

Coastal waters present some unique problems for organisms, however. Rough, surging waters can shred or bash organisms against rocks. Many coastal animals have adaptations (both physical and behavioral) that help protect them from the surging waters. For instance, some animals remain in burrows or possess tough, protective shells. Since surging waters can quickly wash organisms onto the shore or pull them out to sea, a number of coastal organisms have adaptations for clinging to stationary objects. For example, algae use holdfasts to cling to rocks; mussels anchor themselves with powerful cords; and abalone hold tight with their large, muscular foot. Remaining attached to a solid object in churning waters is clearly advantageous, but it also creates problems. How do these sedentary organisms forage for food, find mates, or disperse offspring so that they can colonate new, suitable habitats?

A researcher named I. E. Effort studied sand crabs (*Emerita analoga*) along the Pacific Coast of North America and found, not surprisingly, that the main ocean currents greatly affected the crab's dispersal and, ultimately, its geographic distribution. Sedentary adult sand crabs release tiny larvae for dispersal. The larvae drift along with the plankton from current to current. When the California Current turns sharply westward near Baja, California, all larvae drift out to open sea and perish. As a result, the California Current determines the southern limit of the distribution of this

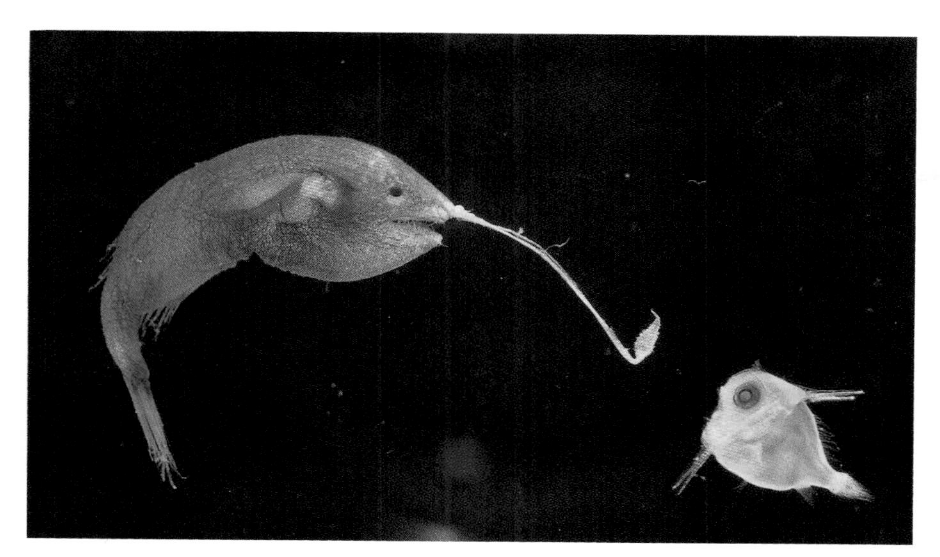

FIGURE 40-4
Generating its own light, the suspended lantern on this anglerfish attracts this larval fangtooth in the black of the ocean's aphotic zone. Backward-pointing teeth prevent even large fishes from escaping this predator's grasp.

sand crab. Likewise, the northern limit of distribution is affected by a north-flowing current that carries larvae from southern Oregon to Alaska. All larvae produced in Oregon can drift only further north to the Gulf of Alaska, where they all perish in the frigid water. Consequently, Oregon populations are colonized only by larvae from southern populations.

Shallow waters are also found along coral reefs. In these shallow, warm, tropical waters, exceptionally diverse communities flourish. For example, F. H. Talbot and his

FIGURE 40-5
The ocean's bounty. Abundant light and the constant circulation of nutrients make coastal waters the ocean's richest habitats.

colleagues identified nearly 800 species of fish around one small island on the southern edge of the Great Barrier Reef in Australia. At the northern edge of the Great Barrier Reef, over 1,500 species of fish were recorded. Depending on location, coral reefs are classified as one of three types: (1) *fringe reefs*, which extend out from the shores of islands and continents; (2) *barrier reefs*, which are separated from shores by channels or lagoons; or (3) *atolls*, islands of coral with a shallow lagoon in the center (Figure 40-6).

Coral reefs develop only in tropical areas (between 30° north and 30° south latitudes), where water temperatures never fall below 16°C (60°F). Persistent cold currents prevent coral reefs from developing along the western edges of continents.

The Intertidal Zone Unlike in coastal waters, where organisms are continually submerged in shallow waters, the organisms that inhabit **intertidal zones** must be able to survive in both water and air, as the flow of tides rhythmically submerge and expose their habitats. As its name suggests, the intertidal zone lies between high and low tides at the interface between the ocean and the land. Recurring tides and the geology of the shoreline create a variety of intertidal habitats: rocky shores that often contain *tide pools* (isolated pools that are left behind in rock depressions when the tide recedes) and *mud flats* or *sandy beaches* that form along nonrocky shorelines (Figure 40-7).

The pattern by which organisms inhabit the intertidal zone is a striking example of the dynamic interplay between physical factors (temperature and dehydration) and interactions between organisms (competition and predation). This interplay often produces four distinct strata of organisms in the intertidal zone (Figure 40-8). The uppermost stratum is the **splash zone,** where organisms receive only sprays of water at high tides. Below the splash zone is the **high intertidal zone,** which is followed by the **middle intertidal zone** and the **low intertidal zone.** Organisms in the high intertidal zone must be able to withstand exposure to air for longer periods than do those that inhabit the middle or low intertidal zones.

Estuaries

Estuaries form where rivers and streams empty into oceans, mixing fresh water with salt water (Figure 40-9). An organism's location ion an estuary depends on its ability to tolerate different concentrations of salt. Salinity changes,

(a)

(b)

(c)

FIGURE 40-6

Coral reefs. ***(a)*** A barrier reef surrounding the island of Taiatea in French Polynesia. ***(b)*** A fringe reef around Society Island, Tahiti. ***(c)*** A circular atoll: a ring of coral built around an island that later submerges, creating a central lagoon.

FIGURE 40-7

Part-time oceans: intertidal habitats. ***(a)*** *Tide pools* are reservoirs of seawater trapped in depressions during low tide. The pools are filled with flowerlike sea anemones, sea stars, crabs, small fishes, and a variety of seaweeds and other algae. ***(b)*** A *mud flat* off Point Reyes, California. ***(c)*** A *sandy beach* below Na Pali sea cliffs, Polihale State Park, Kauai, Hawaii. ***(d)*** *Rocky shores.* Large numbers of animals live in the moist tangle of brown seaweeds that drape the shore's rocks at Elkhead Cove in California.

however, as tides bring in new influxes of salty sea water or as storms increase the flow of fresh water. At any one spot in an estuary, salinity may change within minutes, from very low concentrations (equal to that of fresh water) to very high concentrations (the salinity of sea water), creating an ever-changing environment for the organisms that live there.

Despite such fluctuations in salinity, estuaries are tremendously fertile habitats for organisms; estuaries are continually enriched with nutrients from rivers and from the debris that is washed in by tides. Once nutrients enter the estuary, tidal action and the slow mixture of water prevent their escape. Plankton flourishes in the estuary's warm,

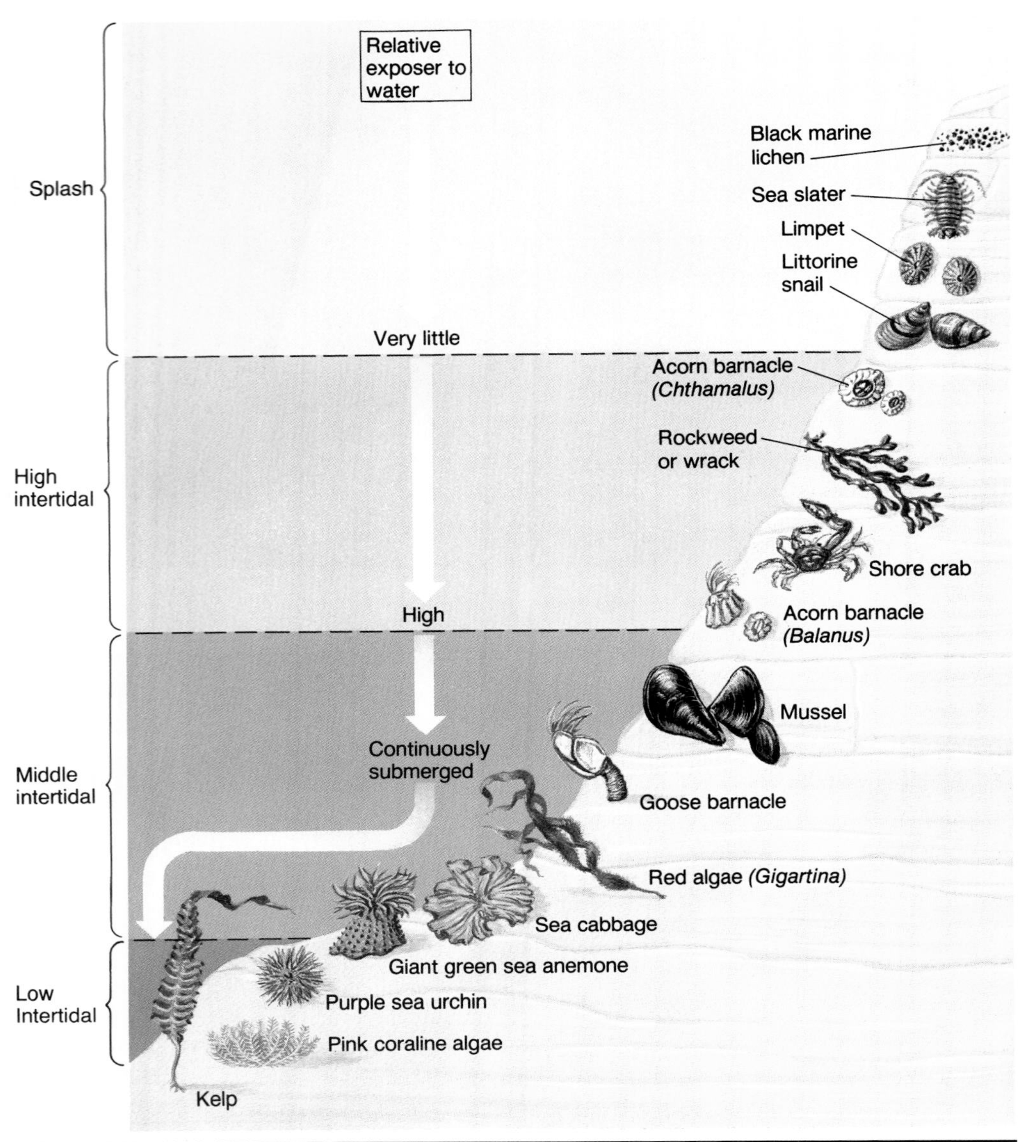

FIGURE 40-8

Intertidal zones. Organisms that live in the *splash zone* must survive with only periodic sprays of moisture, whereas organisms are submerged in water about 10 percent of the time in the *high intertidal zone,* 50 percent of the time in the *middle intertidal zone,* and 90 percent of the time in the *low intertidal zone.*

murky waters. Phytoplankton and rooted plants along the estuary's edges provide food for crustaceans, fishes, shellfishes, and for the young of many open ocean animals that use rich estuaries for spawning.

Freshwater Habitats

Oceans contain 97.2 percent of the earth's total water. The remaining 2.8 percent is fresh water: 2 percent is permanently frozen in ice caps and glaciers; 0.6 percent is groundwater; 0.017 percent is concentrated in lakes and rivers; and 0.001 percent is in the atmosphere as ice and water vapor. Since the quantity of water on earth is so enormous, even a small percentage like 0.017 percent produces more than 52,000 cubic miles of freshwater habitats on earth. The principal freshwater habitats are flowing rivers and streams and standing lakes and ponds.

Rivers and Streams Size determines whether fresh water flowing over the land forms a **stream** or a **river;** smaller streams converge into larger rivers. The constant flow of water erodes the land, sculpting dramatic landscapes and creating unique habitats that support particular groups of organisms.

FIGURE 40-9
Estuaries are shallow inlets, where freshwater mixes with salty seawater. Nearly one-half of the ocean's living matter are found in such nutrient-rich estuaries. This estuary is on Fraser Island in Australia.

The velocity of a current affects not only erosion but also the deposition of sediments and the supply of oxygen, carbon dioxide, and nutrients, factors that determine which organisms inhabit a stream or river and where organisms occur. Furthermore, a current can vary even in one spot. Friction along the edges and bottom slows water velocity, allowing algae to cling to rocky surfaces, plants to take root, and animals to reside safely without being swept away. Calmer water is also found behind rocks, pebbles, and mounds, creating stable habitats for smaller organisms.

In a river's open waters, the never-ending tug of the current presents unique challenges for animals. Some fishes use their fins to force themselves down to the less turbulent bottom waters. Conversely, powerful trout and salmon can swim fast enough to oppose the current or even to swim upstream.

Lakes and Ponds **Lakes** and **ponds** are standing bodies of water that form in depressions of the earth's crust. As in the case of streams and rivers, the difference between a lake and a pond depends on size: Lakes are larger and usually deeper than ponds.

A lake typically has three zones: a shallow **littoral zone** along the water's edge, a **limnetic zone** that encompasses the lake's open, lighted waters, and a dark **profundal zone** at depths below which light is unable to penetrate. Rooted plants and floating algae are the characteristic photosynthesizers of the littoral zone. In the limnetic zone, phytoplankton and photosynthetic plants provide food for zooplankton, fishes, and other animals. And, as in the deep, dark waters of oceans, heterotrophic predators and scavengers inhabit the lake's profundal zone. Since oxygen often becomes scarce near the bottom of a lake, anaerobic bacteria and other species that can survive anaerobically are also abundant in the profundal zone. For example, some midge larvae that live in lakes have specialized adaptations for withstanding long periods of anaerobic conditions. This is not true for *all* midge larvae, however; those that inhabit streams die quickly without oxygen. Thus, stream and lake midges have evolved along two adaptive lines, and neither species is capable of invading the other's habitat.

In many lakes, bursts of biological activity occur in the spring and autumn, when the lake's waters naturally circulate (Figure 40-10), brining a supply of rich nutrients to the surface. In the spring, the sun warms the upper waters, and the winds stir the lake, driving heat and oxygen down into the deep waters and circulating nutrients up to the surface. This *spring overturn* stimulates a sudden increase in bio-

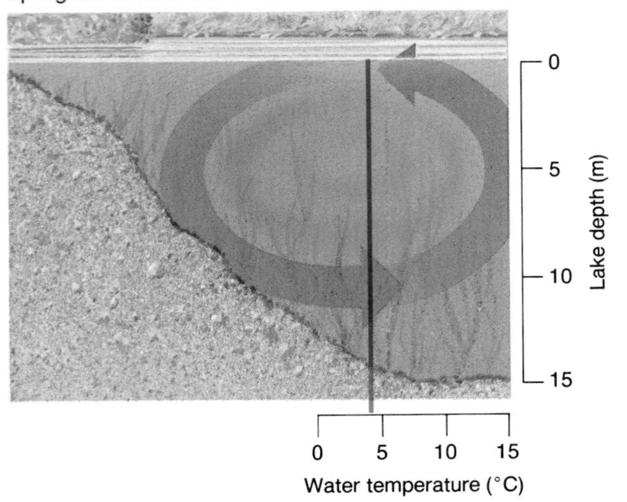

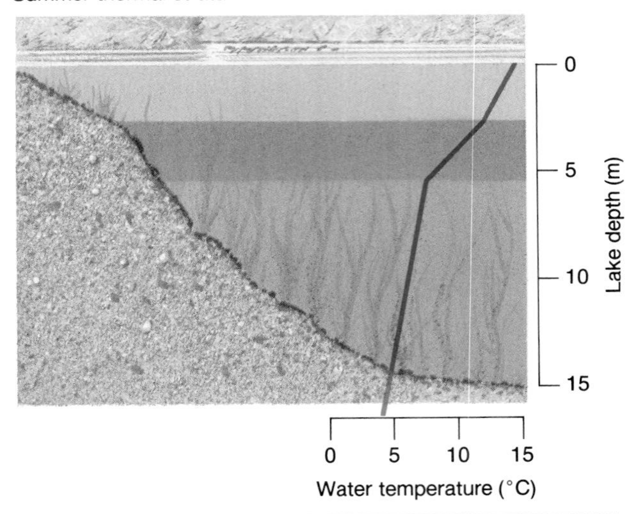

FIGURE 40-10
Seasonal changes in a temperate lake. Water temperature is nearly uniform during spring and autumn overturns. In the summer, warmer surface waters resist mixing, forming distinct temperature zones.

logical activity. As spring passes into summer, heated surface waters become more buoyant and resist mixing. Water circulation stops altogether, producing three temperature layers: a layer of warm surface water; a relatively thin, middle layer called a *thermocline,* in which water temperature declines rapidly; and a bottom layer of cold, relatively still water. Without mixing, oxygen is quickly depleted on the lake's bottom, and nutrients begin to accumulate, slowing down biological activity in the lake throughout the summer.

In the autumn, the lake's thermal layers are disrupted, as heat is dissipated from surface waters to the cold air. As the colder surface waters begin to sink and the autumn winds increase, circulation in the lake is restored, producing a more uniform range of temperature, nutrients, and oxygen. Rising warmer water from the bottom brings nutrients upward and warms the surface waters, producing another burst of biological activity during this *autumn overturn.*

Unlike rivers and streams, lakes and ponds may eventually fill in with sediment and organic matter and become dry land (Figure 40-11). This process may take months, or it may take thousands of years, depending on the size of the lake or pond, the amount of biological activity, and the rate of filling and draining. Thus, biologists speak of a lake's "life span" or a lake's "natural aging process."

A recently formed young lake is termed *oligotrophic* (little nourished) because it contains relatively few nutrients. Oligotrophic lakes support very little life; as a result, they are usually crystal clear. Middle-aged lakes are *mesotrophic* (moderately nourished). Unlike oligotrophic lakes, mesotrophic lakes support large populations of organisms. Finally, the nutrient-rich waters of old, *eutrophic* (fully nourished) lakes are rich in nutrients and support the largest populations and the greatest diversity of species. As the organisms of eutrophic lakes continually add increasing supplies of organic matter to the water, the filling rate is accelerated. Outside sources of nutrients, such as human sewage or runoff from nutrient-soaked agricultural lands, also promote lake filling.

BIOMES: PATTERNS OF LIFE ON LAND

Large terrestrial (land) ecosystems are called **biomes.** Plants from the bulk of the living mass in a biome; as a result, each biome is characterized by the predominant type of plant that grows there. Dense, tall trees form forest biomes; short, woody plants form shrublands; and grasses and herbs form grasslands. Since the prevailing climate (especially temperature and moisture) is the primary factor in determining the types of plants that grow in an area, the earth's terrestrial biomes tend to follow global climate patterns (Figure 40-2). Climate also changes with elevation, producing successive layers of biomes on a single mountainside (Figure 40-12). Climate affects terrestrial biomes in another, unexpected way: Prevailing climate patterns are spreading pollutants from human activities over very wide distances, causing disruptions in biomes several hundred and thousands of miles away (see The Human Perspective: Acid Rain and Acid Snow: Global Consequences of Industrial Pollution).

(a)

(b)

(c)

FIGURE 40-11

A lake's life cycle. ***(a)*** Newly formed *oligotrophic* lakes are clear. Oligotrophic lakes eventually begin to fill with sediment, providing nutrients for greater numbers of organisms. ***(b)*** As sediment and life increases, oligotrophic lakes become *mesotrophic* lakes, ***(c)*** which then become *eutrophic* lakes.

Studies reveal a pattern in species diversity in terrestrial biomes. The largest number of species are supported in tropical habitats; progressively fewer species are found in temperate and polar areas. For example, in a Malaysian rain forest, biologists counted 227 species of trees in an area of 5 acres (2 hectares), whereas a deciduous forest in Michigan contained only 10 to 15 tree species in an area of the same size. In tropical Mexico, 293 species of snakes were found, compared to 126 in the United States, and only 22 in Canada. Such "global gradients" are found for all organisms, from plants to invertebrates to mammals. In one study, over 150 species of mammals were recorded in Central America, compared to only 15 species in northern Canada.

Forests

Over 30 percent of the earth's land surfaces are covered by forests, or dense patches of trees. Biologists recognize three main forest biomes: (1) lush **tropical rain forests** that grow in a broad belt around the equator; (2) **deciduous forests,** whose trees drop their leaves during unfavorable seasons; and (3) **coniferous forests** that are dominated by evergreen conifers.

Tropical Rain Forests Warm temperature, abundant rainfall (over 250 millimeters, or 100 inches, a year), and roughly constant daylength throughout the year produce rich, tropical rain forests in Central and South America, Africa, India, Asia, and Australia (Figure 40-2). Over half of the earth's forests are tropical rain forests. They contain more species of plants and animals than do all other biomes combined. One hectare (10,000 square meters, or 2.5 acres) of tropical rain forest may contain more than 100 species of trees, 300 species of orchids, and thousands of species of animals, most of them insects.

Trees towering over 50 meters (160 feet) form the upper story *(overstory)* of the tropical rain forest (Figure 40-13). Below these giants is a layer of tall trees (the *understory*) so densely packed that very little sunlight penetrates through their canopy. Since much of the light is blocked, plant growth is greatly suppressed below these two layers.

◁ THE HUMAN PERSPECTIVE ▷

Acid Rain and Acid Snow: Global Consequences of Industrial Pollution

Two alarming trends have been recently documented in the biomes of North America and Europe. The first trend is the change in the color of several lakes, from murky green to crystal clear. Sounds good, right? Not really, for a green lake is a biologically active lake, teaming with microscopic algae, which are eaten by small aquatic animals, which, in turn, are eaten by fish. A clear lake is biologically sterile, devoid of aquatic life. How widespread is lake sterility? In eastern Canada, nearly 100 lakes have become sterile, as have more than 1,000 lakes in the northeastern United States and approximately 20,000 lakes in Sweden.

The second trend is the premature death of an excessive number of trees, especially those found on high slopes that face prevailing winds. Huge patches of dead trees, totaling more than 17 million acres (7 million hectares) of trees in North America and Europe, look as if they've been burned, but there have been no fires. More than 50 percent of the forests in Germany alone are affected in this way, impacting more than 1.2 million acres (500,000 hectares).

Just how are dead trees and dead lakes related? The destruction of both trees and lakes is caused by acid deposition from **acid rain** or runoff from **acid snow.** Acid deposition not only destroys forests and kills lakes, it also damages crops, alters soil fertility, and erodes statues and buildings.

The chemicals that create acid rain and snow (sulfur oxides and nitrogen oxides) come primarily from human activities. Although sulfur oxides are released during volcanic eruptions, forest fires, and from bacterial decay, quantities of sulfur oxides from human activities far exceed those that come from natural sources. Nearly 70 percent of sulfur oxides comes from electrical generating plants, most of which burn coal. Most nitrogen oxides come from motor vehicles and industries, including electrical generation. When sulfur and nitrogen oxides mix with the water in the air, they form acids:

$$SO_2 + H_2O \rightarrow H_2SO_4 \text{ (sulfuric acid)}$$
$$NO_2 + H_2O \rightarrow HNO_3^- \text{ (nitric acid)}$$

Acid rain or acid snow has a pH below 5.7, the pH of unpolluted rain. Over the past 25 years, rains in the northeastern United States maintained an average pH of 4.0. The lowest recorded pH for rainfall was 2.0, reported in Wheeling, West Virginia. The rain in Wheeling was more acidic than lemon juice!

Acids that create acid rain and snow remain airborne for up to 5 days, during which time they can travel over great distances. For example, the acid rainfall that killed many of the lakes in Sweden was caused by pollutants that were released in England. The acid rain that is damaging trees and lakes in the Adirondack Mountains of New York originated in the upper Mississippi and Ohio River Valleys. Since these acids circulate in large air masses, acid deposition is widespread, spreading from Japan to Alaska, from New Jersey to Canada, to name just a few places.

The rate of destruction caused by acid rain and snow is increasing. A 1988 survey of U.S. lakes lists 1,700 lakes as having high acidity. Another 14,000 lakes were identified as becoming acidified. Scientists estimate that by the turn of the century, over half of the 48,000 lakes in Quebec, Canada, will have been destroyed.

In 1979, the U.S. Congress passed the "Acid Precipitation Act" to *identify* sources of acid deposition, but so far very little has been done to curtail the release of these pollutants. Congress is considering a plan to cut sulfur oxide emissions by nearly 50 percent, and nitrogen oxides by 10 percent by the year 2000. Most other industrialized nations have already taken steps to curb acid deposition; the United States remains of one of a few that has not. Such steps might include: (1) installing scrubbers on power plant smoke stacks; (2) using coal that is low in sulfur; (3) using coal that has been pretreated to remove sulfur; or (4) reducing auto and truck use.

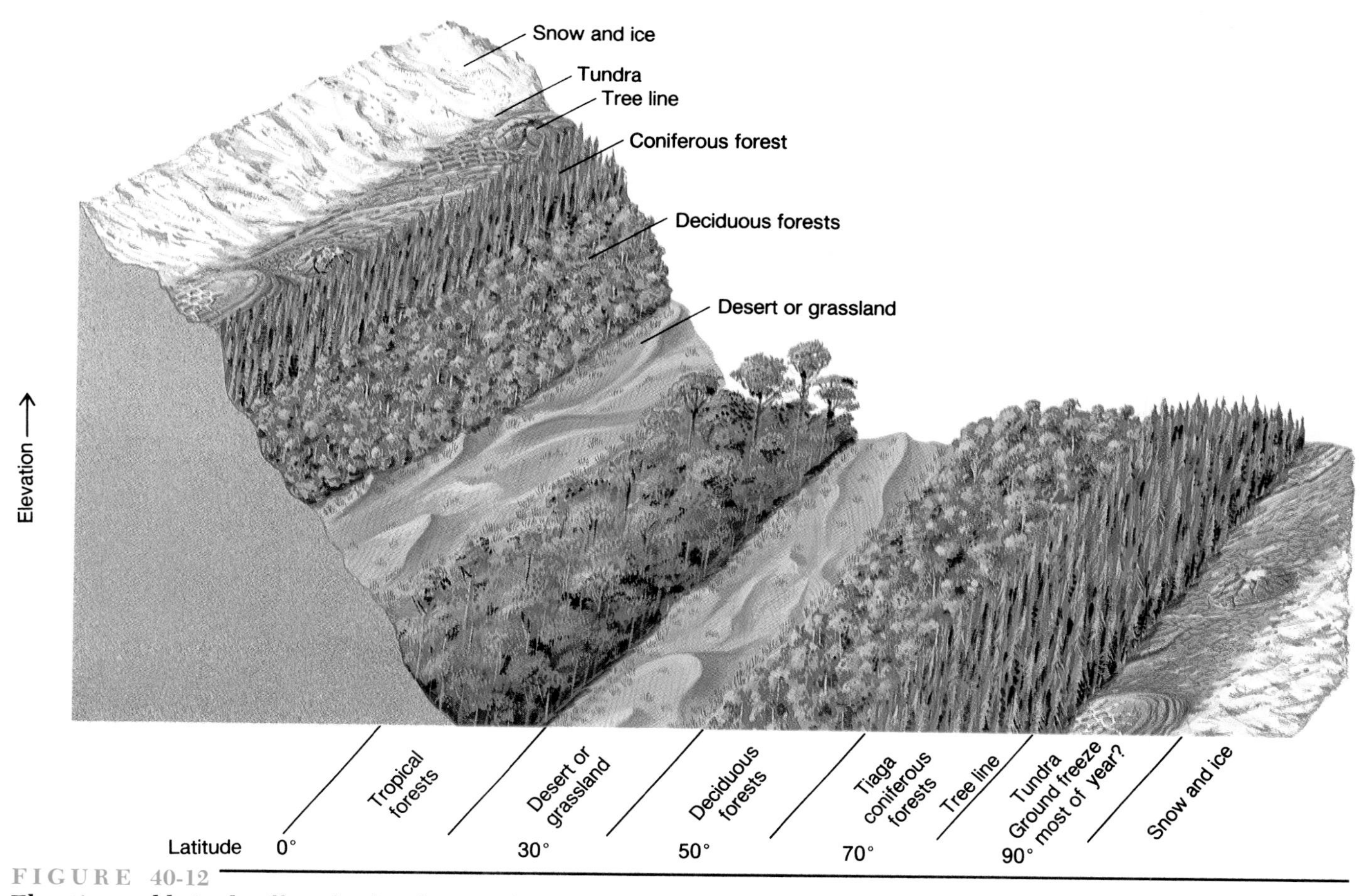

FIGURE 40-12
Elevation and latitude affect the distribution of biomes. Terrestrial biomes change according to elevation, as well as with distance from the equator.

Where a glimmer of sunlight does penetrate, ferns, shrubs, and mosses crowd the forest floor.

Of all the biomes, the rate of decomposition is fastest in the tropical rain forest. A dead animal or fallen tree can swiftly be cleared from the forest floor by hordes of fungi and bacteria that promptly carry out decomposition. Despite such rapid decomposition, virtually no nutrients accumulate in the soil; the nutrients are either absorbed immediately by plants or washed away by steady rains. Surprisingly, the earth's lushest forests have poor soils, a condition that has contributed to the rapid destruction of the earth's tropical rain forests by humans (see Chapter 21, The Human Perspective, page 428.).

Deciduous Forests In those areas of the earth that have distinct seasons, biological activity follows a seasonal pattern (Figure 40-14). Unfavorable growing conditions during one or more seasons during the year produce deciduous forests with trees, such as maple, beech, hickory, and oak, that produce leaves during warm, wet periods and lose their leaves at the onset of the dry season (in tropical deciduous forests) or the cold season (in temperate deciduous forests).

The trees found in deciduous forests are less dense than are those found in tropical rain forests, enabling sunlight to reach the forest floor. In the presence of adequate light, many layers of plants develop beneath the top layer of deciduous trees. (In both aquatic and terrestrial biomes, vertical layering of plants is associated with a decrease in light.) Plants growing below the upper layer receive varying amounts of sunlight throughout the year, as overstory deciduous trees gradually produce leaf buds, bear fully expanded leaves, and then drop their leaves at the beginning of the unfavorable season. The growth and reproductive cycles of understory plants coincide with brief periods of maximum sunlight and favorable conditions.

In one study, E. Lucy Braun of the University of Cincinnati sampled the deciduous forests of North America and found that tree diversity (the number of different kinds of trees) decreases moving north to colder regions or further west to drier regions. Within a localized area, however,

FIGURE 40-13
A picture of unrivaled diversity. Life in the tropical rain forest is more varied and more abundant than is that found in any other biome on earth. Immense, broad-leafed evergreen trees drip moisture onto masses of clinging epiphytes and enormously long vines, some over 200 meters long. These giant trees have widespreading buttresses that keep them upright, even in shallow tropical soil.

tree diversity was found to be primarily affected by soil moisture and soil calcium levels; drier sites had fewer kinds of trees, as did areas with less calcium.

Coniferous Forests Some of the most extensive forests in the world are populated by evergreen conifers (cone-bearing trees), such as pines, firs, and spruces (Figure 40-15). Belts of coniferous forests girdle the huge continental land masses in the Northern Hemisphere (Figure 40-2) and blanket higher elevations of mountain ranges in North, South, and Central America, and Europe.

The transition from deciduous forests to coniferous forests is a result of colder winters. During the spring and summer, melting snow fills the lakes, forming watery bogs and marshes within many coniferous forests. This is why the coniferous forests of northern latitudes are called **taiga,** which is Russian for "swamp forest." The growing season is relatively short; overall biological activity is restricted to a period of only 3 to 4 months. As a result, decomposition by fungi and bacteria is limited to these relatively brief warm periods. Consequently, the forest floor accumulates a thick layer of needles, producing acidic and relatively infertile soils. The mean temperature is below 0°C (32°F) for more than 6 months of the year.

Coniferous forests typically have two stories of plants: a dense overstory of trees and an understory of shrubs, ferns, and mosses. Unlike the tropical rain forests, coniferous forests are usually not made up of a mixture of many tree species but of expanses of many individuals of a few species.

Tundra

The timberline is a zone in which trees thin and eventually disappear; it marks the boundary between the coniferous forest and the tundra biome (Figures 40-2 and 40-12). In Russian, **tundra** means a marshy, unforested area, describing the frigid, treeless landscape. There are two types of

(a) (b)

FIGURE 40-14

The yearly rhythm of a deciduous forest. Deciduous trees in temperate areas produce leaves in the spring, when temperatures are warm. During the summer, the overstory trees have fully expanded leaves that shade the plants below. ***(a)*** A shortened photoperiod and the cool temperatures of autumn trigger the breakdown of chlorophyll, revealing pigments that turn leaves brilliant shades of orange, red, and purple. ***(b)*** Protected terminal buds enable dormant deciduous trees to withstand the bite of freezing winds and snows during the winter.

FIGURE 40-15

A coniferous forest in the icy grip of winter. Many animals hibernate or migrate to less severe habitats. Only a few, like this moose, remain active during harsh winter months, forced to travel over large areas in search of scarce food.

FIGURE 40-16

Life is fleeting in the tundra. During the brief summer growing period, ground-hugging tundra plants erupt with new growth, produce flowers, and set seeds, before severe weather returns. In the arctic tundra shown here, permafrost prevents drainage of water from melting snow, producing multitudes of shallow ponds.

tundra: *alpine tundra,* which are found at high elevations of mountain ranges; and *arctic tundra,* which are found to the north at high latitudes in Alaska, Canada, and in northern Europe and Asia (Figure 40-16). Both types look similar, featuring low-growing plants (often only 10 centimeters, or about 4 inches tall) that often belong to the same species. Tundra plants include mosses, lichens, perennial forbs, grasses, sedges, and dwarf shrubs.

The climate that produces a tundra is brutal. The growing season lasts only 50 to 90 days per year, with an average temperature in the warmest month of less than 10°C (50°F). Thick snow covers the ground, and icy winds blow often during winter. During the brief summer period, melting snows create frigid marshes and ponds. In the arctic tundra, only the top 0.5 meters (1.5 feet) thaw, leaving permanently frozen soil, or *permafrost,* that halts root growth, restricts drainage, and impairs decomposition. Under such harsh conditions, it is not surprising that relatively few plants and animals have evolved adaptations to help them survive the rigorous tundra climate. In fact, there are only about 600 plant species in the entire arctic tundra region of North America, which covers thousands of square miles. This is a smaller number than you would find in a single square mile of tropical rain forest.

Grasslands

Not too long ago, over 30 percent of the earth's lands were covered by **grasslands** of densely packed grasses and herbaceous plants (Figure 40-2 and 40-17). In North America, grasslands were once more widespread than any other biome. But the combination of rich soil and favorable growing (and living) conditions made grassland habitats prime

FIGURE 40-17
Grassland in New South Wales, Australia.

targets for agriculture, livestock grazing, and urbanization. Today, most of the earth's grasslands have been cleared for farming and human development or have been damaged as a result of overgrazing.

Grasslands naturally develop in regions with cold winters, hot summers, and seasonal rainfall (more annual rainfall than in the desert, but not enough to support a forest) and are found worldwide, in places like South Africa, Australia, South America, and the former Soviet Union. Natural fires periodically clear grasslands, opening up space and releasing nutrients for new growth. As in other terrestrial biomes, precipitation patterns greatly affect the nature of the grassland. For example, as precipitation decreases from east to west, *tallgrass prairies,* with plants reaching 2 meters (6.5 feet) in height, gradually give way to *mixed grass prairies,* with grasses growing no more than 1 meter (3.28 feet) tall, which, in turn, give way to *shortgrass prairies,* with bunch grasses less than 0.5 meter (1 foot) in height.

In 1979, a team of researchers from Colorado State University, J. A. Scott, N. R. French, and J. W. Leetham, analyzed herbivory (animals eating plants) in the three types of prairies found in the western United States. The researchers discovered that only a small fraction (less than 15 percent) of the plants were consumed by animals in shortgrass prairies. Herbivore consumption was higher in the tallgrass prairie (about 40 percent) and even higher in mixed grass prairies (50 percent). Since they measured herbivory both above and below ground, the researchers uncovered something very significant and distinctive about the grassland biome: Between 80 and 90 percent of the herbivory occurred underground, mainly by nematodes (page 873.). These results helped confirm the hypothesis that overall growth of grassland plants is primarily limited by root consumption, followed by soil water and competition among the plants for nutrients and light.

Savannas

A combination of grassland and scattered or clumped trees forms a **savanna** biome. *Tropical savannas* are found in South America, Africa, Southeast Asia, and Australia and cover nearly 8 percent of the earth's land (Figure 40-18). In many temperate areas, pockets of savannas are found sandwiched between grasslands and forests.

Like pure grasslands, savannas are characterized by seasonal rainfall, punctuated with a dry season. During the dry season, the above-ground stems of the grasses and herbs die, providing fuel for fast-moving surface fires. The grasses and herbs recover quickly from fires by resprouting from underground roots and stems. If they are killed, the plants are replaced by fast-growing seedlings. Since ground fires

FIGURE 40-18
The African savanna has flat-topped acacias and dry grassland. Some tropical savannas in central Africa are studded with palm trees.

move rapidly through the savanna, the trees are not usually damaged.

The tropical savannas of Africa support large populations of herbivores, including wildebeest, gazelles, impalas, zebras, and giraffes. Like many grasslands, tropical savannas are now being used for grazing. Overgrazing reduces grass cover and allows trees to invade the area, reducing the number of animals that can inhabit overgrazed savannas.

Shrublands

Woody shrubs predominate in **snrublands.** In regions of the earth with a Mediterranean-type climate (hot, dry summers and cool, wet winters), shrubs grow very close together. The shrubs typically have small, leathery leaves with few stomates and thick cuticles to retard water vapor loss (Chapter 18). Remarkably, similar shrublands develop in all areas that have warm, dry summers and cool, wet winters, from areas around the Mediterranean Sea, to coastal mountains in California and Chile, to the tip of South Africa, to southwestern Australia. In California, this type of shrubland is called a **chaparral** (Figure 40-19).

Organisms that live in Mediterranean-type shrublands must survive many stressful periods. For example, over the long, hot summers, when water is needed most by organisms, there is little or no rainfall. Since the supply and demand for water are directly out of sync, most plant and animal activity is restricted to the spring, when temperatures are warm and the soil is still moist from winter rains.

Fires are common in Mediterranean-type shrublands. The accumulation of dry, woody stems and highly flammable litter greatly increases the chances of fire during the hot summer. Most plants have evolved adaptations that allow them to cope with recurring fires in these environments, however. In the chaparral of California, for example, a burned area often recovers within just a few years. Even 1 month after an intense fire, many chaparral plants resprout from protected, underground stems, or their seeds germinate, stimulated by the fire itself.

In tropical regions that have a short wet season, another type of shrubland develops: the **tropical thornwood.** Most thornwood plants lose their small leaves during the dry season, reducing transpiration and exposing sharp thorns that discourage even the hungriest browser. One type of thornwood, the *Acacia* plant, has coevolved with certain ants, forming a close partnership that helps both species survive in these habitats. The plant provides food and shelter for the ants, and the ants patrol the ground and stems of the plant, aggressively warding off hungry herbivores and removing flammable debris that may accumulate under the plant (Figure 40-20).

FIGURE 40-19

A chaparral is a shrubland that forms in those regions of California that have a Mediterranean-type of climate. In addition to being tolerant to drought, most chaparral plants are adapted to fire and are able to regrow rapidly after a fire. Layers of charcoal testify that fire is a natural component of the chaparral, and of forests, grasslands, savannas, and other shrublands.

FIGURE 40-20

Mutual aid. Swollen thorns of this acacia tree provide a home and brooding place for ants, while nectar and nutrient-rich plant structures produced at the tips of leaves provide the ants with a balanced diet. Ants protect the acacia by attacking, stinging, or biting herbivores (or scientists). Ants also remove flammable debris from around the base of the tree, forming a natural fire break. Any plant seedling that grows within this "bare zone" is quickly clipped and discarded, protecting the acacia from competing plants, including its own offspring.

Deserts

Intense solar radiation, lashing winds, and little moisture (annual rainfall of less than 25 centimeters or 10 inches) create some of the harshest living conditions in the biosphere. These conditions characterize the **deserts,** which cover about 30 percent of the earth's land and occur mainly near 30° north and south latitude, where global air currents create belts of descending dry air. Some deserts are produced in the rainshadows of high mountain ranges. A **rainshadow** is a reduction in rainfall on the leeward slopes of mountains, slopes that face away from incoming storms. The skies over the deserts are generally cloudless. The sun quickly heats the desert by day, producing the highest air temperatures in the biosphere, the record being 57.8°C (138°F) in Death Valley, California. High daytime temperatures and persistent winds accelerate water evaporation and transpiration of water vapor from plants. High evapotranspiration and low annual rainfall characterize all deserts, producing sparse perennial vegetation of widely spaced shrubs (Figure 40-21*a*). Despite such harsh living conditions, only portions of the Sahara Desert in Africa and the Gobi Desert of Asia are totally devoid of organisms, an affirmation of the tenacity of life and the power of evolution to produce a great diversity of plants and animals with adaptations that enable them to survive the desert's severe climate.

Plants have evolved many adaptations for surviving the rigors of the desert (see Bioline: Survival in the Desert, Chapter 19). For example, following brief spring and summer rains, the desert floor often becomes carpeted with masses of small, colorful annual plants (Figure 40-21*b*). These annuals germinate, grow, flower, and release seeds, all within the brief period when water is available and temperatures are warm. By remaining dormant as seeds the remainder of the year, annuals avoid the most severe stresses of the desert.

The evolution of succulent tissues with cells devoted to storing water allows cacti and other succulents to escape dehydration by accumulating water in specialized roots, stems, or leaves. During dry periods, succulents frugally tap their internal water reserves. Other adaptations of desert plants include long tap roots that siphon deep groundwater supplies. Other plants simply endure extreme heat and dryness as a result of adaptations that enable them to survive various degrees of dehydration.

Many desert animals rely on learned and instinctive behavior to avoid the desert heat and dryness. Some, such as kangaroo rats and ground squirrels, remain in cool, humid underground burrows during the day and search for food at night or in the early morning or late afternoon, when temperatures are lower. If these animals venture out during the day, their underground burrows act as heat sinks, quickly

(a) *(b)*

FIGURE 40-21

The desert. ***(a)*** During dry periods, a few hardy perennial bushes, cacti, and Joshua trees are scattered over the scorching desert floor. Many animals remain underground, in moist, cool burrows. ***(b)*** Following a spring rain, the desert floor blooms with colorful annual growth.

removing the heat that they acquired while scurrying across the desert in search of food. Many birds avoid stressful desert conditions by simply flying to less hostile areas, while other animals *aestivate;* that is, they sleep through the driest part of the year. With watertight skins, desert snakes and lizards are ale to conserve water even during the heat of the day. The desert toad *(Bufo punctatus)* uses a survival strategy similar to that employed by succulent plants. It stores water in its urinary bladder, carrying its own version of an internal well.

Larger animals, like humans, can survive extreme desert heat by sweating. Body heat causes the water in sweat to evaporate, cooling the sweating person. On a very hot day, a human sweats about 1 liter of water per hour, helping to maintain body temperature within a tolerable range. In contrast, camels do not sweat, and, contrary to popular opinion, they do not store water in their humps. With a body size almost five times that of a human, camels absorb, and therefore must dissipate, more heat than a human. How do they do it, if they neither sweat nor store water? In 1964, a well-respected animal physiologist, K. Schmidt-Nielsen, discovered the answer: First, camels are able to tolerate a wide variation in body temperature, from 35°C (97°F) at night to 40°C (104°F) during the day. Excess heat from the day is stored in the body and then dissipated at night. Second, the camel's dense fur reduces evaporation and cuts down the flow of heat from the hot environment to the skin. Schmidt-Nielsen proved this second point by shearing all the fur off a camel; the animal's evaporation of water increased 50 percent.

There is greater diversity of plants and animals (numbers of species) in land habitats than in aquatic ones, even though water is necessary for all life and is clearly more available in aquatic habitats. Greater numbers of organisms live in aquatic habitats than in terrestrial ones, however. (See CTQ #5.)

REEXAMINING THE THEMES

Relationship between Form and Function

Plants establish the basic structure and dynamics of terrestrial ecosystems or biomes. The growth forms of plants (height, leaf size and shape, and so on) create the vertical and horizontal structure in which all other organisms live. In addition, the rhythm of plant activity affects the functions of all organisms, even other plants. Plant structure and function also modifies the nonliving environment, affecting such factors as light penetration, nutrient cycling, water availability and retention, temperature, and humidity.

Biological Order, Regulation, and Homeostasis

Ecosystems and biomes are levels of biological order that comprise a community of organisms and their physical environment. Both components are permanently interlinked; any change in one will have an affect on the other. A reduction in available water, for instance, will reduce the number of individuals in the community, bringing the size of the community into balance with water availability. Similarly, changes in a community of organisms can alter the physical environment. For example, habitat destruction by humans in tropical rain forest ecosystems is changing the world weather patterns. Weather changes, in turn, will affect the communities of many ecosystems.

Evolution and Adaptation

Natural selection is ecology in action. Not all offspring are equally equipped for acquiring energy and nutrients from their environment, for finding a mate and reproducing, or for surviving under the unique set of environmental conditions present in an ecosystem or biome. Those individuals that are better fit to the existing ecological situations tend to produce more offspring than do those that are less fit. In this way, the more adaptive characteristics of the more fit individuals are passed on to the next generation, increasing the frequency of genes that code for more favorable traits. Natural selection results in adaptation and, under appropriate conditions, in the formation of new species.

SYNOPSIS

The biosphere is composed of the earth's lands, waters, and air, which support all life as we know it. Although organisms are scarce in extreme environments, the biosphere contains all of the resources and conditions necessary for life. Since resources for life are limited, recycling of matter within the biosphere is necessary for the continuation of life throughout time.

The earth's oceans support the greatest quantity of life. Most aquatic life is concentrated in shallow coastal waters along the fringes of land and coral reefs.

Biomes cover vast expanses of the earth's land and are characterized by particular plants. Since prevailing climate is the primary factor that controls plant distributions, terrestrial biomes follow global climate patterns and elevational gradients.

Forests cover more than 30 percent of the earth's land. The major forest biomes include tropical rain forests, deciduous forests, and coniferous forests. Tropical rain forests contain more species of organisms than do all other biomes combined. These rich forests are rapidly being destroyed by humans.

Originally, grasslands covered more than 30 percent of the earth's land. Today, most grasslands are used for agriculture, grazing, and other types of development.

Nearly 8 percent of the earth's lands are covered by savannas of grass and scattered trees. Overgrazing is changing the savannas, reducing their ability to support animal life.

Dense shrublands develop in land areas that have a relatively long dry season. Because fires are frequent in shrublands, many organisms have evolved adaptations for quick recovery from fire.

Nearly 30 percent of the land is covered with hot, dry deserts. Harsh living conditions in the desert have produced many anatomic, physiological, and behavioral adaptations in desert organisms.

Key Terms

biosphere (p. 904)
hydrosphere (p. 904)
lithosphere (p. 904)
atmosphere (p. 904)
ecology (p. 905)
climate (p. 905)
pelagic zone (p. 908)
epipelagic zone (p. 908)
mesopelagic zone (p. 908)
bathypelagic zone (p. 908)
benthic zone (p. 908)
coastal waters (p. 908)
neritic zone (p. 908)
intertidal zone (p. 912)
splash zone (p. 912)
high intertidal zone (p. 912)
middle intertidal zone (p. 912)
low intertidal zone (p. 912)
estuary (p. 912)
stream (p. 914)
river (p. 914)
lake (p. 915)
pond (p. 915)
littoral zone (p. 915)
limnetic zone (p. 915)
profundal zone (p. 915)
biome (p. 916)
acid rain (p. 918)
acid snow (p. 918)
tropical rain forest (p. 917)
deciduous forest (p. 917)
coniferous forest (p. 917)
taiga (p. 920)
tundra (p. 920)
grassland (p. 922)
savanna (p. 923)
shrubland (p. 924)
chaparral (p. 924)
tropical thornwood (p. 924)
desert (p. 926)
rainshadow (p. 926)

Review Questions

1. Complete the chart on aquatic ecosystems by placing the following terms in the appropriate space

pelagic zone	littoral	limnetic
standing	freshwater	aphotic
mesopelagic	ponds	profundal
moving	streams	epipelagic

2. Rank the earth's terrestrial biomes, starting with that which covers the greatest land surface and ending with that which covers the smallest area. Why do some biomes cover more land than others?
3. Rank the earth's terrestrial biomes, starting with that which has the greatest number of species (the highest diversity) and ending with that which has the smallest number of species.
4. Explain why tall trees do not grow in the following biomes:
   a. desert b. tundra c. grassland
5. Explain how lakes have alternating periods of low biological activity and high biological activity.
6. List the strata of organisms found in the intertidal zone. What kinds of adaptations are necessary to help organisms withstand increasing lengths of exposure to dry air?
7. List the two most important environmental factors that determine the distribution and abundance of organisms in the ocean.
8. In what ways are the following similar?
   a. lakes and rivers.
   b. profundal zone and desert.
   c. forest understory and bathypelagic zone.
   d. coastal waters and littoral zone.
9. Discuss the changes that would occur in the biosphere and the effect on biodiversity (the numbers of kinds of organisms) of the total destruction of all tropical rain forests. Do you think total destruction is likely?
10. For each of the following, list the major adaptations needed for survival. Are there distinct differences for plants and animals? Make a list of different plant and animal adaptations for each habitat and briefly explain why such differences exist.
    a. prolonged hot and dry season.
    b. surging waters against a rocky coastline.
    c. calm, warm fresh water with abundant nutrients and many species of organisms.
    d. dramatic fluctuations in water salinity.

Critical Thinking Questions

1. Scientists are warning that without the filtering of ultraviolet sunlight by the ozone layer, incidents of skin cancer and crop damage will increase. Explain how increased amounts of UV can increase skin cancer and crop damage. If by the year 2050, the ozone layer is reduced by another 60 percent, as some scientists predict, what other problems may develop?
2. Life cannot be naturally supported outside the biosphere, yet we send astronauts into space. What fundamental requirements must be met for establishing a space station that will maintain life indefinitely? Design a space station capable of supporting life, including humans. Include a list of all types of organisms that would populate the station.
3. Prepare a chart, similar to the one below, showing the ways in which each organism affects its environment and how it is affected by the environment.

Organism	How It Affects Its Environment	How It Is Affected By Its Environment
earthworm		
moss growing on a rock		
maple tree		
elephant		
reef building coral		
bacterium of decay		

4. The earth's diverse climates create the myriad habitats that support a tremendous variety of organisms. In what ways would the diversity of climate (and therefore the diversity of life) change if conditions were different? For each of the following conditions describe the impact on climate change and then the impact on life:
   a. the earth does not rotate on its axis once every 24 hours.
   b. the earth is not tilted 23.5°.
   c. all the earth's land mass is contained in a single, large continent. (Is the location of this land mass significant? If so, be sure to state its location and effects.)
5. What environmental and evolutionary factors account for the fact that species diversity is greater in terrestrial biomes, yet the abundance of organisms is greater in aquatic habitats?
6. For each of the human activities listed, describe its effects on the biosphere: deforestation, automobile travel, agricultural use of fertilizers, use of pesticides, heavy industry, destruction of wetlands, building large cities.

Additional Readings

Brewer, R. 1988. *The science of ecology.* New York: Saunders College Publishing. (Intermediate)

Carson, R. 1962. *Silent spring.* Boston: Houghton Mifflin. (Introductory)

Cloud, P. 1983. The biosphere. *Sci. Amer.* SEPTEMBER:176–189. (Introductory)

Disilvestro, R. 1989. *The endangered kingdom: The struggle to save America's wildlife.* New York: Wiley Science Editions. (Introductory)

Newell, N. 1972. The evolution of reefs. *Sci. Amer.* JUNE:54–65. (Introductory)

Schmidt-Nielsen, K. 1964. Desert animals: Physiological problems of heat and water. Oxford, England: Oxford Univ. Press. (Introductory)

Sutton, A., and M. Sutton, 1966. *The life of the desert.* New York: McGraw-Hill. (Introductory)

Sutton, A., and M. Sutton. 1979. *Wildlife of the forests.* New York: Harry N. Abrams. (Introductory)

CHAPTER

41

Ecosystems and Communities

STEPS TO DISCOVERY
The Nature of Communities

FROM BIOMES TO MICROCOSMS

THE STRUCTURE OF ECOSYSTEMS

Components of the Biotic and Abiotic Environments

Tolerance Range

Limiting Factors

ECOLOGICAL NICHES AND GUILDS

Niches

Guilds

ENERGY FLOW THROUGH ECOSYSTEMS

Trophic Levels

Food Chains, Multichannel Food Chains, and Food Webs

Ecological Pyramids

BIOGEOCHEMICAL CYCLES: RECYCLING NUTRIENTS IN ECOSYSTEMS

Gaseous Nutrient Cycles

The Phosphorus Cycle: A Sedimentary Nutrient Cycle

Important Lessons from Biogeochemical Cycles

SUCCESSION: ECOSYSTEM CHANGE AND STABILITY

The Climax Community

Primary Succession

Secondary Succession

BIOLINE

Reverberations Felt Throughout an Ecosystem

THE HUMAN PERSPECTIVE

The Greenhouse Effect: Global Warming

STEPS TO DISCOVERY
The Nature of Communities

Conflicting views may arise even among brilliant scientists in the same field. Each unresolved conflict heralds a research opportunity for an alert and eager student. When Robert H. Whittaker began graduate school in 1946 at the University of Illinois at Urbana to study insect ecology, he soon found himself embroiled in a 30-year-old conflict among plant ecologists regarding the nature of communities of organisms.

Before Whittaker's time, many scientists had already observed that groups of plants and animals tended to form repeatable, discrete communities of organisms. Indeed, distinct communities have been described at least as far back as 300 B.C. by Theophrastus, the Greek philosopher who studied with Aristotle and succeeded him as head of the Peripatetic school. Theophrastus' two books on botany, *History of Plants* and *Etiology of Plants,* remained the definitive works on the subject until the Middle Ages. In these books, Theophrastus noted that particular plants tended to group together, forming repeating patterns in similar environments. Between 1800 and the early 1900s, a great deal of research and philosophical discussion centered on the causes of such plant communities.

Carl L. Willdenow, one of the first botanists to study the distributions of plants, noted that similar climates tended to produce similar plant communities, even in regions that are widely separated. Willdenow's findings in-

Hundreds of random samples (small circles) of insects and plants in the Great Smoky Mountains showed that both plants and

trigued Friedrich H. A. von Humboldt, a wealthy Prussian student who was studying botany under Willdenow (along with mathematics and chemistry) at the University of Gottingen in Germany. Upon graduating, von Humboldt traveled to South America, where he collected over 60,000 plant specimens and documented the relationship between climate and plant communities. Von Humboldt eventually coined the term **association** to describe a community that is consistent in the composition of plant species and general appearance, and whose distribution is correlated with a specific climate and other physical factors. Von Humboldt was one of the first scientists to recognize that communities formed as a result of many causes, including elevation, latitude, temperature, rainfall, soils, and so on.

Not everyone shared Von Humbolt's view, however. By the turn of this century, botanists had still not agreed on the principal cause of plant communities. For example, two eminent botanists, Frederick E. Clements of the University of Nebraska and Carnegie Institution of Washington, D.C., and Henry A. Gleason of the New York Botanical Gardens, formulated opposing views on the nature of plant communities. Clements considered a plant community a "super organism"; that is, a well-defined association of plant species that are always found together, much as the cells, tissues, and organs of an organism are always found together. According to Clements, the groups of species in a community all have identical distribution limits, so they recur together in distinct communities. In contrast, Gleason argued that each plant species was distributed individualistically; that is, instead of all species in a community having identical distribution limits, the limits of each species were determined by the species' own genetics that determine its physical and physiological characteristics and tolerances. As a result, each species' distribution changed gradually, making it difficult to divide vegetation into discrete associations or communities. Although Clements proposed his view in 1916, and Gleason proposed his view in 1926, this conflict remained unresolved until the mid-1940s. At that time, Robert Whittaker entered graduate school.

Whittaker referred to the conflict between Clements and Gleason as "exciting confusion' for a new graduate student. Whittaker decided to test these conflicting views on insects. Whittaker sampled the distribution of insects at different elevations in the Great Smoky Mountains National Park in Tennessee and North Carolina. The hypothesis Whittaker formulated was in line with Clements' view. That is, Whittaker proposed that he would find distinct groups of insect species along the continuous elevational gradient, each group more or less separated by abrupt transitions. He discovered that insect populations changed *irregularly* with elevation, however, disproving his hypothesis.

Whittaker decided to retest the same hypothesis, this time on plant species in the same forest. He took 300 *random* samples of plants over the mountain range to see whether natural units of plant species emerged. Again, he failed to find definite groups of plant species. Whittaker finally rejected his hypothesis and reanalyzed his results to see whether Gleason's individualistic view applied to the vegetation in the Great Smoky Mountains. When he plotted plant distributions on a chart with elevation and moisture gradients as axes (an analysis technique he invented), Whittaker verified that plants distributed themselves individualistically and that the vegetation found did not form a mosaic of discrete associations. In his words, Whittaker found "a subtly wrought tapestry of differently distributed species populations variously combining to form the mantle of plant communities covering the mountains."

Whittaker's results resolved the conflict between Clements and Gleason: Clements' group view was rejected, and Gleason's individualistic view was scientifically supported. As often occurs in science, at the same time that Whittaker was completing his studies, other investigators in the United States and in the former Soviet Union had also discovered the validity of the individualistic view of species distributions.

insects distribute themselves independently. Data from different samples are shown by histograms.

Organisms are continually affected by other organisms as well as by the surrounding physical environment.

- A long drought destroys vast stretches of grasses and trees in the African savanna, causing thousands of animals to starve and leaving others frail and malnourished.
- The cold runoff from exceptionally heavy winter snows keeps streams cool throughout most of the summer, slowing hatching larval development of blackflies, mayflies, and caddisflies, which, in turn, reduce the number and size of some fishes that normally eat these larvae.
- Acrid Los Angeles smog blows into mountain forests, where it reduces photosynthesis in pines by as much as 80 percent. Unable to manufacture enough resin (sap) to protect themselves against burrowing insects, thousands of pines are dying from bark beetle infestations in the forests.

These examples illustrate a fundamental ecological principle. Because of this constant interplay, the organisms and the physical environment in a particular area function as an organized unit known as an **ecosystem.** This level of biological organization includes both the community of living organisms and the physical environment.

Biologists have discovered a great deal about many of the levels of biological organization, particularly the cell, tissue, organ, organ system, and organism levels, but relatively little is known about the levels of organization studied in ecology (populations, communities, and ecosystems). This is not surprising because, by definition, ecological levels of organization include all former levels, making them the most complex of all.

In addition to enormous complexity, the timing of ecological events is often difficult to fathom and measure, ranging from a few nanoseconds (a predator capturing its prey) to millions or billions of years (gradual climatic change). Add to these considerations the fact that it is virtually impossible to study whole populations, communities, or ecosystems under controlled conditions, and it is easy to see why ecological studies often require very long periods of time and the collaboration of a number of scientists from many fields.

▼ ▼ ▼

FROM BIOMES TO MICROCOSMS

The variety of ecosystems in the biosphere is enormous. Each expansive biome contains countless ecosystems (Figure 41-1). The boundaries that separate ecosystems are not always sharp delineations, however, since all ecosystems are linked with other ecosystems, to some degree. For example, consider a lake and a cave, two ecosystems that at first seem clearly independent. However, the rate at which the lake fills with sediment may affect the number of beetles that live in a cave, even if the lake and the cave are separated by several miles.

This is possible because bats link the two "separate" ecosystems as these animals forage at night for insects around the lake, where insects are plentiful, and return to the cave during the day to rest. As the lake fills with sediment and organic material, more plants grow along the margins of the lake, providing more food for greater numbers of plant-eating insects, the dietary mainstay for the bats. When well-fed bats return to the cave, they defecate large amounts of feces onto the cave floor. The bat feces, in turn, nourish the growth of mold, a staple of the cave beetles' diet. Thus, the amount of sediment in the lake eventually affects the number of beetles in the cave: More lake sediment means more plants; more plants mean more insects; more insects mean more bat feces; more bat feces means more mold; and more mold means more beetles. Furthermore, since beetles, crickets, flies, springtails, moths, spiders, centipedes—indeed, nearly every organism that makes up the entire cave community—depend on the supply of bat droppings, the lake ecosystem affects many aspects of the cave ecosystem, and the cave affects the lake, as bats regulate the number of insects (Figure 41-2).

Furthermore, a lake may not be linked only to cave ecosystems but to other ecosystems as well. For example, runoff from a nearby forest ecosystem provides nutrients and debris to the lake, while the river ecosystem that drains the lake removes sediments and nutrients. In fact, many biologists view the entire biosphere as one giant global "ecosystem" of tremendous complexity and order.

The earth's vast biomes are subdivided into ecosystems, unique communities of organisms that interact with one another and with the surrounding environment. Although a separate unit of organization, each ecosystem is linked to and affected by other ecosystems. (See CTQ #2.)

THE STRUCTURE OF ECOSYSTEMS

Each ecosystem is a functional unit in which energy and nutrients flow between the physical, nonliving abiotic

FIGURE 41-1

Ecosystems within ecosystems. ***(a)*** A South American tropical rain forest contains many ecosystems, each with a unique community of organisms and a unique set of physical factors. ***(b)*** The ecosystem supports colonies of leaf cutter ants that harvest leaves to cultivate a fungus garden ecosystem for food. ***(c)*** Perched high on a tree branch, the top of a bromeliad collects water, forming a tiny pond ecosystem, the home of this red-eyed tree frog.

FIGURE 41-2

Bats interconnect a cave ecosystem with a lake ecosystem. The more insects found around a lake, the more food available to cave-dwelling bats such as these Mexican free-tail bats. When bats return to a cave to rest, they defecate, forming the substrate for the growth of mold. Beetles and a number of other cave-dwelling organisms in turn consume the mold.

◁ BIOLINE ▷

Reverberations Felt Throughout an Ecosystem

Some great lessons in ecology are stumbled upon accidentally. For example, in the 1950s, the raising of ducks on Long Island, New York, produced some unexpected and economically devastating consequences. No one foresaw that allowing duck farms near the Great South Bay of Long Island would totally destroy the lucrative oyster industry in the bay, but that is just what happened. Before the ducks arrived, the famous blue-point oysters had thrived in the bay, eating a normal mixture of diatoms, unicellular green algae, and dinoflagellates. Shortly after the duck farms were established along the tributaries to the bay, oysters and some other shellfish were found starving to death. Eventually, all the oysters disappeared.

How could such a disaster have happened? The answer lay in the large duck farms that produced enormous quantities of duck droppings that were washed into the tributaries and eventually into the bay. As a consequence of the bay's slow circulation rate, the duck droppings quickly changed the nutrient balance in the bay, drastically altering the prevailing mixture of algae. One species of green algae that was ordinarily present in very small numbers prospered under these new nutrient-rich conditions, while the populations of other phytoplankton species plummeted. Unable to digest the green algae that proliferated, the oysters quickly starved to death. Since the disaster, several attempts have been made to reestablish a normal nutrient balance, in the hopes of reintroducing oysters to the Great South Bay, but all have failed.

The ecological story of "the ducks and the oysters" illustrates how a change in the biotic environment (increasing the number of ducks) can alter the abiotic environment (the levels of inorganic nutrients in the bay), which, in turn, changes the biotic environment (the proportion of algae species), causing another change in the biotic environment (the death of oysters), which produces yet another change in the biotic environment (the loss of food for humans and the bay's other consumers of oysters). The story also illustrates how one seemingly minor alteration can send reverberations throughout an ecosystem, usually with unexpected and destructive results.

environment, and a community of living organisms that make up the biotic environment. The abiotic and biotic environments continually affect each other, producing interdependent connections. Even a slight modification in one factor can disrupt an entire ecosystem (see Bioline: Reverberations throughout an Ecosystem).

It is often easier to envision how the abiotic environment changes the biotic community than vice versa. For instance, a long period of freezing temperatures or a devastating fire may kill many organisms outright. Extremely strong winds can uproot plants and blow flying insects and birds far sway from their natural habitat. The relationship between the abiotic and the biotic environment is two-way, however; organisms change the abiotic environment as well. For instance, recall from Chapter 35 that the earth's original atmosphere was devoid of oxygen. Ancient organisms released oxygen during photosynthesis, gradually adding this important gas to the atmosphere. In other words, organisms dramatically altered the abiotic environment, building up the oxygen of the earth's atmosphere to its current level. The biotic environment affects the abiotic environment in various other ways as well: Plants contribute to the formation and fertility of soils; coral reefs change the flow and temperature of oceans; dense forests modify humidity, temperature, and the amount of light and rain that reaches the forest floor; and today, human activities produce pollutants that are changing the earth's climate.

COMPONENTS OF THE BIOTIC AND ABIOTIC ENVIRONMENTS

Biologists divide ecosystems into five subcomponents, two of which comprise the abiotic environment, and three that comprise the biotic environment (Figure 41-3). The two abiotic subcomponents are:

1. *abiotic resources:* the energy and inorganic substances (nitrogen, carbon dioxide, water, phosphorus, potassium, and so on) needed by organisms for the construction of organic compounds; and
2. *abiotic conditions:* the substrate and/or medium (air, water, and soil) in which organisms live, and the surrounding conditions, such as temperature and water currents.

FIGURE 41-3

The basic components of an ecosystem are the abiotic environment and the biotic community. The abiotic environment includes energy, inorganic substances (essential elements), and the substrate and/or medium in which the community lives. Within the biotic community, primary producers convert energy and inorganic nutrients into organic food molecules, which, in turn, are passed on to consumers and decomposers. Respiration, excretion, and decomposition return essential elements to the abiotic environment. Each year, the earth's primary producers convert an amount of energy equivalent to the output of about 2 billion nuclear power plants, enough to power all life in the biosphere.

The three subcomponents of the biotic environment are:

1. **primary producers:** autotrophs (algae, bacteria, and plants) that use sunlight or chemical energy to manufacture food from inorganic substances;
2. **consumers:** heterotrophs that feed on other organisms or organic wastes; and
3. **decomposers** and **detritovores:** heterotrophs that get their nutrition by breaking down the organic compounds found in waste organic matter and dead organisms **(detritus).** Decomposers are primarily microscopic bacteria and fungi, whereas detritovores are typically larger animals, such as some worms, nematodes, insects, lobsters, shrimp, and birds that feed on detritus.

TOLERANCE RANGE

For each abiotic resource or condition, an organism is able to survive and reproduce only within a certain maximum and minimum limit. For example, the lethal temperature limits for many land animals are a minimum of 0°C (32°F) and a maximum of about 42°C (107°F), whereas the lethal temperature range is often smaller for aquatic animals. This range is known as an organism's **tolerance range** (Figure 41-4). In addition to identifying the tolerance range for a single organism, tolerance ranges can be determined for an entire species by determining the tolerance ranges of all members of the species and then setting the tolerance range limits at the maximum and minimum levels found. The **Theory of Tolerance,** as this concept came to be called, was proposed in 1913 by Victor Shelford, an animal ecologist at the University of Chicago. Since the time the theory was proposed, studies have revealed that tolerance ranges

- differ for each abiotic factor (a marine organism may have a broad range of tolerance for temperature and a narrow range for salinity);
- have an optimum at which conditions are best for the organism (Figure 41-4);
- have some factors that may affect the tolerance range of other factors, especially when conditions approach the maximum or minimum limits for one factor (grasses grown in nitrogen-deficient soils need more water than do those grown in nitrogen-rich soils);
- change as an organism passes through different phases of its life cycle (eggs, gametes, embryos, and young organisms usually have narrower ranges than do adults); and
- may differ slightly from population to population within a species (populations with different genetically fixed tolerance ranges are called **ecotypes** of a species).

LIMITING FACTORS

When the limits of tolerance for one or a few factors are approached, those factors usually take on greater importance than do all the others in determining where or how well an organism will survive. These more critical factors become **limiting factors** because they alone impose restraints on the distribution, health, or activities of the organism. Since each species has a multitude of tolerance ranges (one for each abiotic factor), the ability to identify a few key limiting factors helps ecologists understand otherwise extremely complex ecosystems.

This concept of limiting factors was first proposed in 1840 by Justus von Liebig of the University of Heidelberg and has become known as Liebig's **Law of the Minimum** (Figure 41-5). An agriculturist and physiologist, Liebig discovered that the yield of a crop was restricted by the soil nutrient most limited in amount. His Law of the Minimum states that the growth and/or distribution of a species is dependent on the one environmental factor that is available in the shortest supply.

Examples of Liebig's law are common: Low amounts of phosphate in a lake drastically reduce algae growth, which, in turn, limits the growth of all consumers; the amount of

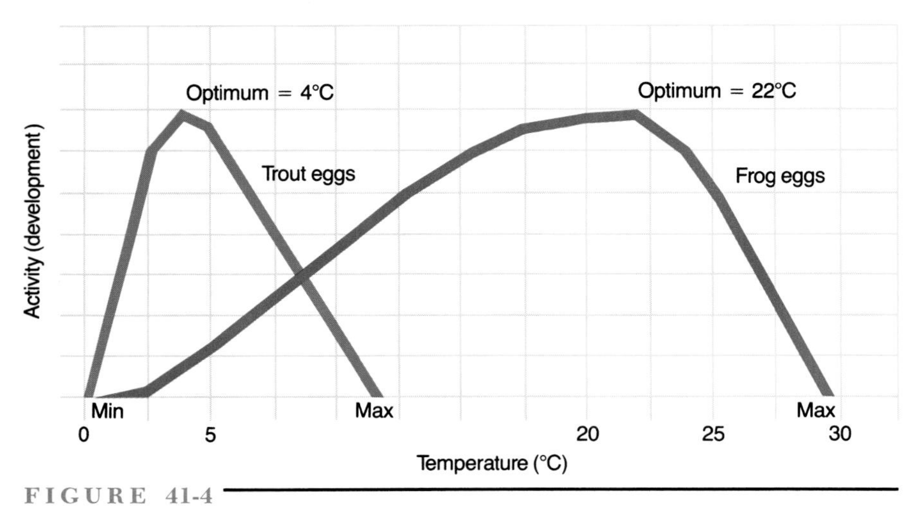

FIGURE 41-4
The limits of life. For each physical factor, organisms have minimum and maximum tolerance limits. Within most tolerance ranges, such as the temperature ranges for the development of trout and frogs, lies an optimum, at which growth is greatest.

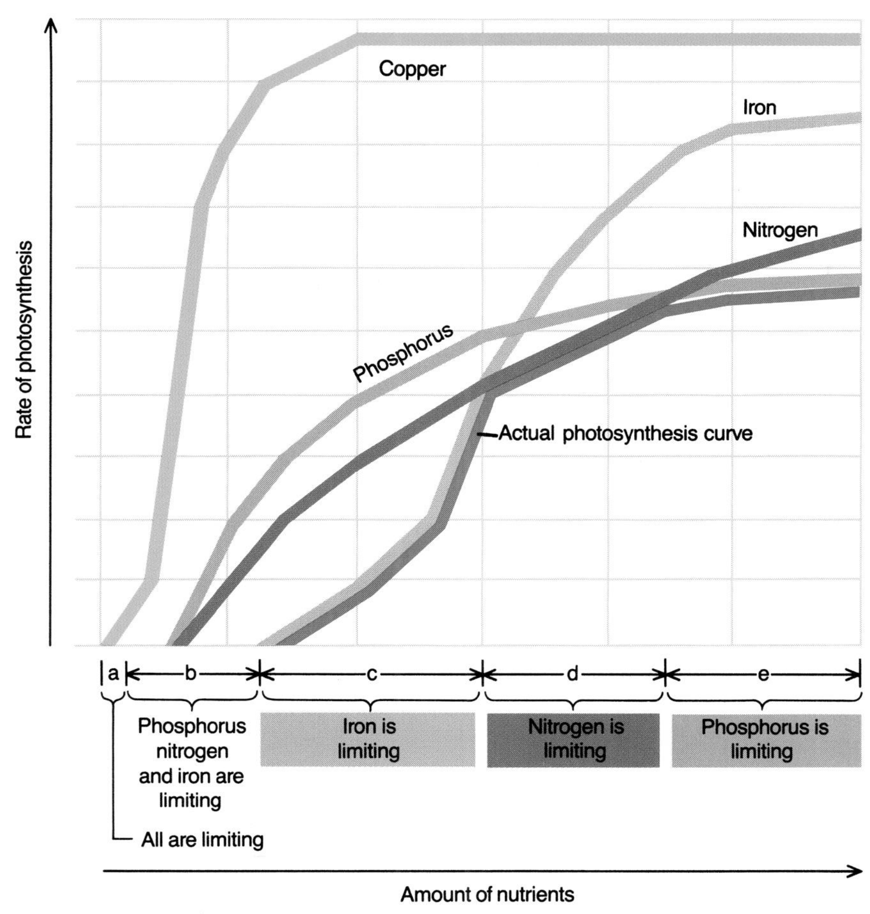

FIGURE 41-5

Nutrient availability and photosynthesis. These graphs illustrate the Law of the Minimum. The rate of photosynthesis in phytoplankton (black line) is limited by the nutrient in the shortest supply. In areas ***(a)*** and ***(b)*** on the graph, one or more essential nutrients are lacking, so there is no photosynthesis. Low supplies of iron limit photosynthesis in area ***(c)***; of nitrogen in area ***(d)***; and of phosphorus in area ***(e)***.

zinc in soil is usually so scarce that it alone limits the yield of many agricultural crops; the limited rainfall in deserts severely restricts plant growth; low light intensity on the floor of a tropical rain forest limits photosynthesis in understory plants.

Exceptions to the Law of the Minimum occur when one factor changes the tolerance for another. For example, changes in the oxygen content of water can change the temperature tolerance of American lobsters. When the oxygen content of water is low, lobsters tolerate temperatures only as high as 29°C (84°F), whereas when the oxygen content is high, the temperature tolerance rises to 32°C (89°F). Another exception to the Law of the Minimum is found when one factor substitutes for another, such as when mollusks use strontium to build their shells when calcium is limiting.

Organisms and the nonliving physical environment are inseparably linked; each continually affects the other. Organisms in an ecosystem are primary producers, consumers, or decomposers. Organisms tolerate a certain range of conditions for each abiotic factor; one or a few key limiting factors often determine its growth and distribution. (See CTQ #3).

ECOLOGICAL NICHES AND GUILDS

The organisms that comprise a community possess structural and behavioral adaptations that enable them to grow, reproduce, and survive in a particular ecosystem. Each organism in an ecosystem occupies a specific **habitat,** the physical location in which the organism lives and reproduces (the bottom of a lake, under a rock, inside another organism, and so on). In addition to needing space, organisms also require energy and nutrients. The processes by which organisms acquire these resources partly define their "role(s)" in an ecosystem. Having identified the habitats and roles of organisms that make up ecosystem and investigated what happens when habitats or roles overlap, ecologists have proposed two fundamental concepts to help evaluate and describe the total life history pattern of individuals that occupy the same ecosystem.

NICHES

Together, an organism's habitat, role, requirements for environmental resources, and tolerance ranges for each abiotic factor comprise its **ecological niche.** Since the ecological niche includes all aspects of an organism's existence—its residence, activities, requirements, and effects—it too is the outcome of evolution through natural selection. The full scope of adaptations an organism acquires through natural selection establishes the range and boundaries of an organism's—or a species'—ecological niche.

If only two or three components are considered, it is possible to plot an ecological niche on a graph, each axis representing a different factor (Figure 41-6). It is impossible to draw a graph that includes *all* the components of the ecological niche (all tolerance ranges, the total range of habitats, and all functional roles), however, because such a representation would circumscribe a multidimensional area, or what ecologists refer to as a **hypervolume.** The hypervolume represents the *potential* niche, or the **fundamental niche.** Most organisms never realize their fundamental niche because interactions with other organisms (such as competition for limited resources or mates) often reduce the range of available habitats or possible functional roles. (We will discuss the range of organism interactions in the next chapter.) The remaining portion of the hypervolume, in which an organism actually exists and functions, is its **realized niche.** In other words, the fundamental niche represents all *possible* ranges under which an organism or species can exist, while the realized niche is the range of conditions in which the organism *actually* lives and reproduces.

Ecologists speak of **niche breadth** as a measure of the relative dimensions of an organism's ecological niche (Figure 41-7). Some species have very broad niches. Hawks, for example, have a relatively broad niche because they visit many habitats over large areas and have evolved adaptations that enable them to feed on several kinds of organisms. Conversely, the cotton boll weevil has a narrow niche because it lives, feeds, and reproduces on cotton plants alone. Species with narrow niches possess adaptations for very specific habitat and environmental requirements. For example, the cotton boll weevil has a long, tubular mouth specifically adapted for feeding on a cotton plant; female boll weevils have adaptations for laying eggs in cotton buds (bolls); and boll weevil larvae can only survive by eating the internal tissues of flower buds and developing cotton bolls.

Although they may not always be apparent, practical benefits sometimes arise from studying an organism's niche breadth. For example, consider the constant battle between farmers and weeds. All farmers want to stop weeds from

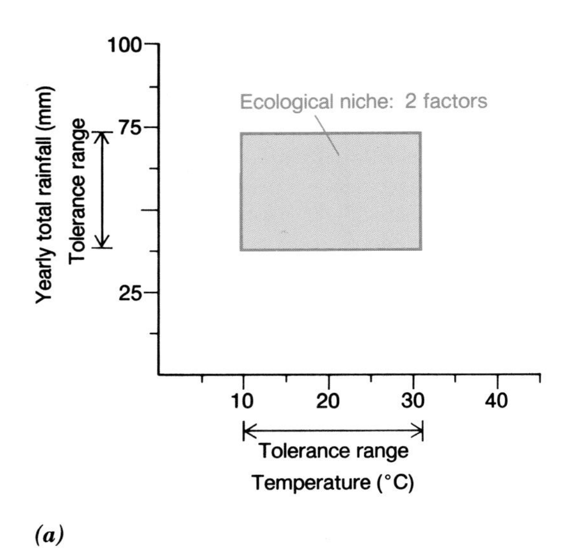

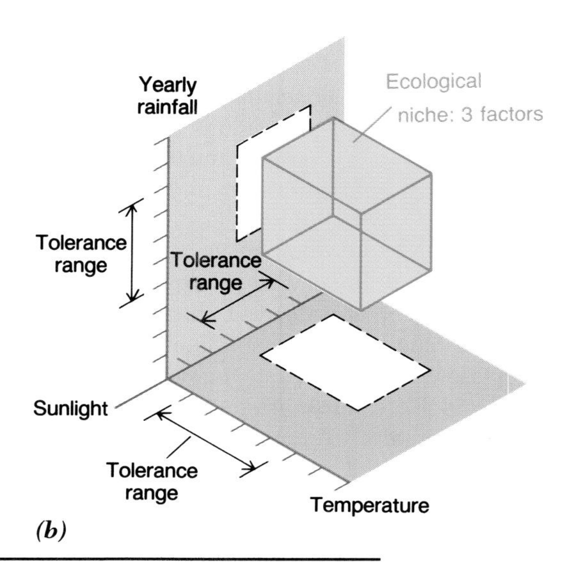

FIGURE 41-6
An ecological niche based on the tolerance range of two ***(a)*** and three ***(b)*** factors.

(a) (b)

FIGURE 41-7

Opposite ends of the spectrum. ***(a)*** The boll weevil's specific adaptations for living and reproducing only on cotton plants results in a narrow niche breadth, whereas ***(b)*** adaptations that enable a hawk to harvest a variety of animals for food from a number of habitats result in a broad niche breadth.

invading and overgrowing their agricultural fields, since weeds contaminate harvests and cut down yields by competing with the crops for space and limited nutrients. One way to stop weed invasion is to apply a chemical weed killer that destroys only the weeds, not the crops. Unfortunately, such "selective chemical killers" are not available for every kind of weed. Alternatively, weeds could be stopped by using a **biological control,** whereby another organism destroys the weeds. Of course, it is important that the organism kill only the weed and not the crop plants; in other words, the weed-controlling organism must have a narrow niche.

Many examples of biological weed control exist. One in particular involves the unlikely pairing of the klamath weed (*Hypericum perforatum*) and a leaf-feeding beetle *Chrysolina quadrigemina*). After studying the "natural history" of *C. quadrigemina,* scientists discovered that the feeding habits of the beetle were very specific. These beetles fed only on plants that had leaves of the same surface texture and margin as that of the klamath weed. Fortunately, the beetle was active during the same period that the klamath week germinates and enters its period of rapid growth. The beetle's life history pattern was ideal for controlling the klamath weed. Unfortunately, the two organisms did not naturally occur together; they didn't even exist on the same continent. Once the leaf-feeding beetle was introduced into the United States, however, it became a very effective biological control agent against the klamath weed. This example reaffirms the importance of "basic research" in science. If separate scientists had not been studying the "natural history" of both of these organisms, and if they had not shared this information with each other, effective control of the klamath weed would still be a fantasy.

Niche overlap occurs when organisms share the same habitat, have the same functional roles, or have identical environment requirements in some other way. Niche overlap leads to competition for the same needed resource; the greater the overlap, the more intense the competition. For example, consider two germinating acorns of different species of oak growing side by side in an oak woodland. Both young plants share the same habitat, require similar amounts of space, light, water, and nutrients, and play essentially the same roles in the ecosystem as do mature oak trees. Because of such close similarities in niche requirements (high niche overlap), these plants will likely compete for needed resources.

When two species have *identical* niches, competition can become so intense that both species are unable to coexist in the same ecosystem; one species eventually excludes the other. This phenomenon is called the competitive exclusion principle and is discussed in more detail in Chapter 42.

GUILDS

Although species with identical niches cannot coexist in the same ecosystem for long periods, species with *similar* niches often can. In a forest, for example, all insect-eating bird species have similar habitats, roles, and nutrient requirements, yet they are able to coexist. Such groups of species with similar ecological niches form a **guild.** Not surprisingly, members of a guild interact frequently, often with intense competition when resources become scarce.

Species that have similar ecological requirements, yet live in different ecosystems, are called **ecological equivalents.** (The niches of ecological equivalents do not overlap because the species do not share the same habitat.) Ecological equivalents usually look very similar, the product of convergent evolution (page 735). For example, placental mammals in North America and their marsupial counterparts in Australia are ecological equivalents (see Figure 33-16). Although these animals have completely different ancestry and are separated by thousands of miles of ocean, North American mammals and Australian marsupials have evolved strikingly similar adaptations in response to very similar ecosystem characteristics: Remember, evolution takes place on the ecological stage. These animals may be ecologically equivalent, but they are not the same species. With distinctively different ancestry, ecological equivalents contribute to the great diversity of life on earth.

Each organism lives and reproduces in a specific habitat, requires specific environmental resources, has fixed ecological tolerances, and plays a specific role in an ecosystem. Together, these factors define an organism's ecological niche. (See CTQ #4.)

ENERGY FLOW THROUGH ECOSYSTEMS

All organisms must secure a supply of energy and nutrients from their environment in order to remain alive and reproduce. Biologists categorize organisms according to their energy-acquiring strategy: Primary producers harvest energy through photosynthesis or chemosynthesis, forming a direct link between the abiotic and biotic environments, whereas consumers and decomposers obtain energy and nutrients from other organisms.

TROPHIC LEVELS

Tracking the transfer of food (energy and nutrients) among the organisms in an ecosystem is relatively easy if you follow a single feeding path. You simply count how many leaves a caterpillar eats, how much phytoplankton goes into a copepod, or how many prairie dogs a hawk consumes. Energy and nutrients are transferred step by step between organisms. For example, in a streamside community, a hawk eats a snake that ate a frog that ate a moth that sipped nectar from the flower of a periwinkle plant that converted the sun's energy into chemical energy during photosynthesis.

This linear feeding pathway has the same organization as do those of all ecosystems: It begins with a primary producer (the periwinkles in the streamside community), which provides food for a **primary consumer** (the moth), which is eaten by a **secondary consumer** (the frog), which is eaten by a **tertiary consumer** (the snake), and so on, until the final consumer dies and is disassembled by decomposers. Each step along a feeding pathway is referred to as a **trophic level** (*trophic* = feeding).

The sequence of trophic levels maps out the course of energy and nutrient (food) flow between functional groups of organisms (primary producers, primary consumers, secondary consumers, and so on). As Figure 41–8 illustrates, trophic levels are numbered consecutively to indicate the order of energy flow. Trophic level 1 is always populated by primary producers, and Trophic level 2 is always populated by primary consumers. Levels of consumers beyond the primary consumer are then numbered sequentially. The final carnivore, called the **ultimate,** or **top carnivore,** and those organisms that escape being eaten (such as humans), eventually die and are consumed by decomposers.

FOOD CHAINS, MULTICHANNEL FOOD CHAINS, AND FOOD WEBS

The transfer of food from organism to organism forms a **food chain** (Figure 41-8). Like a trophic level diagram, a food chain is a flowchart that follows the course of energy and nutrients through an ecosystem. Unlike trophic level diagrams, food chains name the organisms (rather than the group) that occupy each link. The example we gave earlier of periwinkle plants, moths, frogs, snakes, and hawks is a food chain because the organism that occupies each step is identified.

Most ecosystems contain many food chains. When more than one food chain originates from the same primary producer, a **multichannel food chain** is formed. In Figure 41-9, for example, different parts of a single manzanita shrub provide energy and nutrients for at least five independent food chains.

Often, linkages form between food chains, creating networks of connections. In the streamside ecosystem, for example, frogs may sometimes eat flies instead of moths, and snakes may eat small rodents instead of frogs (the flies and rodents originating from other food chains). When all interconnections between food chains are mapped out for an ecosystem, they form a **food web** (Figure 41-10). A food web illustrates all possible transfers of energy and nutrients among the organisms in an ecosystem, whereas a food chain traces only one pathway in the food web.

Food webs may intersect within ecosystems. For example, in a grassland, energy and nutrients may flow between a *grazing food web* and a *detritus food web.* In the grazing food web, living tissues of photosynthetic grasses are consumed by a variety of herbivores, which may then be consumed by a variety of carnivores. But not all plants, or all herbivores and carnivores are eaten, so when these organisms die, their bodies enter the detritus web. Even before they die, animals excrete organic wastes that also enter the

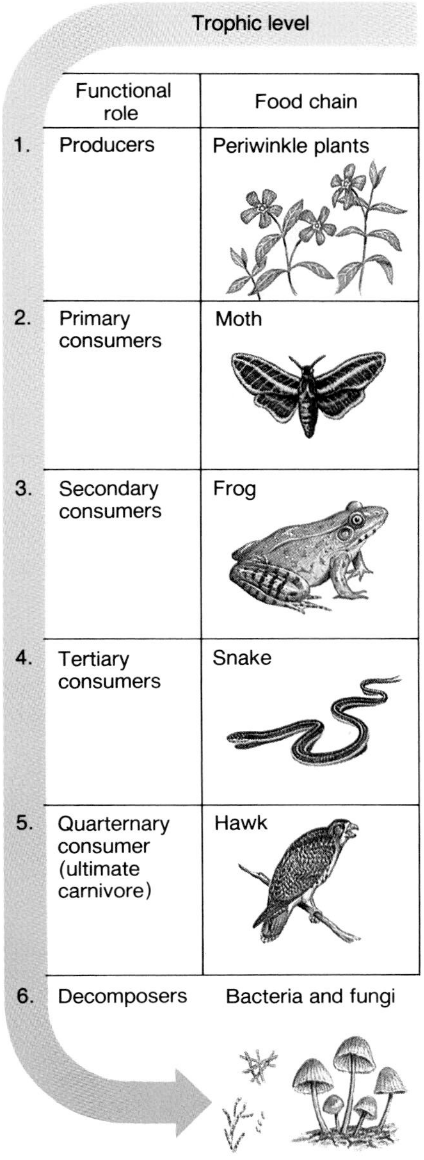

Trophic level

	Functional role	Food chain
1.	Producers	Periwinkle plants
2.	Primary consumers	Moth
3.	Secondary consumers	Frog
4.	Tertiary consumers	Snake
5.	Quarternary consumer (ultimate carnivore)	Hawk
6.	Decomposers	Bacteria and fungi

FIGURE 41-8

A feeding path in a streamside ecosystem. Energy and nutrients flow through the biotic community, as food passes from trophic level to trophic level. Trophic level numbers indicate the order of flow.

detritus web, where earthworms, insects, millipeds, fungi, and bacteria eventually break down organic matter to simple inorganic molecules.

ECOLOGICAL PYRAMIDS

Feeding relationships are graphically represented by plotting the energy content, number of organisms, or **biomass**—the total weight of organic material—at each trophic level. Such graphs are called **ecological pyramids** because of their triangular shape. Each trophic level of a food chain forms a tier on the pyramid; that is, each successive trophic level is stacked on top of the level that represents its food source.

Pyramid of Energy

Energy always flow one way through a food chain; it never recycles. All the energy that enters a food chain is eventually dissipated as unusable heat energy. Since energy does not recycle, there must be an extraterrestrial resource of energy to refuel life continually. For more than 99 percent of the earth's ecosystems, this extraterrestrial energy source is radiant energy from the sun.

A **pyramid of energy** illustrates the rate at which the energy in food moves through each trophic level of an ecosystem (measured in kilocalories of energy transferred per square meter of area per year—kcal/m^2/year). A typical pyramid of energy for a food chain in a grassland ecosystem during the spring (excluding decomposers) looks like this:

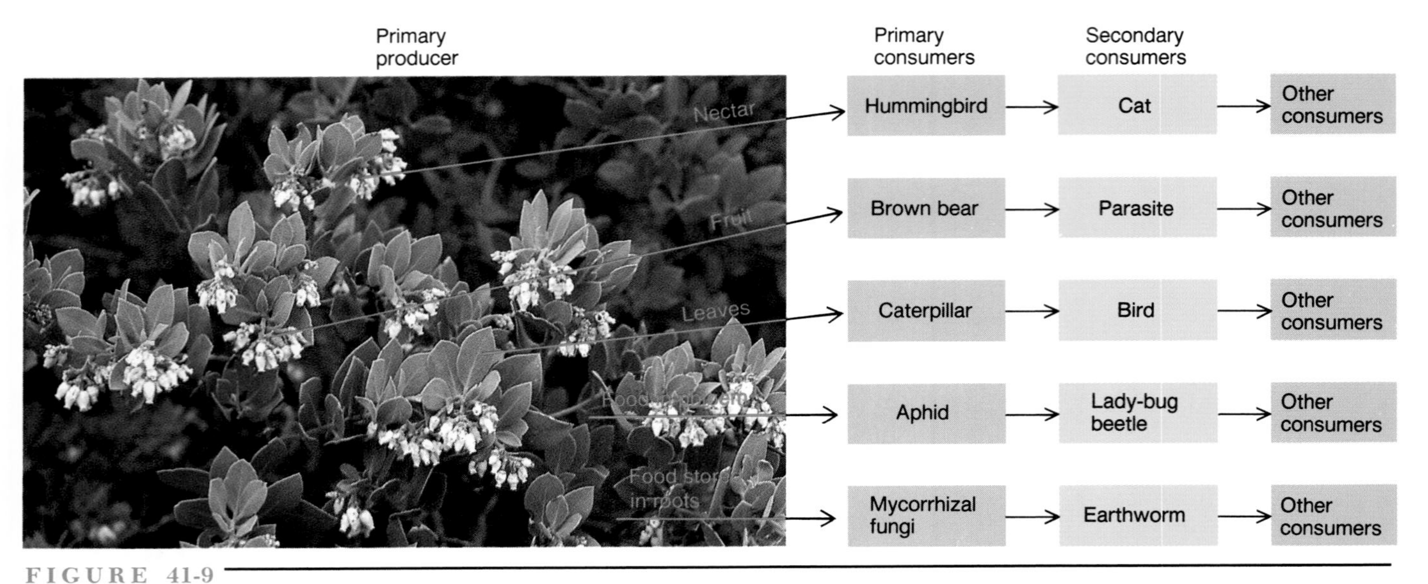

FIGURE 41-9

Multichannel food chains. This manzanita *(Arctostaphylos pringlei* var. *drupacea)* the primary producer for many food chains, including the five drawn here.

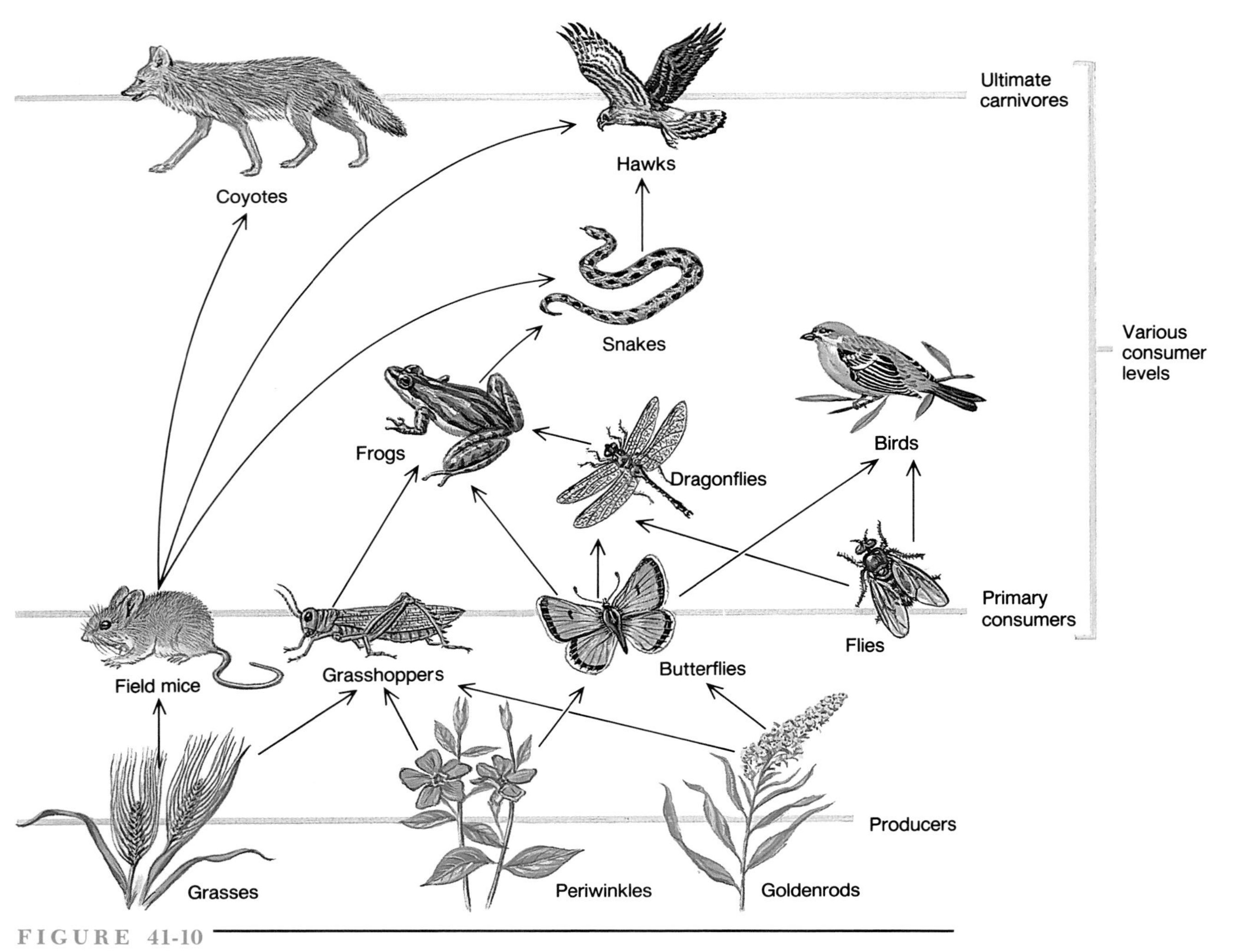

FIGURE 41-10

A simplified food web. A complete food web charts all possible feeding transfers within an ecosystem.

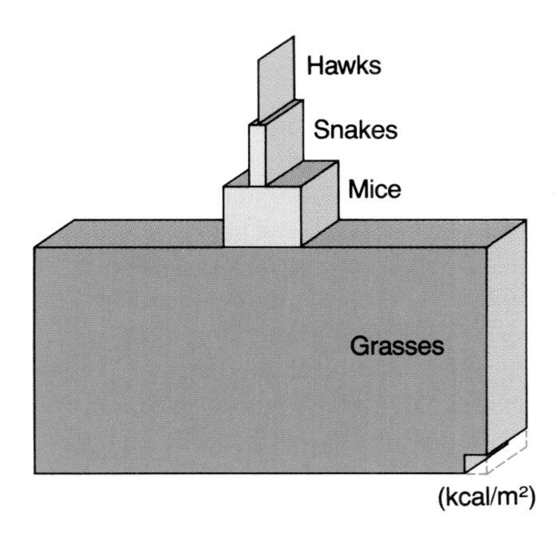

As you can see, there is a dramatic reduction in usable energy as food is transferred from one trophic level to the next. On average, only about 10 percent of the energy in any trophic level is converted into biomass in the next level. The least efficient transfer is between trophic levels 1 and 2; as much as 99 percent of the energy available in the producers is not transferred to the primary consumer level.

What happens to this huge amount of lost energy? Some energy is lost as heat during chemical conversions (the second law of thermodynamics); some energy is used by the organisms in each trophic level for their own metabolism and biological processes; not all food in one trophic level is eaten by organisms in the next trophic level; not all of the food that is eaten is usable (some energy is excreted or defecated as waste), and some energy powers nonbiological phenomena, such as fire. Since such a small fraction of energy gets passed on to successive trophic levels, the number of links in a food chain is limited and rarely exceeds more than four or five. Supporting a food chain with more than five trophic levels would require an enormous energy base at trophic level 1, a situation that does not occur very frequently.

Pyramid of Numbers

The rate of energy flow through a trophic level determines how many individuals can be supported in an ecosystem as well as the overall biomass of the organisms. Because energy decreases with each successive trophic level, the number of organisms and their biomass also decreases. The following **pyramid of numbers** illustrates this principle for a grassland ecosystem:

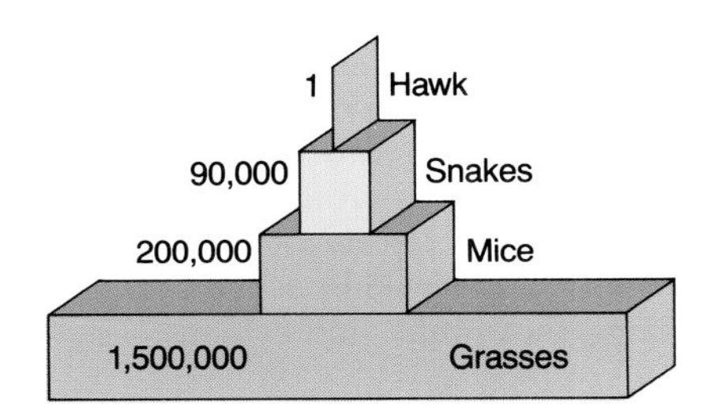

In grasslands, all primary producers are relatively small, compared to primary consumers, so it takes a large number of plants to provide enough food for big consumers, such as wildebeests or bison. In ecosystems where *larger* producers support *smaller* consumers, the pyramid of numbers becomes inverted. In a forest ecosystem, for example, a few large trees provide enough food to support many insects and insect-eating birds. The pyramid of numbers for such a food chain look like this:

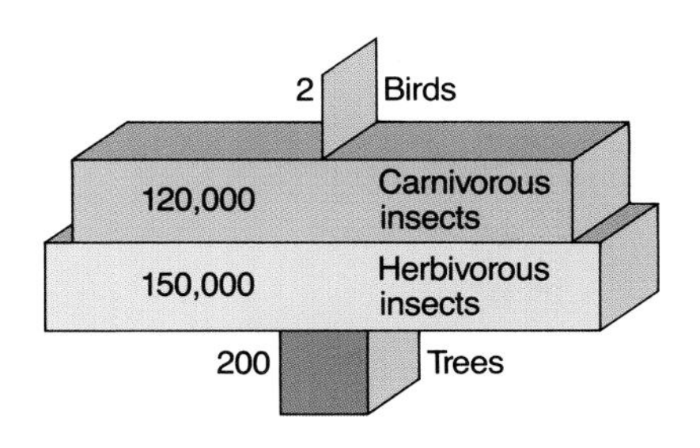

Pyramid of Biomass

A **pyramid of biomass** represents the total dry weight (in grams of dry weight per square meter of area—gm/m²) of the organisms in each trophic level at a particular time. Although most pyramids of biomass are "upright," they may become inverted, depending on when the samples are taken. For example, in the open oceans, where producers are microscopic phytoplankton, and consumers range all the way up to massive blue whales, the biomass of consumers may temporarily exceed that of the primary producers if data are taken when the number of phytoplankton is low. During such sampling periods, the pyramid of biomass could look like this:

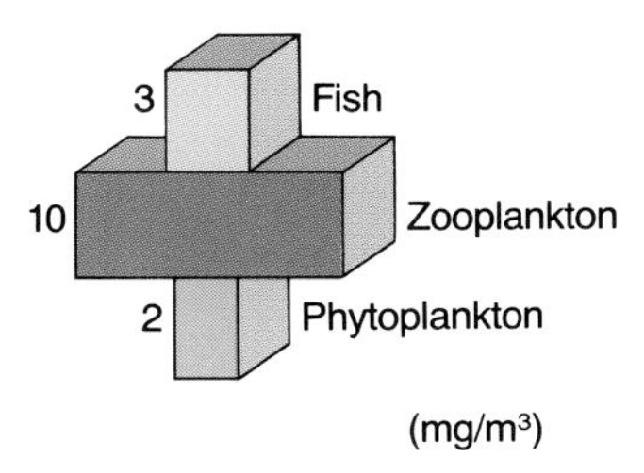

If samples are taken during the spring, however, when phytoplankton populations are immensely large, or if multiple generations of phytoplankton are included, the pyramid of biomass assumes the upright pyramid shape:

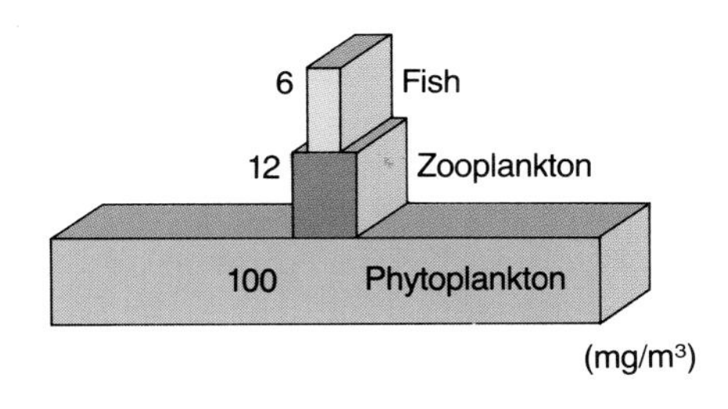

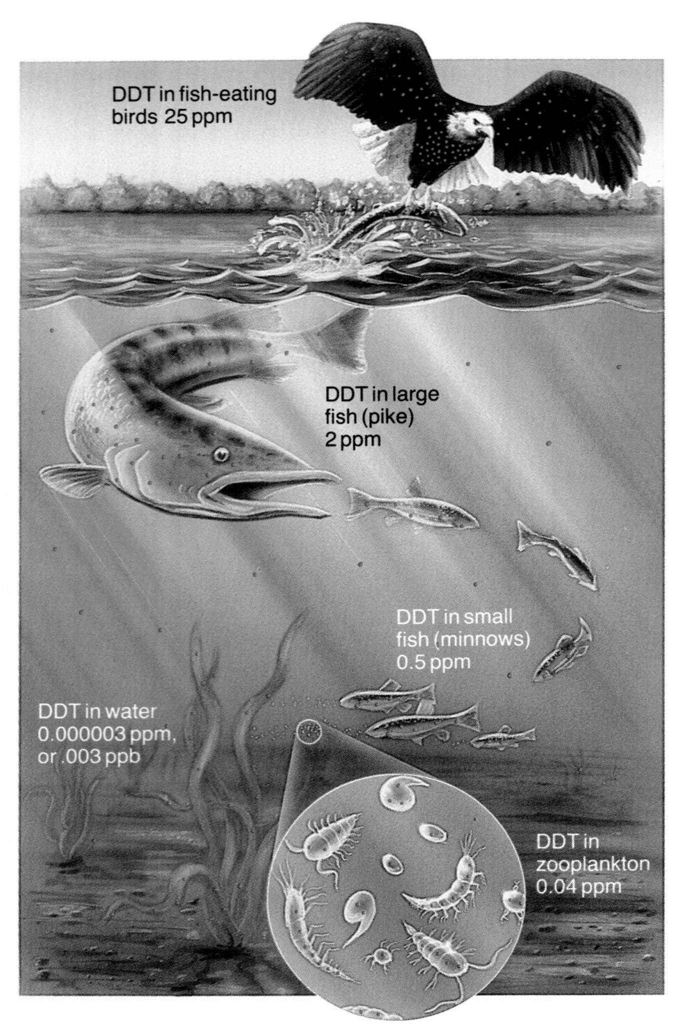

FIGURE 41-11

DDT, bioconcentration, and biological magnification. DDT was so effective in killing pests in agricultural fields, it became widely used. Unfortunately, runoff from these fields contaminated streams, rivers, and oceans with DDT. The algae growing in these aquatic ecosystems absorbed the DDT. As the algae were eaten by small aquatic animals, the DDT was passed on. As these animals were eaten by progressively larger fish, the DDT continued to pass along the entire food chain. DDT accumulates at each feeding level because organisms cannot metabolize DDT (such *bioconcentration* is represented by the red dots). When birds eat larger fish, the high doses of DDT interfere with calcium deposition in the birds' eggs, making the egg shells so thin and weak that they can no longer protect developing birds. Because of its harmful effects on wildlife, DDT was banned in the United States in 1968. The chemical is still widely used in other nations, however, and continues to contaminate much of the earth's waters.

Although it is possible for pyramids of biomass and numbers to be inverted, an energy pyramid can never be inverted; there must always be more energy at lower levels to maintain life at higher trophic levels.

Bioconcentration and Biological Magnification

Ecologists often use ecological pyramids to illustrate how toxic chemicals in the environment can gradually accumulate in the bodies of the organisms in an ecosystem, a phenomenon known as **bioconcentration.** Bioconcentration of a toxic chemical may build to a high enough level to kill the organism. Bioconcentration may also lead to **biological magnification,** the buildup of chemicals in the organisms that form a food chain. Biological magnification exposes organisms toward the end of a food chain (at higher trophic levels) to potentially dangerous levels of chemicals.

The use of DDT as a pesticide provides a clear example of both bioconcentration and biological magnification (Figure 41-11). DDT is fat-soluble, so it accumulates in the body fat of organisms (bioconcentration). Runoff from the use of DDT on agricultural lands mixes the DDT with local water supplies. Small, single-celled animals, like zooplankton, bioconcentrate the DDT in their fat-storage regions. The zooplankton are then eaten by small fish, which are eaten by larger fish, which are eventually eaten by fish-eating birds, each of which also accumulates DDT. The combination of bioconcentration and biological magnification results in DDT concentrations that are several million times greater in the fish-eating birds than in the water. Such magnification has restricted the reproduction of many birds, including peregrine falcons, ospreys, brown pelicans, American eagles, and the California condor, by reducing calcium deposition in their eggshells. Without a hard, protective eggshell, the birds' eggs break easily, and few embryos survive.

The energy for all life in an ecosystem originates from the abiotic environment. The primary producers introduce energy from the sun as well as nutrients from the abiotic environment into the biotic environment. To understand how ecosystems function, ecologists often map and measure the flow of energy and nutrients between organisms, generating food chains, food webs, and ecological pyramids. (See CTQ #5.)

BIOGEOCHEMICAL CYCLES: RECYCLING NUTRIENTS IN ECOSYSTEMS

All organisms are composed of chemical elements (Chapter 4). Of the more than 90 naturally occurring elements found in the biosphere, only about 30 are used by organisms. Some elements, such as carbon, hydrogen, oxygen, phosphorus, sulfur, and nitrogen, are needed in large supplies, whereas others, including sodium, manganese, iron, zinc, copper, and boron, are required in small, or even minute, amounts.

Unlike radiant energy, which showers on the biosphere daily, there is no outside source to supply the elements essential for life. Since there is a finite amount of each, essential elements must be recycled (reused) over and over again for life to continue. Organisms today are using the same atoms that were present in the primitive earth. Some of the atoms that comprise your body may have been part of an ancient bacterium, a long-extinct tree fern, or a dinosaur. During recycling, elements pass back and forth between the biotic and abiotic environments, forming **biogeochemical cycles.** Primary producers typically introduce elements into the biotic environment by incorporating them into organic compounds, and consumers and decomposers release the elements back into the abiotic environment by breaking down complex organic molecules into simple inorganic forms. Without such biological order and regulation, nutrient recycling could not occur. The rate of nutrient recycling depends primarily on where the element is found in the abiotic environment. For example, elements that cycle through the atmosphere or hydrosphere, forming *gaseous nutrient cycles,* recycle much faster than do those of *sedimentary nutrient cycles,* which cycle through the earth's soil and rocks.

GASEOUS NUTRIENT CYCLES

Gaseous nutrient cycles are those in which the element occurs as a gas at some phase in its cycle, and a large proportion of the element resides in the earth's atmosphere. Although several elements have gaseous cycles, we will discuss only the water (hydrologic), carbon, and nitrogen cycles here because of their central role in living organisms.

The Hydrologic Cycle

Nutrient cycles may involve more than one element. For example, in the **hydrologic cycle,** both hydrogen and oxygen cycle together in the form of water molecules, which often change from one phase to another (liquid water to water vapor, or vice versa) as they move between the earth's oceans, land, organisms, and atmosphere (Figure 41-12).

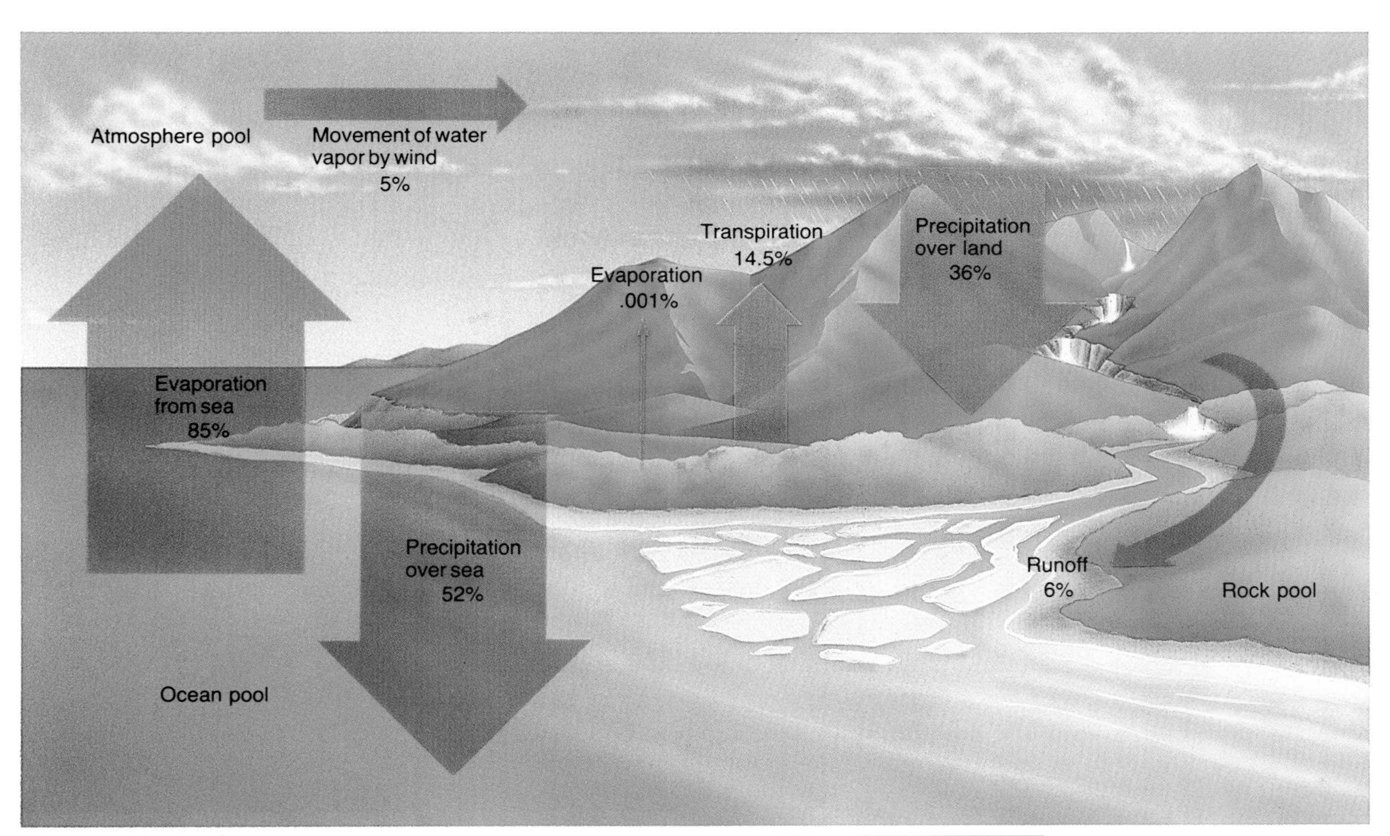

FIGURE 41-12
The hydrologic (water) cycle.

The largest reservoir of water, about 97 percent of the earth's available water, is in the oceans; only about 0.001 percent is present as water vapor in the atmosphere. Yet, despite this huge difference, the greatest quantities of water are exchanged between the oceans and the atmosphere via evaporation and precipitation. Over 80 percent of the liquid water that is evaporated during the hydrologic cycle comes from the oceans and moves into the atmosphere. About 75 percent of water in the atmosphere falls back into the oceans as precipitation, or rain. The remainder either stays in the atmosphere in the form of clouds, ice crystals, and water vapor, or falls back to earth as rain over the land.

Although some of the rain that falls on the land is intercepted by vegetation generally rain reaches ground level and either percolates into the soil or runs off the soil surface into streams and rivers, where it may eventually flow back to the ocean. Some water is pulled downward into the ground by gravity, contributing to ground water supplies, most of which eventually flows back to the ocean. Of course, some of the water in the ground is absorbed by plants. For many plants, more than 90 percent of the water taken in by roots is released back into the atmosphere as water vapor through transpiration. The remainder hydrates plant cells and tissues or is used in biochemical reactions, particularly photosynthesis. In comparison with other biogeochemical cycles, organisms seem to play a relatively minor role in the hydrologic cycle.

The Carbon Cycle

Large amounts of carbon are continually exchanged between the atmosphere and the community of organisms. As a result, the recycling rate of carbon is especially rapid (Figure 41-13). In the **carbon cycle,** primary producers constantly extract carbon dioxide from the atmosphere and use the carbon to form the chemical backbone for building organic molecules. Organic molecules are then disassembled by all organisms during cellular respiration, releasing carbon as carbon dioxide.

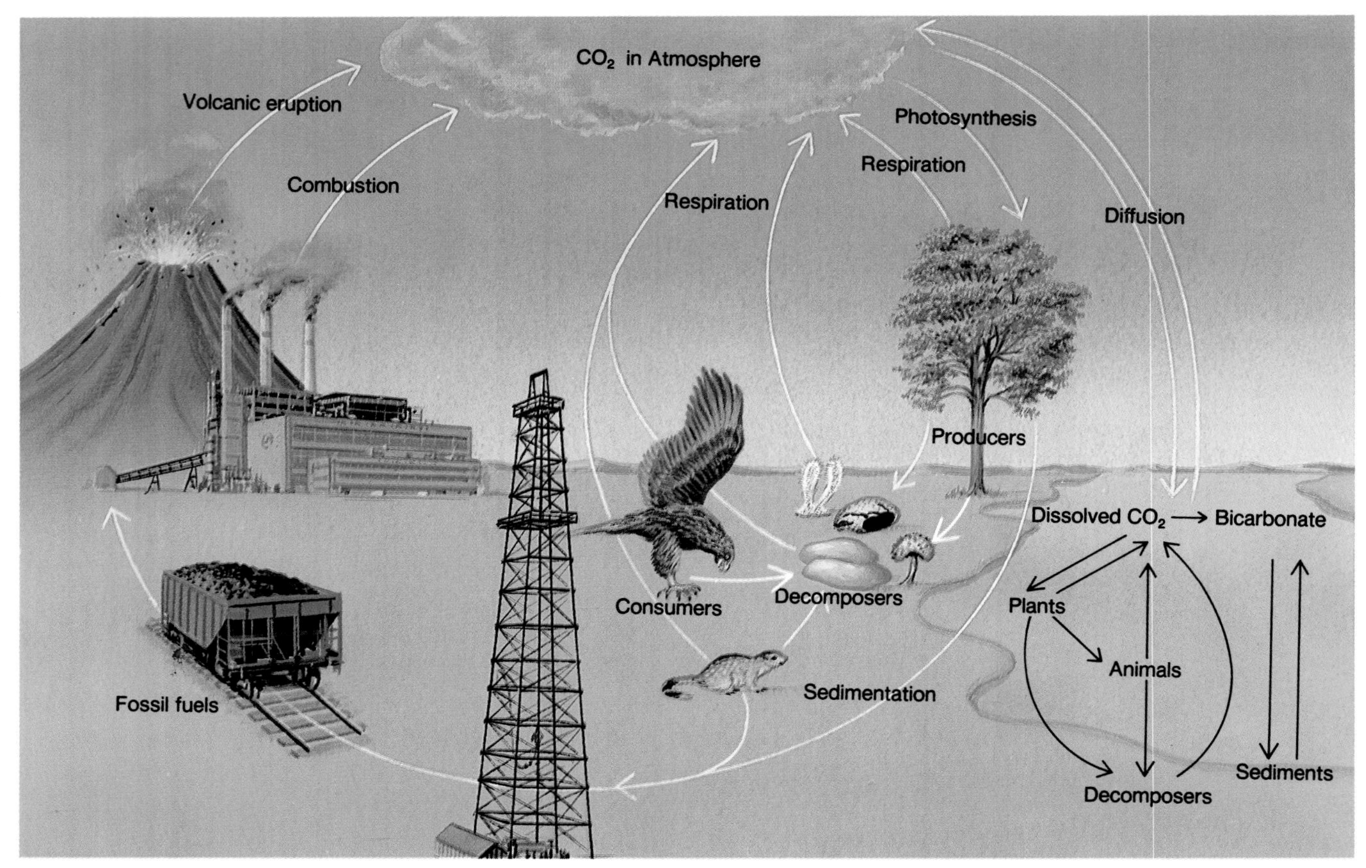

FIGURE 41-13

The carbon cycle. Atmospheric and dissolved carbon dioxide are used by primary producers to make energy-rich organic compounds during photosynthesis. When producers are eaten, carbon-containing organic compounds are passed to consumers and, eventually, to the decomposers. All organisms (producers, consumers, and decomposers) release carbon as carbon dioxide during respiration, most of which is returned to the atmosphere. Organisms that are not decomposed may eventually form fossil fuels. Combustion of fossil fuels also releases carbon as carbon dioxide back into the atmosphere.

On land, producers extract carbon dioxide gas directly from the atmosphere. In water, gaseous carbon dioxide must dissolve before it can be incorporated into organic compounds by aquatic autotrophs. Not all of the carbon dioxide dissolved in water is used in photosynthesis. Resulting carbon-containing bicarbonates and carbonates, compounds with low solubility, eventually settle to the bottom of oceans, streams, and lakes. Additional carbon sediments form as the skeletons and shells of organisms accumulate. The settling of carbonate, shells, and skeletons can tie up carbon for long periods of time in sediment, limestone, and several other forms of rock.

Even on land, dead organisms and organic matter may not be decayed by decomposition, a process that would ordinarily release carbon dioxide back into the atmosphere. Nondecayed organic material may build up deposits that turn into fossil fuels, producing a carbon reservoir within the earth. Most of the world's fossil fuels were formed during the Carboniferous period, between 285 million and 375 million years ago, when shallow seas repeatedly covered vast forests, preventing decomposition. As we burn fossil fuels, we return this carbon to the atmosphere (see The Human Perspective: The Greenhouse Effect: Global Warming).

The Nitrogen Cycle

Although the earth's atmosphere is 79 percent nitrogen gas (N_2), only a few microorganisms are able to tap this huge reservoir, initiating a series of conversions and transfers that produces the **nitrogen cycle** (Figure 41-14). These "nitrogen-fixing" microorganisms include bacteria and cyanobacteria.

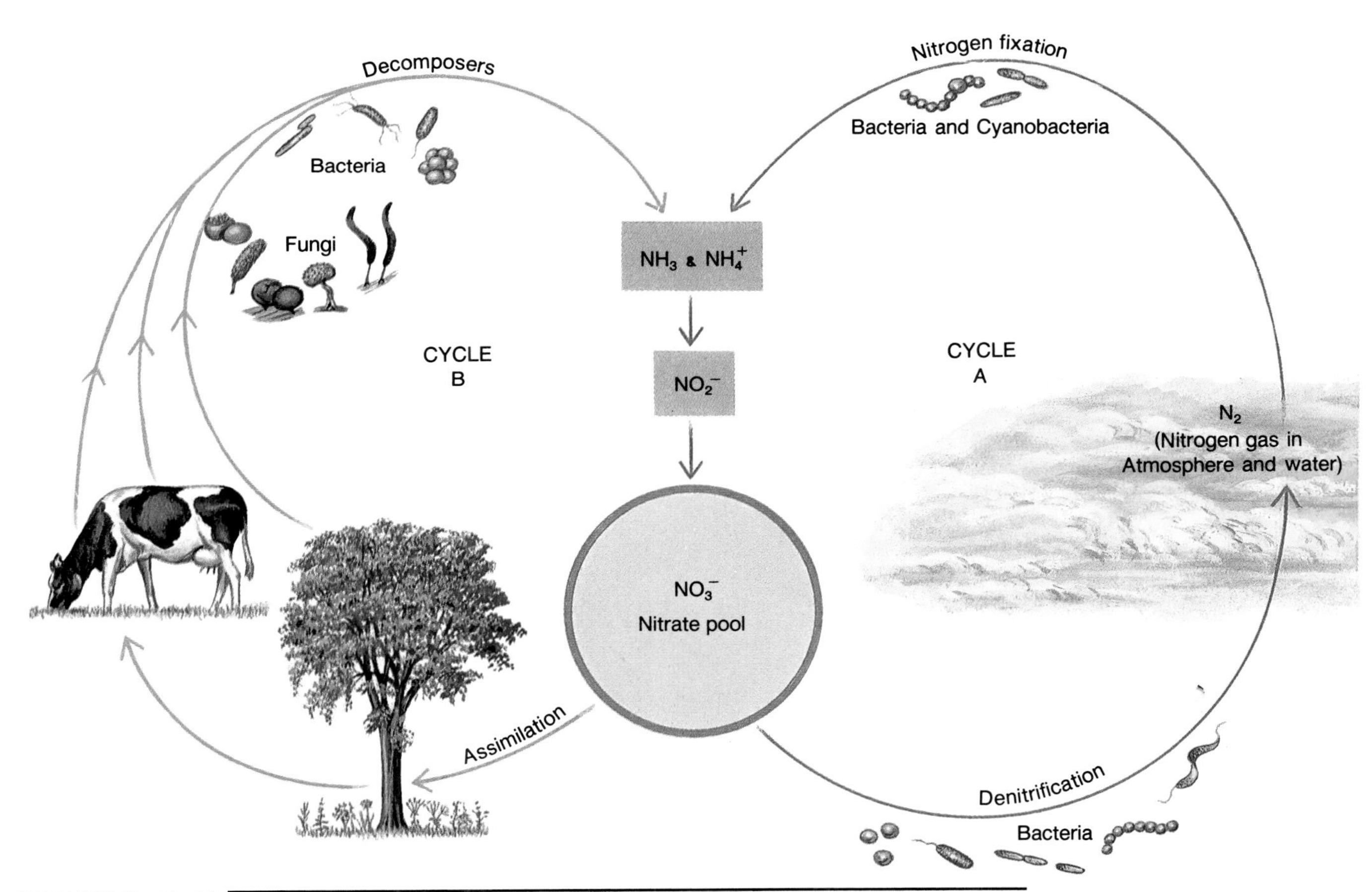

FIGURE 41-14
The nitrogen cycle. A pool of nitrates joins together two cycles that exchange nitrogen between the biotic community and the abiotic environment. In cycle A, nitrogen gas from the atmosphere is converted to ammonia or organic compounds by bacteria during *nitrogen fixation*. The ammonia is converted to nitrites and then nitrates by nitrifying bacteria. The nitrates are either converted by other bacteria into nitrogen gas during *denitrification* (completing Cycle A), or they are absorbed by primary producers and the nitrogen incorporated into nitrogen-containing organic compounds (beginning Cycle B). The ammonia that is excreted by consumers or released during decay is converted into nitrites and then into nitrates, returning nitrogen to the nitrate pool (completing Cycle B).

◁ THE HUMAN PERSPECTIVE ▷

The Greenhouse Effect: Global Warming

The **greenhouse effect** is a term that has been coined to describe the rise in global temperature that is triggered by increased amounts of gaseous pollutants, mainly carbon dioxide, that trap heat within the atmosphere. Gaseous pollutants absorb heat-generating infrared radiation from the earth, preventing it from escaping into space at night. The greenhouse effect in the atmosphere is similar to what happens when a car is left parked in the sun with its windows closed. The car's windows are like the gases in the atmosphere: They allow the sun's radiant energy to enter the car, heating the interior but preventing the inside heat from escaping; the car gets hotter and hotter.

As levels of carbon dioxide and pollutants increase in the atmosphere, they trap more and more heat, causing global temperatures to rise gradually, a phenomenon referred to as "global warming." By analyzing air bubbles that formed in the ice of Antarctica and Greenland over the past 1,000 years, scientists have documented an increase in atmospheric carbon dioxide of over 25 percent; 85 percent of this increase occurred between 1870 and 1989, mostly from the burning of fossil fuels. At the current rate, the global carbon dioxide level is expected to double by 2050, raising the average global temperature between 2°C and 5°C (3.5°F and 9°F). The major consequences of global warming include the following:

- The sea level will rise during the next century. Using computer models, scientists predict that rising sea levels may cover the homes of more than 20 million Americans who live on the East Coast and ruin rice production in Asia. Most of the rice grows in low-lying regions that would be flooded with salt water. Increases in sea temperature will also begin melting the polar ice caps, contributing to rising sea levels. Satellite photographs show that the polar ice caps have shrunk by 6 percent over the past 15 years.
- Reduced rainfall and rising temperatures will amplify the world's hunger crisis. Most of the world's food supply is currently produced in the band of agricultural land found in North America, Europe, and the Soviet Union. Climatic changes in these regions will cause crop production to fall, shifting farming northward to mountainous areas that are more difficult to farm.
- Many marine organisms breed in coastal estuaries, marshes, and swamps. As seas rise, these vital areas will be flooded, and many plant and animal species will be lost, as will much of our seafood supplies.

Since carbon dioxide has the biggest impact on global warming, reducing activities that produce carbon dioxide will help curb the problem. Much of the carbon dioxide that is released into the atmosphere originates from burning fossil fuels to produce electricity and deliberate burning of huge expanses of tropical rain forests. Thus, reducing energy consumption and saving tropical rain forests (see the Human Perspective, Chapter 21) would make a significant difference.

Nitrogen fixers convert atmospheric nitrogen gas to ammonia (NH_3) in a process called **nitrogen fixation.** Once nitrogen is converted, other groups of soil bacteria, collectively known as the *nitrifying bacteria,* convert ammonia into nitrites (NO_2) and then nitrates (NO_3). Plants absorb nitrates and incorporate the inorganic nitrogen into organic molecules: nucleotides and amino acids, the fundamental building blocks of DNA, RNA, and proteins. When the producers are eaten or die, nitrogen is passed on to the consumers and decomposers in a food chain. The decomposers then convert the nitrogen-containing organic molecules into inorganic ammonia, which is also released directly from the consumers (in urea) as a means of eliminating the excess nitrogen that would otherwise accumulate and poison the organism.

Once again, ammonia is converted into nitrites and then into nitrates by nitrifying bacteria, making nitrogen available for absorption by plants. Some nitrogen is converted to nitrogen gas during **denitrification** by other bacteria, called **denitrifying bacteria;** nitrogen fixation is therefore needed to renew usable nitrogen resources. Some plants form nodules on their roots, housing nitrogen-reducing bacteria, thereby receiving a direct source of nitrates (refer to Figure 20-1*b*).

The nitrogen cycle illustrates the critical role microorganisms play in the biosphere. Without bacteria, there

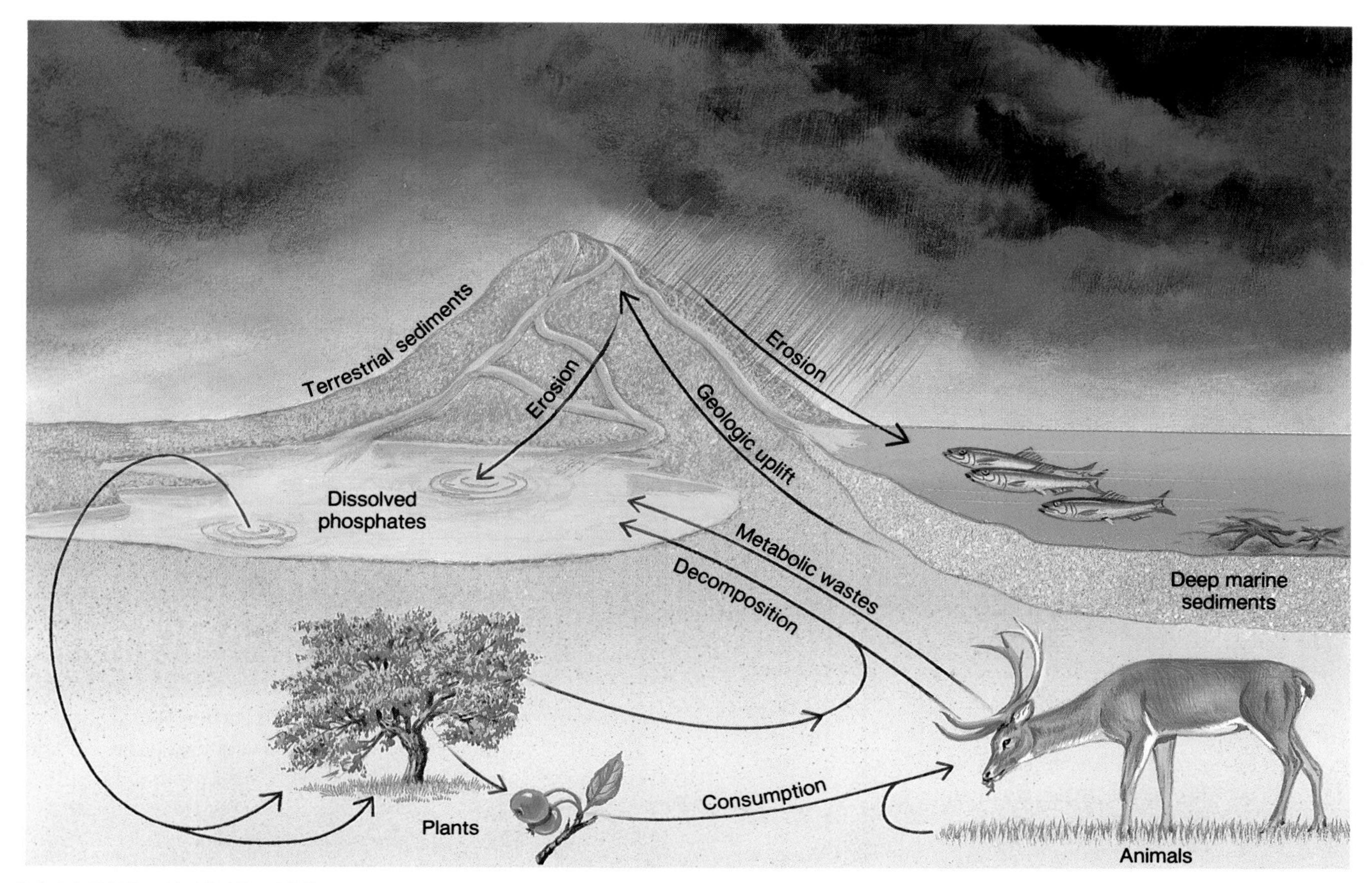

FIGURE 41-15

The phosphorus cycle. Phosphorus has a *sedimentary nutrient cycle* because most of the element is found in sedimentary rocks. Geologic uplift raises phosphate sediment, and erosion caused by waves and rain dissolves the phosphorus. Some dissolved phosphates are absorbed by primary producers, incorporated into organic compounds, and then passed to the other organisms in a food chain. However, the majority of dissolved phosphates runs off and accumulates as precipitated solids at the bottom of streams, lakes, and oceans, where it becomes part of new sediments. Decomposition and wastes from some animals, such as birds, also add to the dissolved phosphate pool.

wouldn't be sufficient nitrogen recycled to support the diversity of life on earth. Without bacteria-dependent nitrogen fixation, there wouldn't be any source of usable nitrogen for producers or the consumers they support. In other words, without bacteria, life on earth would cease.

THE PHOSPHORUS CYCLE: A SEDIMENTARY NUTRIENT CYCLE

Some elements never, or only rarely, exist as a gas. Such elements accumulate in the soil or rocks and, as a result, have a sedimentary nutrient cycle. Elements with sedimentary cycles include calcium, iron, magnesium, sodium, and phosphorus. We use the phosphorus cycle (Figure 41-15) to illustrate the general features of a sedimentary nutrient cycle.

All organisms require large amounts of phosphorus to construct ATP, DNA, RNA, and cellular membranes. Most phosphorus is contained in rock deposits. Erosion and runoff from rain dissolve the phosphorus in rocks and form phosphates (PO_4^{-2}). Plants and other primary producers absorb phosphates and use the phosphorus to build organic molecules. When these primary producers are eaten, phosphorus is passed to the primary consumer and then to the other organisms in a food chain. Decomposers break down phosphorus-containing compounds and release it back into the environment as phosphates, which are either reabsorbed by plants or leached out of the soil, where they accumulate in sediments. Since phosphorus is easily leached from soil, it is one of the least available essential elements in the biosphere. The large quantities of phosphorus needed by organisms often makes this element a major limiting factor in many ecosystems.

IMPORTANT LESSONS FROM BIOGEOCHEMICAL CYCLES

All nutrient cycles share certain characteristics:

- The abiotic environment provides the principal reservoir of elements in air, water, or earth.
- The introduction of elements into the biotic environment almost always requires primary producers (mainly plants and algae).
- Microorganisms play a crucial role in nutrient recycling.

Nutrient cycles illustrate a very important and fundamental ecological principle: organisms depend on other organisms. Even technologically advanced human societies ultimately depend on plants and simple microorganisms to supply their essential nutrients. Without them, nutrients could not recycle, and all life on earth would quickly grind to a halt. This is one of the reasons why many biologists are so concerned about how human activities have accelerated the extinction rate of many organisms (page 997). Because of their crucial role in nutrient recycling, we must pay just as much attention to the extinction of plants and microorganisms as we do to large, familiar animals.

Unlike energy, nutrients recycle through ecosystems over and over again, enabling life to persist on earth over immense periods of time. Nutrients are exchanged back and forth between the abiotic and biotic environments through the activities of primary producers and decomposers. (See CTQ #6.)

SUCCESSION: ECOSYSTEM CHANGE AND STABILITY

Ecosystems are constantly changing as energy and elements flow from the abiotic environment to the biotic community and as organisms interact within the biotic community. Ecosystems also change as seasonal shifts generate fluctuations in both the abiotic and biotic environments. Changes triggered by regular, seasonal fluctuations make an ecosystem dynamic, but they do not cause *permanent* changes in the composition and organization of organisms in the biotic community. However, permanent changes in the biotic community do occur in newly formed habitats and in areas disturbed by fire, floods, hurricanes, drought, or the activities of humans. Such large-scale changes trigger the process of **succession,** a progression of distinct communities that eventually leads to a community that remains stable and perpetuates itself over time.

THE CLIMAX COMMUNITY

Communities that remain more or less the same over long periods of time, such as an area of mature forest or grassland, are called **climax communities.** The populations of organisms that make up a climax community are in equilibrium with their abiotic environment. Thus, the kinds of organisms and their abundance remain relatively constant over long periods.

Climax communities tend to contain many species in a highly organized trophic structure. Generally, large amounts of organic compounds are manufactured by producers of climax communities, but consumers and decomposers use nearly all of the excess, so the total biomass does not increase. In addition, the majority of plant species that make up the climax community are long-lived, and a comparatively small portion of their energy is used for reproduction; most is diverted for growth.

Severe or long-term changes in either the abiotic or biotic environments can permanently change the organization and composition of the biotic community, however. For example, cycles of declining temperatures during the Pleistocene Epoch produced a series of ice ages that destroyed and permanently altered most communities in the biosphere. Another large-scale factor that often causes permanent changes in a community is fire. Fires not only kill many organisms outright, they also modify conditions so much that different organisms predominate in the affected area.

When a community has been permanently changed, or when an entirely new habitat is formed (such as following a volcanic eruption or when a glacier retreats), a variety of species invade the area, forming a *pioneer community.* As pioneer species take hold, they modify the environment by changing the soil, the temperature of the ground, the amount of light that penetrates to ground level, and many other environmental characteristics. Eventually, the pioneer community changes conditions so much that new species invade the community. The new group gradually displaces the pioneer community and forms its own community. The process continues—one community replacing another—until a stable, climax community develops. This orderly, directional sequence of communities that leads to a climax community is succession. The entire series of successional communities, from pioneer to climax, forms a successional *sere.* Each community in a sere is called a *stage.*

Ecologists recognize two types of succession: 1. **Primary succession** occurs in areas where no community existed before (new volcanic islands, deltas, dunes, bare rocks, or lakes); and 2. **secondary succession** occurs in disturbed habitats where some soil, and perhaps some organisms, still remain after the disturbance. Fires, floods, drought, and many human practices (such as clearing forests for agriculture and construction projects) would prompt secondary succession. Secondary succession also occurs on abandoned farmlands, in overgrazed areas, and in forests cleared for lumber.

PRIMARY SUCCESSION

New habitats do not remain bare for very long. Even a dry rock is soon colonized by lichens, a "compound organism" composed of a fungus and resident algae (Figure 41-16). The hyphae of the fungus are able to grow into even the tiniest rock fissures, prying the rock open. At the same time, hyphae secrete chemicals that help erode the rock. The combination of intrusive growth and chemical erosion, together with abrasion from the wind and water and repeated heating and cooling, gradually crumbles the rock into small fragments, forming sand. When the lichens die, they mix with the sand, initiating the process of soil formation. As sand and organic matter accumulate, moss spores and grass seeds are eventually able to germinate and grow. These organisms continue the process of disassembling and chemically dissolving the rock. When they die, their remains add even more organic matter to the developing soil. Eventually, the soil is rich enough to support the growth of large plants that outcompete the lichens, and the pioneer lichen community becomes replaced by a new community of mosses, grasses, ferns, and other plants. After hundreds of years, a forest climax community may grow in an area that was once nothing but bare rock.

The transformation from bare rock to forest is only one example of primary succession. Primary succession also takes place in lakes (see Figure 40-11), on hardened lava flows, and in any other newly exposed substrate. Primary succession in the sand dunes along the shores of Lake

FIGURE 41-16
Primary succession on a bare rock begins with lichens. Eventually, the lichens crumble the rock, initiating soil formation. The process continues over many years, until what was once bare rock may eventually support a desert, woodland, or even a forest, depending on the climate.

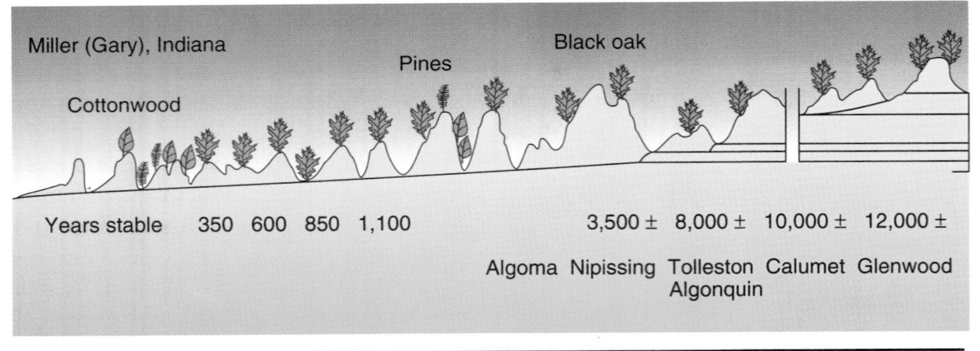

FIGURE 41-17
Primary succession on dunes along Lake Michigan. This diagram illustrates a profile across the sand dunes of Lake Michigan. The level of the lake is gradually falling, uncovering new sand substrate, on which primary succession occurs. Since regions along the shore have emerged most recently, they demonstrate the earliest stages of dune succession. Farther away from the shoreline, successively older stages of dune succession occur. These older dune systems originated along earlier and higher beaches.

FIGURE 41-18
The eruption of Mount St. Helens in Washington on May 18, 1980, initiated both primary and secondary succession. Near the center of eruption, species of bacteria quickly populated warm, standing pools that formed from rains and melting snows, beginning primary succession in each pool. Away from the eruption center, the force of the eruption severed trees at their base, seared off branches, and blew over burnt trees like toothpicks. Soil remained in both these outer areas, where secondary succession is now occurring. This series of photographs were taken from the same spot ***(a)*** 3 months, ***(b)*** 2 years, ***(c)*** 4 years, and ***(d)*** 9 years after the May, 1980 eruption.

Michigan has been investigated extensively. Since the retreat of the glaciers that formed the Great Lakes area during the Pleistocene Epoch, the level of Lake Michigan has gradually been falling, creating a nearly ideal situation for studying sand-dune succession. The initial substrate is dune sand (nearest the lake shore), but areas progressively further from the shore are successively older, each the product of an increasing length of time it took for primary succession to occur (Figure 41–17). Bare sand is colonized by grasses, particularly marram grass *(Ammonphila breviligulata),* which spreads quickly asexually by rhizomes, stabilizing the dune surface. After 6 years or so, marram grass dies out and other grasses become predominant. Cottonwoods *(Populus deltoides)* are the first trees to appear. After 50 to 100 years, pines invade the dunes but are then replaced by black oaks after about 100 to 150 years. The oldest dunes, dated at 12,000 years, still had black oak associations, which were most likely the climax community.

SECONDARY SUCCESSION

Primary succession generally continues over hundreds or thousands of years, but secondary succession is often completed in less time, for two reasons. First, soil is already present, so the process that takes the longest is already completed. Second, in the presence of fully developed soil, plants from the surrounding communities, along with any surviving plants or seeds, quickly colonize the area. As a result, the pioneer community may include individuals of species that are part of the climax community, giving these species an early foothold in the disturbed area.

Of course, rates of secondary succession vary widely, depending on the degree of disturbance, the climate, and the kind of climax community. For example, a cleared area in a tropical rain forest may recover within 10 years, a burned chaparral in 20 years, and a grassland in 40 years; an abandoned farm may revert back to a deciduous forest in 150 years; and a disturbed area in the desert may take several thousands of years to recover, if it recovers at all.

The first eruption of Mount St. Helens in 1980 initiated both primary and secondary succession (Figure 41-18). Near the center of the eruption, primary succession began almost immediately as bacteria populated warm pools. Farther away from the eruption center, secondary succession is currently taking place in areas where tree trunks were severed at their base by the blast and where trees remain standing but were burned during the eruption. The rate of secondary succession is different in each zone because of the varying degree of disturbance.

Although ecosystems bustle with activity as organisms continually interact with each other and with the physical environment, only large-scale changes in the biotic or abiotic environments trigger permanent change in the biotic community. Such change produces a succession of replacement communities that eventually leads to a community that remains stable over long periods. (See CTQ #7.)

REEXAMINING THE THEMES

Biological Order, Regulation, and Homeostasis

Nutrient cycles take place within ecosystems. These critical biogeochemical cycles require order and regulation for the transfer of chemical elements between the abiotic and biotic environments and within the biotic community. Generally, a number of organisms are critical to recycling elements, including some of the simplest microorganisms. Without the participation of essential organisms, particularly primary producers and microorganisms, chemical elements will not recycle. Without recycling of finite amounts of elements, life would eventually grind to a halt.

Acquiring and Using Energy

Unlike chemical nutrients, energy does not recycle through ecosystems. Primary producers, primarily photosynthesizing plants, capture and convert radiant energy from the sun into chemical energy. Only primary producers are capable of acquiring energy from the abiotic environment. All other organisms acquire their energy from the primary producers or from some other organism, forming food chains or food webs if all possible transfers within an ecosystem are considered. In a food chain, organisms incorporate and pass on only a small percentage of the energy they acquire. If the energy transfers between organisms are presented graphically, they form a pyramid shape that illustrates the accentuated loss of energy at each transfer.

Unity within Diversity

Despite the tremendous diversity of organisms found on earth, all are dependent on the physical environment to supply the resources they require for survival and reproduction. All organisms need a habitat in which to live; all organisms have tolerance ranges for each environmental factor; and the distribution and survival of an organism is often determined by a single limiting factor. Despite the tremendous diversity of organisms on earth, however, all organisms are similar in that they depend on one another to secure the resources they need for survival and reproduction.

Evolution and Adaptation

An organism's ecological niche is the result of natural selection. Within a specific ecosystem, those traits that enable an organism to acquire more nutrients, to outrun a predator, to find a mate or any other feature that increases survivorship and reproduction are naturally selected. In this way, natural selection leads to adaptations for a particular habitat, with defined role(s) and tolerance ranges for each abiotic factor in a particular ecosystem. For example, organisms with narrow niches have very specific adaptations for specific habitats and roles, whereas the adaptations of organisms with broad niches enable these organisms to secure resources in a variety of ways and, generally, over a wide range of habitats.

SYNOPSIS

Ecosystems are dynamic, self-sustaining units. They are composed of a community of organisms and the surrounding physical environment. Ecosystems are connected to other ecosystems to varying degrees.

Organisms (or species) have a tolerance range for each physical factor. When the maximum or minimum tolerance is approached or exceeded for any given factor, that factor limits the distribution, health, or activities of the organism. Each organism has a suite of adaptations that defines its ecological niche—the combination of an organism's habitat, functional role(s), and total environmental requirements and tolerances.

Energy flows through an ecosystem. Trophic levels, food chains, and food webs track the flow of energy and nutrients between the members of the biotic community.

Essential nutrients cycle between the abiotic and biotic environment, forming biogeochemical cycles. Producers and decomposers are required for exchanging nutrients between the biotic and abiotic environments.

Ecosystems inevitably change. The changes that occur in an ecosystem are triggered by permanent changes that occur in the abiotic or biotic environments.

Key Terms

association (p. 933)
ecosystem (p. 934)
primary producer (p. 937)
consumer (p. 937)
decomposer (p. 937)
detrivore (p. 937)
detritus (p. 937)
tolerance range (p. 938)
Theory of Tolerance (p. 938)
ecotype (p. 938)
limiting factor (p. 938)
Law of the Minimum (p. 938)
habitat (p. 940)
ecological niche (p. 940)
hypervolume (p. 940)
fundamental niche (p. 940)
realized niche (p. 940)
niche breadth (p. 940)
biological control (p. 941)
niche overlap (p. 941)
guild (p. 941)
ecological equivalent (p. 942)
primary consumer (p. 942)
secondary consumer (p. 942)
tertiary consumer (p. 942)
trophic level (p. 942)
ultimate (top) carnivore (p. 942)
food chain (p. 942)
multichannel food chain (p. 942)
food web (p. 942)
biomass (p. 943)
ecological pyramid (p. 943)
pyramid of energy (p. 943)
pyramid of numbers (p. 944)
pyramid of biomass (p. 944)
bioconcentration (p. 946)
biological magnification (p. 946)
biogeochemical cycle (p. 947)
hydrologic cycle (p. 947)
carbon cycle (p. 948)
greenhouse effect (p. 950)
nitrogen fixation (p. 950)
denitrification (p. 950)
denitrifying bacteria (p. 950)
succession (p. 952)
climax community (p. 952)
primary succession (p. 952)
secondary succession (p. 952)

Review Questions

1. Match the example with the term:

Term	*Example*
____1. guild	A. fungi and bacteria
____2. primary producers	B. fruit-eating birds and grass-eating antelopes
____3. primary consumers	C. algae and plants
____4. decomposers	D. two insect-eating bird species

2. List examples, other than those given in the text, of how the biotic community causes changes in the abiotic environment.
3. Give an example of each of the following:
   a. realized niche
   b. Law of the Minimum
   c. secondary succession
   d. food web
   e. ecological equivalents
4. List as many components as you can think of for the fundamental niche and realized niche of a whale (a relatively broad niche) and a tapeworm (a narrow niche). Now try it for humans.
5. Using your experience with house and garden plants, list some effects of limiting factors that you have observed for yourself.
6. Describe some of the ways humans have triggered secondary succession in ecosystems near where you live. Will the original ecosystems eventually return, or will new ecosystems develop?
7. Refer to the nitrogen cycle on page 949. Describe at least three separate pathways that would enable nitrogen to be recycled through the biotic community.
8. Compare the carbon cycle (a gaseous nutrient cycle) and the phosphorus cycle (a sedimentary nutrient cycle). In what ways are they similar? How do these similarities affect the relative rates of recycling for each?
9. Describe how food (energy and nutrients) is transferred from the abiotic environment to the biotic community of an ecosystem. What eventually happens to the energy? What eventually happens to all of the nutrients?
10. Relative to what you have learned in this chapter, why are biologists so concerned about the increasingly rapid extinction of life on earth?

Critical Thinking Questions

1. Had Robert Whittaker decided to test whether groups of organisms have identical distribution ranges in the Great Smoky Mountains by analyzing bird species instead of insects or plants, would you expect the outcome to be the same or different? What about mammals? Would the lifestyle of the organism being investigated (such as whether the organism is stationary like a plant, has a small range like a rodent, or a wide range like a hawk) have any impact on the results? Explain.
2. Linkage between two ecosystems, even if far apart, is illustrated by the forests of the tropics and North America, which provide winter and summer homes, respectively, for many of our most common songbirds. Read "Why American Songbirds Are Vanishing" (*Scientific American,* May 1992) and list all of the factors that are affecting the songbirds. Are there other ways in which these two ecosystems are linked? (HINT: Think globally!)
3. If organisms are so dependent on their physical environment, how do you explain the following? (1) Many plants and animals thrive when introduced into new areas; for example, many European wild flowers thrive as weeds in North America. (2) When environments undergo change, some plants and animals survive, while others are wiped out; for example, removing maple and beech trees from eastern forests promotes the growth of birch and aspen.
4. Like all biomes, a desert encompasses several ecosystems. One type of desert ecosystem is a "wash." Although dry most of the year or often over several years, a desert wash forms as water from rain is channeled, producing a "river," sometimes only the size of only a small trickle, and other times the size of a large flood. Describe as many components of the ecological niches of five organisms (two plants, two animals, and one protist or fungus) that you would expect to find in a desert wash. Are there similarities in niche breadth between the different types of organisms? In which ways do niches overlap between the plants? Between the animals? Does niche breadth and overlap change

over time as water alternates between abundance and scarcity?

5. All ecosystems depend on a flow of energy through the living system and cycling of material elements. Prepare a diagram showing these characteristics of a generalized ecosystem and the role of producers, consumers, and decomposers in the system.
6. Explain why nutrients must be recycled in an ecosystem. What are the natural recyclers? How are human activities affecting natural cycles? Give two specific examples.
7. The following graph presents data taken over 50 years in a small but unique community of organisms that is completely surrounded by a dense forest. The graph illustrates the change in the number of new species and the change in the average height of primary producers. From these data alone, could you determine whether this unique community was stable or undergoing succession? What are the limitations of these data? If you were conducting this research, what other factors would you investigate? How would you survey these factors, and how would you analyze the data you collect to discern whether changes were cyclic or permanent?

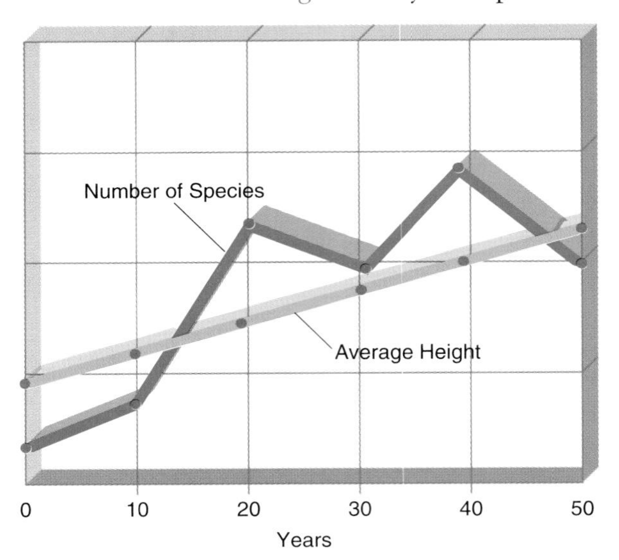

Additional Readings

Brewer, R. 1979. *Principles of ecology.* Philadelphia: Saunders. (Intermediate)

Flannagan, D., and F. Bello (Editors). 1970. *The biosphere.* A Scientific American Book, Scientific American, Inc. San Francisco: W. H. Freeman. (Introductory)

Lewin, R. 1986. In ecology, change brings stability. *Science* 34:1071–1073. (Intermediate)

Odum, E. P. 1983. *Basic ecology.* Philadelphia: Saunders College Publishing. (Intermediate)

Thoreau, H. D. 1937. *Walden.* New York: The Modern Library. (Classic, Introductory)

CHAPTER

42

Community Ecology: Interactions between Organisms

STEPS TO DISCOVERY
Species Coexistence: The Unpeaceable Kingdom

SYMBIOSIS

COMPETITION: INTERACTIONS THAT HARM BOTH ORGANISMS

Exploitative and Interference Competition

Competitive Exclusion: Winner Takes All

Resource Partitioning and Character Displacement

INTERACTIONS THAT HARM ONE ORGANISM AND BENEFIT THE OTHER

Predation and Herbivory

Parasitism

Allelopathy

COMMENSALISM: INTERACTIONS THAT BENEFIT ONE ORGANISM AND HAVE NO AFFECT ON THE OTHER

INTERACTIONS THAT BENEFIT BOTH ORGANISMS

Protocooperation

Mutualism

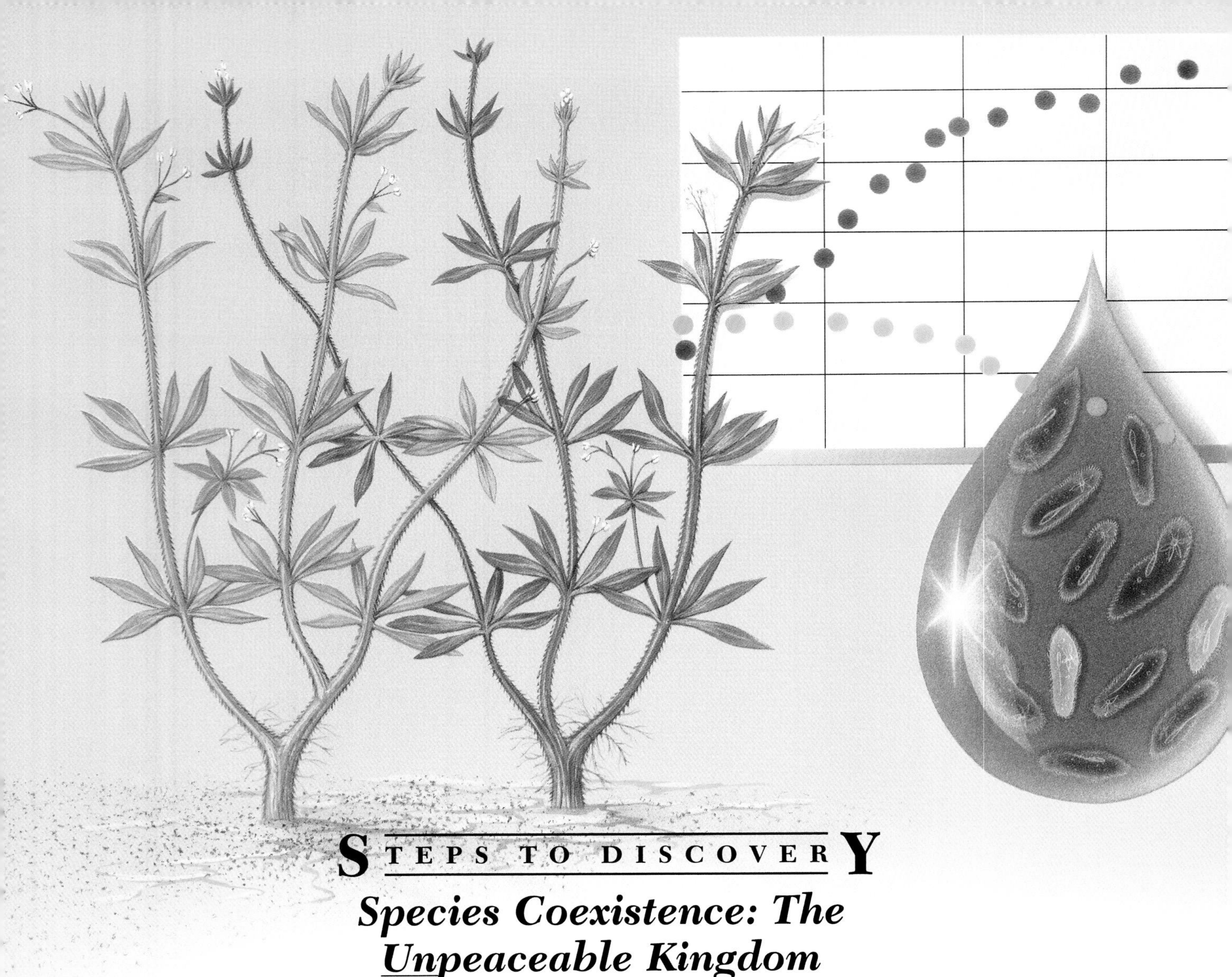

STEPS TO DISCOVERY

Species Coexistence: The Unpeaceable Kingdom

Charles Darwin was an inexhaustible thinker and worker. Not only did he explore the process of evolution, he pursued research in several other areas of biology, including orchid pollination, selective breeding, animal taxonomy, plant movements, and competition. Darwin observed that when two individuals lived in the same area and required the same limited resources—space, food, or whatever—those individuals would compete with one another for that resource. He also noted:

> *As species of the same genus have usually, though by no means invariably, some similarity in habits and constitution, and always in structure, the struggle will generally be more severe between species of the same genus, when they come into competition with one another, than between species of distinct genera.*

In other words, the more closely related the competing individuals, the more similar their needs and the more intense the competition for limited resources.

Since Darwin's time, many scientists have studied competition. These studies eventually led to the formulation of a mathematical principle regarding competition. At the center of this discovery was G. Gause, a Russian microbiologist at the University of Moscow. As in many scientific investigations, Gause acquired insight not only from his own research efforts but from the research of other biologists as well as two mathematicians and a physicist, L. Boltzmann. In 1905, Boltzmann wrote:

As the graphs illustrate, competition between similar species having identical requirements, whether between bedstraw plants

... plants spread under the rays of the sun the immense surface of their leaves, and cause the solar energy before reaching the temperature level of the earth to make syntheses of which as yet we have no idea in our laboratories. The products of this chemical kitchen are the object of the struggle in the animal world.

Although Boltzman refers only to animal competition in this statement, plants also enter into competition with one another, often for the "rays of the sun" to which they expose the "immense surface of their leaves." Indeed, Gause later acknowledged his debt to botanists (plant ecologists, in particular) because their work provided much of the insight and data he needed to formulate his principle. Gause wrote:

Botanists have already recognized the necessity of having recourse to experiment in the investigation of competition phenomena.

One of the first plant ecologists to investigate plant competition was Sir Arthur G. Tansley, who founded the British Ecological Society. In his presidential address to the First Annual General Meeting of the Society, held in 1914, Tansley emphasized the importance of competition among plants in community dynamics and urged fellow ecologists to initiate research on plant competition. Three years later, Tansley reported the results of his own studies on the competition between two species of bedstraw plants, *Galium saxatile* and *G. sylvestre*. Each species of bedstraw is more abundant in different soil types: *G. saxatile* grows best in silica-rich soils, while *G. sylvestre* thrives in limestone soils.

Tansley grew both plants in various soils, including silica-rich soil and limestone, and then monitored germination, seedling survival, and competition. He found that *G. sylvestre* had a higher germination rate, grew more vigorously, and outcompeted *G. saxatile* in lime-rich soils. Eventually, all the *G. saxatile* seedlings in the lime-rich soils died. In contrast, in lime-poor soils, although *G. sylvestre* had a greater germination rate, *G. saxatile* eventually outcompeted *G. sylvestre*. Tansley was one of the first to show that some plants were better able than others to compete in certain soils, suggesting that competition in natural communities influences the distribution and abundance of plants (organisms) in an ecosystem.

From Tansley's work, and that of other plant ecologists, Gause concluded that light, nutrients, water, and pollinators were common limiting resources for which plants compete. Sources of animal competition include water, food, mates, nesting sites, wintering sites, and sites that are safe from predators.

During the 1920s, researchers began formulating mathematical models to account for what happens when two species live together and require the same limited resource or when one species preys on or parasitizes another. One model, the *Lotka-Volterra equation*, was derived independently by Alfred J. Lotka at Johns Hopkins University in 1925 and by V. Volterra in Italy in 1926. This model described competition between organisms for food or space by comparing changes in population growth (increases or decreases in numbers of individuals) as competing species affect each other. According to the Lotka-Volterra equation, one possible outcome of competition is for one competitor to displace the other completely, causing the weaker species to become extinct.

To test whether this "winner-takes-all" outcome really occurs in nature, Gause initiated a number of studies in 1932 to test competition between microorganisms, first between competing species of yeast and then between competing species of protozoa. In Gause's best-known experiment, he monitored two species of *Paramecium* (*P. caudatum* and *P. aurelia*). Each species was first grown in a separate culture and then in a mixed culture, where the species competed for a limited food supply. When grown separately, the number of individuals of both paramecia increased rapidly and then leveled off and remained constant. When cultured together, however, competition for limited food supplies resulted in the elimination of *P. caudatum*, which were outcompeted by the more rapidly reproducing "winner," *P. aurelia*.

Gause concluded from this and from other similar experiments that "the process of competition under our conditions has always resulted in one species being entirely displaced by another." Gause's experiments supported the Lotka-Volterra equation and the "winner-take-all" outcome of competition and eventually became known as *Gause's Principle of Competitive Exclusion*.

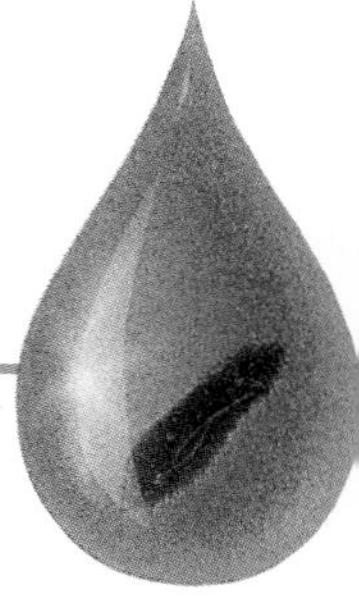

or paramecia (in droplet), results in one species completely outcompeting the other.

Although seemingly calm, ecosystems actually bustle with activity. During warmer months in a forest, for example, the soil teems with bacteria, fungi, nematodes, springtails, amoebas, mites, slugs, worms, beetles, spiders, and scores of other organisms that churn the ground as they move about, grow, and reproduce. As they erupt through the soil surface, delicate plant seedlings absorb the nutrients recycled by microorganisms and fungi. These seedlings eventually grow into herbs, shrubs, and trees that create habitats and manufacture food for countless herbivores, which are, in turn, devoured by an assortment of carnivores.

Although the participants vary from one ecosystem to the next, all ecosystems are similar to the forest described above in that the organisms that live together often interact with one another. Some of these interactions benefit one or both participants; some have neutral consequences; and some harm either or both participants. The general categories of interactions and the eventual outcome for the participants are previewed in Table 42-1.

▼ ▼ ▼

SYMBIOSIS

Some organisms interact because they physically live together or because they live in very close association with one another. A close, long-term relationship between two individuals of different species is called **symbiosis,** which literally means "to live together" (Figure 42-1). Symbiotic interactions include forms of parasitism, commensalism, protocooperation, and mutualism (Table 42–1).

Many biologists now believe that symbiosis played a critical role in the early stages of the evolution of eukaryotic cells. According to the *endosymbiont theory,* the organelles of eukaryotic cells—mitochondria, chloroplasts, and flagella—are descended from once free-living prokaryotes that developed symbiotic relationships with primitive eukaryotic cells (see Chapters 5 and 35). A great deal of evidence exists to support this theory.

There are many modern examples of symbiotic relationships, involving very different types of organisms, such as between fungi and plants (mycorrhizae, page 386), plants and bacteria (root nodules, page 389), fungi and algae (lichens, page 821), sea anemones and fish, and jellyfish and algae. One interesting example involves an oyster and a crab. The larvae of the crab enters the mantle cavity of the oyster. The crab then grows and resides inside the oysters for its entire life. Since it remains sheltered all its life by

(a)

(b)

FIGURE 42-1

Symbiosis: living together. ***(a)*** *Aphids and ants.* Some ants live with groups of aphids. The aphids feed on the sugary juices of plants, often taking in more than their bodies can hold. The excess juice passes through the aphid's digestive system and out its anus, forming a honeydew drop that is lapped up by the ants. The ants protect the aphids by aggressively keeping predators away (such as ladybird beetles and syrphid fly larvae). ***(b)*** *Anemones and algae.* This anemone is green because algae live in its body. The algae conduct photosynthesis, producing food and oxygen for the anemone. In turn, the algae receive carbon dioxide and a safe habitat inside the animal's tissues.

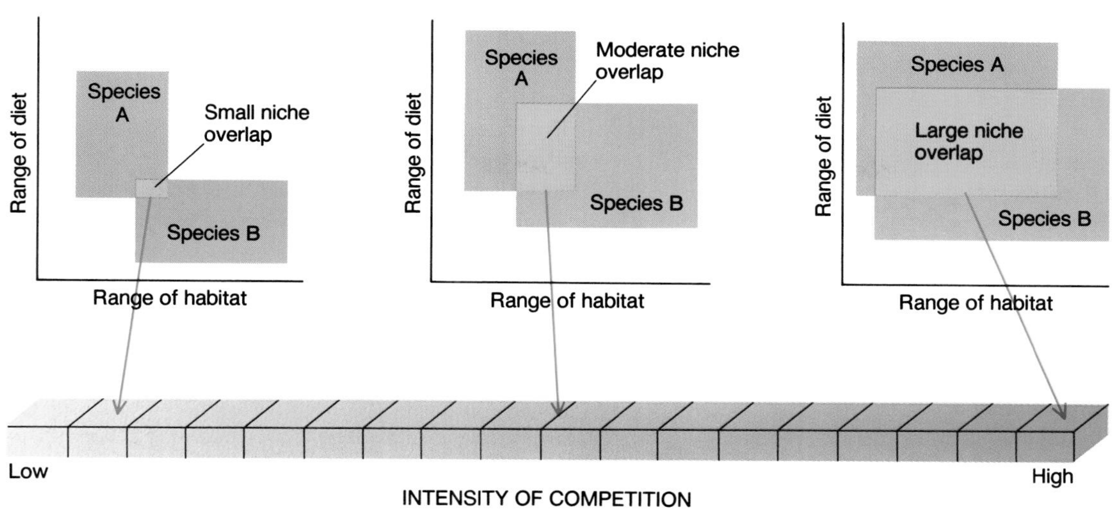

FIGURE 42-2

Niche overlap and competition. In each of the three graphs, habitat and diet requirements are plotted for two species. When niches overlap, species compete for limited resources. The greater the similarity in the requirements of the species, the greater the niche overlap, and the more fierce the competition.

the shell of the oyster, the crab benefits, while the oyster apparently remains unaffected by the crab's presence.

Organisms rely on other organisms. Because of these reliances, some organisms pair up to form a symbiotic relationship, whereby one or both of the organisms provides the needs of the other. (See CTQ #2.)

COMPETITION: INTERACTIONS THAT HARM BOTH ORGANISMS

Recall from Chapter 41 that species can coexist in a community as long as they have slightly different ecological niches, even though their niches may overlap. When a shared resource is abundant, such as oxygen in the air of terrestrial habitats, or water in aquatic habitats, there is more than enough for all. Generally, however, most resources are limited so organisms with overlapping niches enter into **competition.** Competition always harms both participants because each competitor reduces the other's supply of a needed resource. The more similar the requirements of the organisms, the greater their niches overlap; the greater the niches overlap, the more intense the competition (Figure 42-2). Furthermore, since members of the same species require many of the same resources, **intraspecific competition** that occurs between members of one species is often more intense than is **interspecific competition** that occurs between members of different species.

▶ Organisms are powerful agents of natural selection. In fact, they are just as powerful as are environmental factors, if not more so. Competition between organisms is also a powerful natural selection force. As we will see, competition can lead to the extinction of one competitor, to the exclusion of one competitor from an ecosystem, or to rapid evolutionary changes in the characteristics of the competitors.

EXPLOITATIVE AND INTERFERENCE COMPETITION

Organisms compete either *directly* or *indirectly* for a limited resource. Indirect competition occurs when competi-

TABLE 42-1

INTERACTIONS BETWEEN ORGANISMS IN A COMMUNITY

Kind of Interaction	*Organism 1*	*Organism 2*
Competition	Harmed	Harmed
Predation (including herbivory)	Benefited	Harmed
Parasitism[a]	Benefited	Harmed
Allelopathy	Benefited	Harmed
Commensalism[a]	Benefited	Unaffected
Protocooperation	Benefited	Benefited
Mutualism[a]	Benefited	Benefited

[a] May include symbiotic interactions.

tors have equal access to a limited resource but one species manages to get *more* of the resource, reducing the competitor's supplies. This form of indirect competition is called **exploitative competition.** An example of exploitative competition is currently taking place in the California deserts between deep-rooted native plants and newly introduced tamarisk trees. Tamarisk trees were brought to the California deserts from the Middle East to act as windbreaks along freeways and railroad tracks. The rapidly growing and reproducing tamarisk trees are better able to tap groundwater supplies, interfering with water availability to the deserts' native trees, such as mesquite and desert willows, reducing their populations.

In contrast to exploitative competition, **interference competition** is direct: One species directly interferes with the ability of a competing species to gain access to a resource. Interference competition is exemplified by aggressive behavior in animals, as when hyenas drive away vultures from the remains of a zebra, or by **territoriality,** as when male bighorn sheep establish and defend an area against other males of their species (see Chapter 44 on Animal Behavior).

COMPETITIVE EXCLUSION: "WINNER-TAKES-ALL"

Although species with small niche differences are often able to live in the same community, those with *identical* ecological niches cannot do so, even if they share only one scarce resource and many abundant resources. Competition becomes so intense in this case that one species eventually eliminates the other from the community, either by taking over its habitat and displacing the species from the community or by causing the species' extinction. The "winner" species is successful because it possesses some characteristics (adaptations) that give it a slight advantage over its competitor. This advantage enables members of the winner species to capture a greater share of resources, which, in turn, increases the survival and reproduction of the individuals. Consequently, greater numbers of offspring with better-suited traits gradually displace members of the less efficient species, illustrating natural selection in action. As we mentioned in the introduction to this chapter, this "winner-takes-all" outcome is referred to as Gause's principle of **competitive exclusion.**

One example of competitive exclusion involves the day lily *(Hemerocallis),* a popular cultivated plant. Day lilies often escape gardens and begin multiplying along roadsides and in surrounding communities by forming thick clumps of shoots and roots. Once the day lily becomes established outside a garden, few native plants can compete against it; eventually, they become displaced from their original habitat.

The day lily is an example of a plant that becomes established in a surrounding community that does not naturally include day lilies; it is not an example of competitive exclusion between members of the same community. In natural communities, competitive exclusion is not always apparent because there are so many variables to monitor. For instance, six species of leafhoppers *(Erythoneura)* are able to live on a single sycamore tree, feeding side by side on the same leaves. Not only are the habitats and food source of the insects the same, but the species' life cycles are virtually identical as well. In fact, researchers could not find any niche differences among the six species, a seemingly perfect setup for competitive exclusion. Yet, investigations found no evidence whatsoever that these species harm one another, much less that they exhibit competition that results in exclusion. Perhaps competitive exclusion is avoided in this case because shared resources are abundant.

Competitive exclusion is also often not recorded in ecosystems where environmental conditions change frequently. Under continually changing conditions, there simply is not enough time for one species to displace another during the short periods when resources become limited. This occurs in ocean upwellings and in temperate lakes during seasonal overturns because changes develop too quickly for one species of phytoplankton to grow enough to exclude another, despite intense competition between the species for limited nutrients. Similarly, since steady changes occur during primary and secondary succession (page 953), competitive exclusion is not apparent in communities undergoing succession.

A third explanation for why the process of competitive exclusion may go unnoticed in natural communities is the length of time it takes for one species to exclude another. Researchers are often unable to observe a community continuously so they may miss the process of exclusion entirely. For example, goats were introduced on the island of Abingdon in the Galapagos Islands in 1957. The goats browsed on the same low-growing plants as did the island's native tortoises as well as on the leaves found on higher stems and branches. In the absence of predators, the goats reproduced rapidly and consumed all the low-growing food that could be reached by the tortoises. By the time a research team revisited the island in 1962, all of the tortoises were gone. Competitive exclusion had caused the extinction of the Abingdon tortoise over a 5-year period, but the researchers had missed it.

RESOURCE PARTITIONING AND CHARACTER DISPLACEMENT

Competitive exclusion is not the only outcome of competition. Sometimes, a shared resource becomes partitioned in a way that allows competitors to use different portions of the same resource. For example, five species of North American warblers feed in slightly different zones on the same spruce tree, enabling these very similar birds to coexist with minimum competition (Figure 42-3). In addition to such spatial partitioning, a shared resource may be exploited at different times, producing temporal partitioning. An example of temporal partitioning occurs in a grassland ecosystem, where a species of buttercup *(Ranunculus)* grows only

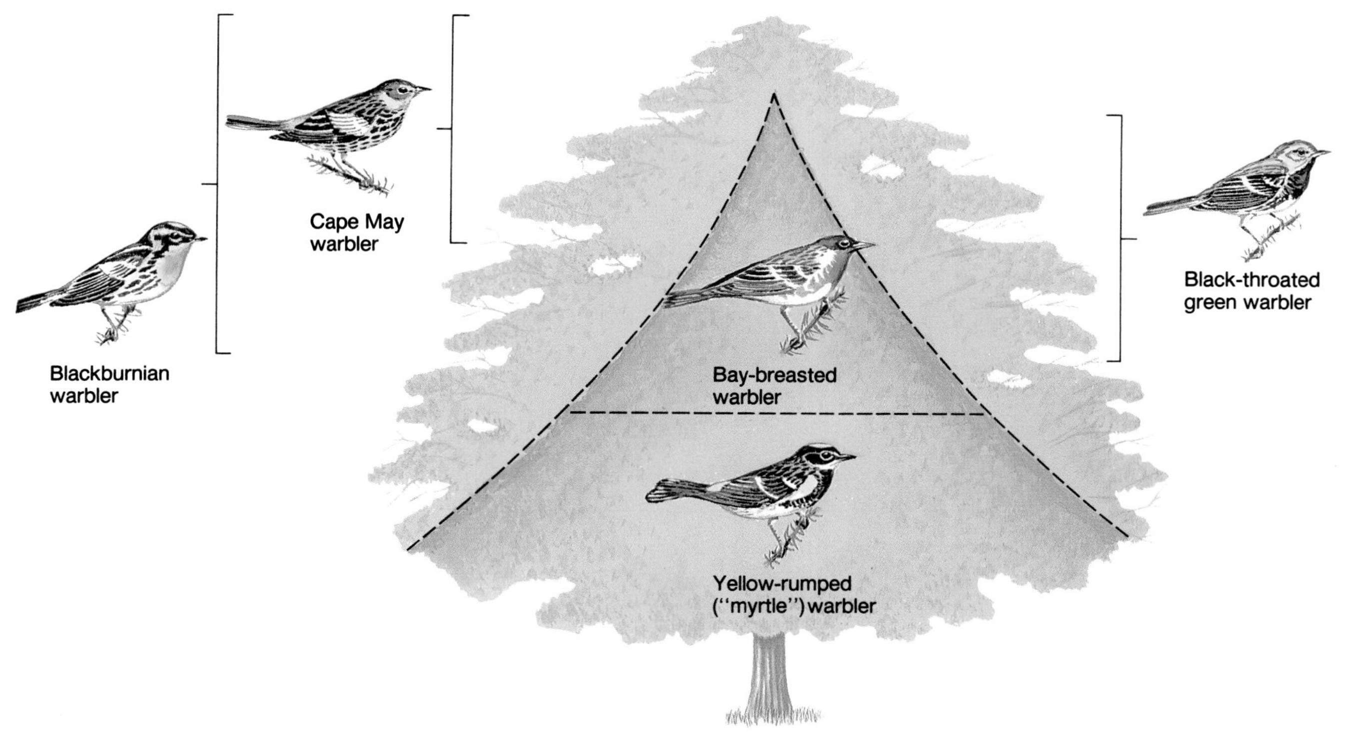

FIGURE 42-3

Resource partitioning: dividing the ecological pie. Five species of North American warblers feed in the same spruce tree, but each feeds in a slightly different zone. Foraging in different areas of a common resource at the same time is one form of resource partitioning. Resource partitioning reduces competition, enabling species with similar ecological niches to coexist in a community.

in early spring, before competing perennial grasses begin to grow. Dividing a resource in time or space is known as **resource partitioning.**

Another alternative outcome to competitive exclusion is **character displacement,** whereby intense competition dramatically affects evolution, leading to changes in one or more characteristics of a species. Imagine two bird species that both harvest the same kind and size of fruit. If, as a result of natural selection, one bird species had evolved a different size bill than the other—a bill that can harvest larger fruits, for example—competition between the species would decrease.

The concept of character displacement was formulated in 1956 by William L. Brown and Edward O. Wilson, both of Harvard University, mainly from the observations of the feeding habits of two insect-eating bird species known as nuthatches. The ranges of these two species of nuthatches overlap only in some places. In these overlapping regions, however, the size and coloration of the bills of the two species are very different. In contrast, in regions where the birds' ranges do not overlap, the size and coloration of the bills of both species in adjacent areas are identical. Scientists postulated that, in overlapping regions, interspecific competition resulted in a selective pressure that resulted in the evolution of different bills. Since there was no such competition in nonoverlapping regions, there was no selective pressure for different bills to evolve.

Competition between organisms is always harmful to both of the individuals involved because each receives less than what it could in the absence of competition. Competition can result in the exclusion of one species from a community or in the species' extinction, or it can lead to behavioral and evolutionary changes that reduce harmful interactions. (See CTQ #3.)

INTERACTIONS THAT HARM ONE ORGANISM AND BENEFIT THE OTHER

All organisms need a source of energy and nutrients to survive and reproduce. When one organism supplies a resource to the other, or is *itself* the resource, the two organisms must interact. Natural selection has produced a bat-

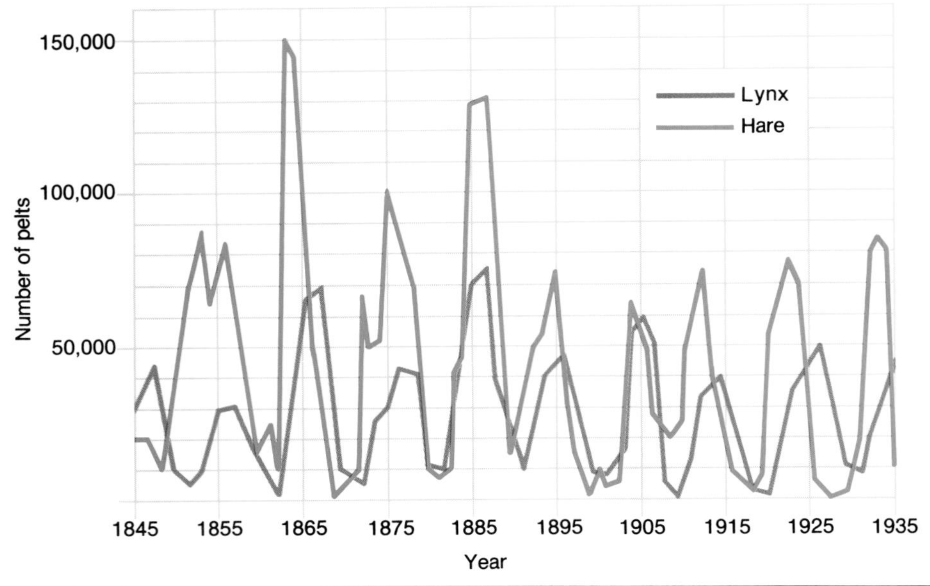

FIGURE 42-4
The ups and downs in numbers of predators and prey. This graph plots the number of lynx and snowshoe hare furs sold by trappers to the Hudson's Bay Company in Canada between 1845 and 1930. As you can see, increases and decreases in the number of prey (the hare) triggered increases and decreases in the number of predators (the lynxes), and vice versa. Other factors may also have affected the lynx and hare populations. Overcrowding and fluctuations in plant growth may have altered the availability of food for the hares, while outbreaks of disease and climate changes could have affected the size of the lynx population.

tery of adaptations that help organisms secure needed resources from others and that help organisms defend themselves from becoming a resource.

PREDATION AND HERBIVORY

During **predation,** one organism (the **predator**) acquires its needed resources by eating another organism (the **prey**). If the prey is a primary producer, the interaction is called **herbivory;** plant-eating animals are called **herbivores.** Organisms that eat other animals for energy and nutrients are **carnivores,** and those, like humans, that eat a mixed diet of plants and animals are **omnivores.**

Predator and Prey Dynamics

Although some predators limit their diets to one type of prey, most rely on more than one species for nourishment. The choice often depends on the abundance and accessibility of prey. During the summer, for example, a red fox mainly eats meadow mice. As the availability of meadow mice dwindles in the cooler seasons, however, the fox shifts to eating the more abundant white-footed mice.

As the availability of prey increases in an area, so does the number of predators; more prey feed more predators. More predators consume greater numbers of prey, however, reducing the availability of prey. In turn, the number of predators drops. This reciprocal interaction generates recurring cycles of increases and decreases in predator and

TABLE 42-2
HOW SOME ORGANISMS AVOID BECOMING PREY

Escape Adaptations	*Effect*
Camouflage	
Cryptic coloration	Hides from predator
Disruptive coloration	Distorts shape and confuses predator
Individual responses	
Startle behavior	Confuses predator
Playing dead	Confuses predator
Shedding body parts	Escapes capture
Outdistancing predators	Escapes capture
Group responses	Warn, protect, and confuse
Defense Adaptations	
Physical defenses	
Armor	Deters an attack
Aposematic	Advertises noxious trait
Mimicry	
Müllerian mimicry	Noxious species avoided
Batesian mimicry	Harmless or palatable species avoided
Chemical Defenses	
Poisons	Kills predators
Hormones	Disrupts predator development
Allelochemicals	Repels predator

prey abundance, resulting in a balance whereby the number of prey remains mostly in balance with the number of predators.

An example of a predator–prey relationship is illustrated in Figure 42-4, where the number of lynxes (the predator) and the number of snowshoe hares (the prey) are plotted for a 90-year period. As you can see, when the number of hares increased, the number of lynxes also increased. As the number of lynxes grew, the lynxes ate more hares, lowering the hare population. A drop in available prey caused a drop in the number of predators, and so on, over the 90-year period. As in other predator–prey relationships, the number of predators (lynxes) lags slightly behind that of the number of prey.

Predator and Prey Adaptations and Defenses

Coevolution between predators and prey over long periods of time has produced an array of remarkable and effective adaptations; some improve the skills of predators in capturing prey, while others improve the prey's chances of escaping predators. Adaptations that aid prey survival can be grouped into two general categories: (1) those that help prey escape being eaten, and (2) those that help a prey defend itself against predator attacks (Table 42-2).

Camouflage Some prey go unnoticed by predators because they blend in with their surroundings or because they appear inanimate (like a dried twig) or inedible (like bird droppings). Such adaptations are called **camouflage** (Figure 42-5) because the color, shape, and behavior of an organism make it difficult to detect, even when in plain sight. Camouflage is not reserved exclusively for prey, however; predators also use camouflage to help conceal themselves while waiting to ambush prey.

FIGURE 42-5

Camouflage: nature's masqueraders. ***(a)*** The "vine" "growing" on this branch is really a grass green whip snake *(Dryophis)*. ***(b)*** This woodcock (directly in the center of the photograph, facing the left) blends in with its surroundings so effectively that is usually escapes detection by predators. ***(c)*** The Malaysian horned frog has adaptations that help the animal blend in with dead leaves lying on the forest floor. Such adaptations include shading (which conceals the frog's eyes) and curly horns that resemble drying leaf tips. These characteristics help make the horned frog virtually invisible to its prey. ***(d)*** Transparent wings help the Costa Rican clearing butterfly *Ithomia* virtually disappear.

(a) *(b)*

FIGURE 42-6
Keeping a "low profile." The fur of the long-tailed weasel *(Mustela frenata)* changes from white in winter ***(a)*** to brown in summer ***(b)***. Both cryptic colors help the weasel integrate into its surroundings and escape detection by predators.

The camouflage of some organisms helps these individuals resemble their background. This type of adaptation is called **cryptic coloration** because the camouflaged organism is hidden from view. The long-tailed weasel exemplifies cryptic coloration. This animal changes color to match the seasonal changes in its surroundings (Figure 42-6). During winter, the weasel's pure white coat helps the animal blend in with the snow (and perhaps conserve heat), whereas during summer, a brown coat helps the weasel blend with the forest floor. The plumage color of several species of grouse and hares also changes seasonally. Laboratory experiments with willow grouse *(Lagopus lagopus)*, for example, reveal that the length of daylight (photoperiod) triggers hormonal changes that coordinate color change.

Disruptive coloration disguises the *shape* of an organism, as in the coloration of the moth shown in Figure 42-7. The color pattern breaks up the outline of the moth when the individual is resting on a dark tree trunk, concealing its shape. Disruptive coloration sometimes means camouflaging vital parts, frequently the organism's eyes, since many predators use eyes as an attack target. With the organism's real eyes camouflaged, the predator's attention is often diverted away from a vital part of the prey (the head). If attacked, an individual that has false eyes in a less vital area of the body (such as a wing) usually escapes with only minor damage (Figure 42-8*a*). Furthermore, false eyes may not only divert a predator's attack; it may also threaten or startle a predator by making the prey appear larger than it really is (Figure 42-8*b*).

Cryptic and disruptive color adaptations are not used solely by animals. Cryptic coloration in some plants helps them resemble less palatable plants, and cryptic and disruptive coloration helps camouflage some plants from herbivores (Figure 42-9).

FIGURE 42-7
Disruptive coloration disguises the shape of this moth as it rests on a tree trunk, creating an image that goes unrecognized by its sharp-eyed predators, the birds.

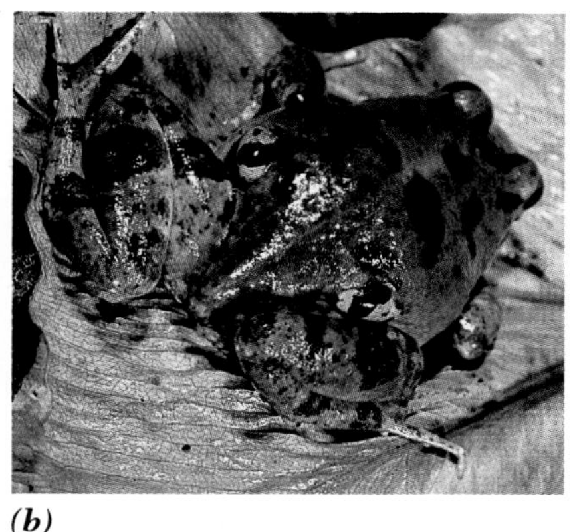

(a) *(b)*

FIGURE 42-8
An eyefull. ***(a)*** In addition to false eyes, the Malaysian back-to-front butterfly *(Zeltus amasa)* deflects predator attacks with false legs and fake antennae on its hind wingtips. An attack on this rear end does not damage vital organs, so the butterfly can dart to safety, minus only a wing fragment. ***(b)*** When confronted with a predator, this South American frog bends over and puffs up its body, revealing large, false eyes. The startled predator is usually frightened away by the intimidating display. If the predator proceeds with the attack, the frog releases an unpleasant secretion from glands located near each false eye.

In addition to coloration, the shape and behavior of an organism also contribute to a successful disguise. The crab spider, African thorn spider, and potoo bird in Figure 42-10 all remain motionless, reducing the chance that they will be detected.

Individual Responses When confronted with a predator, some prey rely on sudden escape responses. Generally, the predator is momentarily stopped by the unexpected response, especially if the escape response seems dangerous. This moment's hesitation may give the prey a chance to escape. Examples of such "last-ditch" responses include the following.

- An owl fluffs its feathers and spreads out its wings, a last-minute bluff that usually startles an attacking hawk.
- A mosquito fish frantically splashes on the surface of a pond when approached by a voracious pickerel (a small fish), making it difficult for the pickerel to launch a pinpoint attack.
- A tiny bombardier beetle sprays hot chemical irritants at a rodent, thwarting the attack.
- An opossum "rolls over and plays dead" when confronted by a coyote; the discouraged predator often searches elsewhere for a fresher meal.

FIGURE 42-9
Appearing more like stones than plants, pebble plants *(Dinteranthus)* usually go unnoticed by passing herbivores.

(a) *(b)* *(c)*

FIGURE 42-10

Nature's impostors. *(a)* The Borneo crab spider resembles bird droppings, a disguise that lures butterflies and other insect prey that eat genuine droppings. At the same time, birds, the predators of the spider, stay clear of what appears to be their own wastes. *(b)* The "twin thorns" on this acacia branch are really parts of an African thorn spider, resting motionless during the day to avoid predators. At night, the spider spins a web to catch insect prey. *(c)* Unwary animals fall easy prey to this dead tree trunk (really a motionless potoo bird), with its eyes half open and neck outstretched. The adult potoo is not the only one with a perfect disguise; its single-spotted egg also blends with the broken tree stump, camouflaging it from predators.

A few organisms may escape a predator's clutches by releasing the seized part of their body. For example, a lizard quickly detaches its tail, which continues to move for several minutes, keeping the predator occupied while the lizard scurries to safety to begin regenerating a new tail.

Finally, some animals escape becoming prey simply by outrunning their predator. A healthy antelope or impala, for instance, can usually outdistance a lion. Like many predators, lions generally capture the young or the weak. By removing the young and weak from the breeding population, predators act as a powerful natural selection agent for the prey population.

Group Responses Schools, packs, colonies, and herds typically defend themselves more effectively than can a single organism. For example, the first smelt fish to notice an approaching predator releases chemicals into the water, which immediately send the school of smelt fleeing in various directions. The confused predator does not know which way to turn. In grazing herds, stronger individuals generally surround the younger and weaker, protecting them from an advancing predator. There is indeed safety in numbers; a predator is less able to pick out a single target among a swarming group.

Physical Defenses Organisms have evolved an arsenal of anatomic features to help protect themselves against direct attacks. Many have protective shells (mollusks and turtles), barbed quills (porcupines), needlelike spines (sea urchins), and piercing thorns, spines, and stinging hairs (plants) that can discourage even the hungriest predator. In fact, your skin is a protective armor against the daily invasions of millions of microbes. Imagine how effective the 9-inch-thick hide and blubber of a whale is as a barrier; whale blubber sometimes even prevents the penetration of a high-velocity harpoon.

Many foul-tasting, poisonous, stinging, smelly, biting, or in other ways obnoxious animals ironically have striking colors, or bold stripes and spots. This type of defense, called **aposematic coloring,** or warning coloration, is the opposite of camouflage; it makes an organism stand out from its surroundings. (*Aposematic* refers to anything that serves to warn off potential attackers.) The distinctive aposematic stripes of a skunk, for example, advertise to potential predators that this animal can yield an obnoxiously smelly counterattack. It usually takes only one encounter for a potential predator to avoid any future entanglements with a skunk. In plants, the red and black fruits of poisonous nightshades are examples of warning coloration.

A bad taste is often of little help for the individuals being attacked because at least part of the animal must be eaten before the predator notices its foul taste. The species as a whole profits, however, because individual predators learn to recognize the characteristics of a vile-tasting species and to avoid the distasteful individuals, sparing other members of the species. In evolutionary terms, the advantage is clear: The species survives.

FIGURE 42-11
Equally distasteful, the *Acrea* butterfly *(left)* and the African monarch *(right)* are Müllerian mimics. These butterflies are not even closely related, yet they resemble each other almost exactly in color, pattern, behavior, and flavor.

Color is not the only aposematic defense. Recently, researchers learned that some foul-tasting moth species make a clicking sound when they are being pursued by bats. Bats associate the clicking sound with the bad-tasting moths and learn to pursue quiet, tasty species. Some tasty moths avoid becoming prey for the bats by making the same clicking sound as do the foul-tasting moths. The clicking of tasty moths is an example of aposematic sound defense, as well as of **mimicry,** whereby one species resembles another in color, shape, behavior, or in this instance, sound, as a mechanism for defense or disguise.

Mimicry If similar-appearing species are equally obnoxious, the resemblance is called **Müllerian mimicry.** In the tropics, for example, many species of beetles have bright orange wingcases with bold, black tips. When attacked, the beetles release drops of their own foul-tasting blood; just a taste deters a predator, sparing the beetle. Consequently, predators learn to stay away from other similarly colored beetle species. Examples of Müllerian mimicry abound in many tropical butterfly species, such as between monarch and *Acraea* butterflies (Figure 42-11).

In some cases of mimicry, such as the moths mentioned in the previous discussion, a harmless or palatable species gets a "free ride" by resembling a vile-tasting or stinging species. When a good-tasting or harmless species (the *mimic*) resembles a species with unpleasant, predator-deterring traits (the *model*), the similarly is called **Batesian mimicry** (Figure 42-12).

FIGURE 42-12
The deadly moray eel (left) and the harmless plesiops fish (right) are an example of Batesian mimicry. When pursued by a predator, the plesiops fish swims head first into a rock crevice. The shape, color, pattern, and false eye of its exposed tail strongly resemble the head of the dangerous Spotted Moray eel, frightening predators away.

Chemical Defenses Some plants and animals release **allelochemicals**—chemicals that deter, kill, or in some other way discourage predators. Such defenses must be swift and effective; otherwise, the predator may inflict fatal damage before the allelochemical has a chance to take effect. To accomplish this defense, many animals rely on noxious propellants or painful, sometimes deadly, bites and stings to blunt a predator's attack.

Some tropical toads and frogs secrete extremely poisonous chemicals. South American tribesmen simply need to touch the tip of their arrows to the skin of poisonous toads to produce a lethal missile that can kill an animal (including a human) within minutes. Other swift-killing poisons are manufactured by the Japanese puffer fish, the Asian goby fish, and the American newt.

Poisonous animals frequently exhibit aposematic coloration, which serves as a blatant signal of the consequences of an attack to experienced predators. For example, the conspicuous stripes on a poisonous monarch butterfly larvae make it easy for predators to recognize this species (Figure 42-13). Interestingly, although both the larva and the adult monarch butterfly are poisonous, neither form manufactures the toxic chemicals itself. Instead, larvae are poisonous only because they eat milkweed plants that synthesize the toxic chemicals. Adult monarchs are poisonous only because the toxic milkweed chemicals are passed on from the larvae to the adults during metamorphosis. Some nudibranchs also derive their defenses secondhand. Ironically, these nudibranchs arm themselves with their prey's stinging cnidocytes (see Figure 39-9) and use them to defend themselves against their own predators.

PARASITISM

Parasitism is another type of interaction that benefits one organism and harms the other. A **parasite** secures its nourishment by living on or inside another organism, called the **host.** Although parasitism is sometimes considered a form of predation, the victim of parasitism usually survives the interaction, whereas the victim of predation is almost always killed. The *larvae* of some insect parasites are lethal, however. Such larvae are called **parasitoids.** For example, after a female tarantula wasp captures and paralyzes a tarantula with her sting, she then lays eggs in the spider's flesh. When the eggs hatch, the larvae gorge themselves on fresh tarantula tissues, killing the helpless spider, who is literally eaten alive.

Most parasites are **host specific;** that is, their anatomy, morphology, metabolism, and life history are adapted specifically to those of their host. For example, the human tapeworm lacks eyes, a digestive tract, and muscular systems. The combination of adaptations it evolved are suited for living inside human intestines, however. They include

- an outer cover that protects the tapeworm from powerful digestive enzymes yet allows the absorption of nutrients;
- a long, flat shape that creates a maximum absorptive surface area yet prevents obstruction of the host's intestine;
- hooks on its "head," which anchor it to the host's intestinal lining;
- a reproductive system with both male *and* female parts, allowing for self-fertilization. Self-fertilization is an important reproductive strategy in a location where contact with another tapeworm is highly unlikely. (Internal parasites are often little more than reproduction "machines," producing millions of offspring, increasing the chances of infecting a new host.)

Some internal parasites require more than one host to complete their reproductive cycle. For example, the fox tapeworm requires not only a fox, as its name indicates, but also a rabbit. Inside the intestines of a fox, the tapeworm produces hundreds of eggs that are released into the environment in the fox's feces. When a rabbit eats a plant that is contaminated with fox feces, the tapeworm eggs enter the rabbit's digestive system. Once inside the rabbit, the eggs

FIGURE 42-13
Secondhand poison obtained from eating milkweed plants makes the monarch butterfly and its larva toxic to predators. The bold stripes on the larva (below), and the distinctive color and pattern of the adult, broadcast danger to predators. After just a few tastes and subsequent episodes of vomiting, predators quickly learn to avoid these bold patterns.

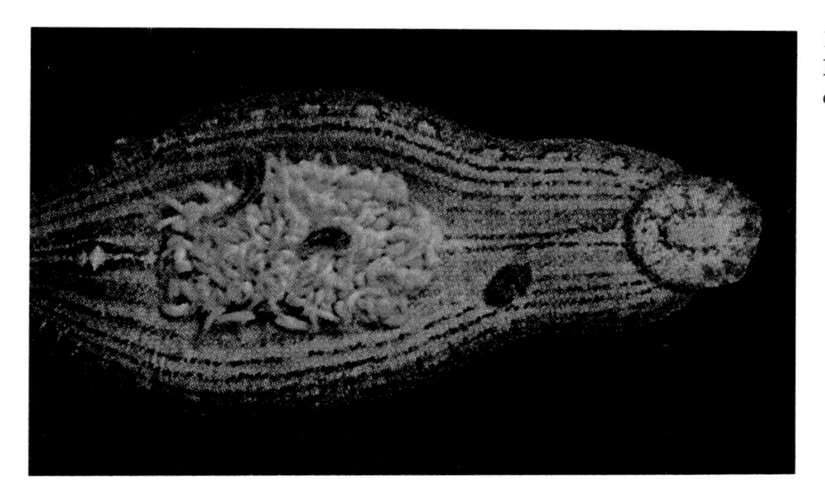

FIGURE 42-14
Even leeches have leeches, illustrating that virtually all organisms—parasites included—have parasites.

hatch, and the larvae bore their way out of the intestine and into the rabbit's muscles. The larvae then form cysts, the resting stage of a parasite. When an infected rabbit is eaten by a fox, the cysts become activated and develop into young tapeworms that attach themselves to the intestines of the new fox host. Within a short time, the fox tapeworm is producing hundreds of eggs a day, and the life cycle begins again.

Apparently, no organism escapes being parasitized. Even parasites can have parasites (Figure 42-14). Animals are hosts to a huge battery of parasites, including viruses, bacteria, fungi, protozoa, and other animals (flatworms, flukes, tapeworms, nematodes, mites, fleas, and lice, for example). The range of plant parasites is equally broad and includes viruses, bacteria, fungi, nematodes, and other plants, such as mistletoe and dodder.

Most organisms are hosts to a number of different kinds of parasites at the same time. A single bird may have 20 different parasites (Figure 42-15), of which there may be hundreds of individuals. For example, researchers counted more than 1,000 feather lice in the plumage of a single curlew.

Although most parasites use their host only as a source of nutrients, some parasites also use their host as a haven for protection from predators. For example, the pearl fish *(Carapus)* develops and lives in a safe, but very unusual place: the anus of a sea cucumber *(Actinopygia)*. As a sea cucumber draws in water through its anus for gas exchange,

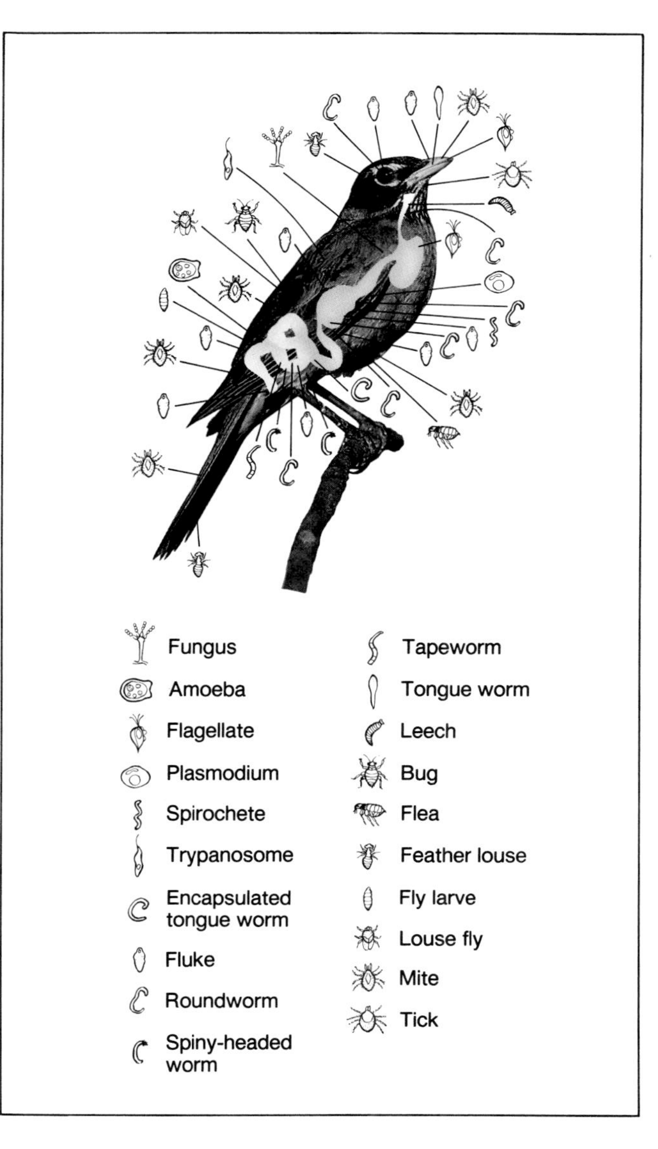

FIGURE 42-15
The invasion begins immediately after hatching. As many as 20 kinds of parasites can infest a single bird.

a newly hatched pearl fish swims in (Figure 42-16). The pearl fish remains inside the sea cucumber, feeding on the host's tissues and taking periodic excursions outside its host to supplement its diet and to reproduce. It returns to the sea cucumber for protection from predators.

Some parasites exploit the behavior of their host, an interaction called **social parasitism.** Examples of social parasitism are provided by European cuckoos, American cowbirds, and African honey guides. After a host bird builds a nest and lays eggs, the parasite bird destroys one of the host's eggs and replaces it with her own. The egg is usually so similar in size and coloration that the host bird fails to recognize it as an alien egg and incubates the parasite's egg as its own. After hatching, the parasitic baby bird instinctively shoves all solid objects out of the nest, including the host's babies and any unhatched eggs. After clearing the nest of its rivals, the parasite snatches up all of the food brought to the nest by its duped "foster parents."

ALLELOPATHY

Some organisms wage chemical warfare on other members of the community. **Allelopathy** is a type of interaction whereby one organism releases allelochemicals that harm another organism. Although some of the animal chemical defenses we described earlier may also be examples of allelopathy, we will confine this discussion to the harmful chemical defenses of plants.

Some plants manufacture allelochemicals that kill herbivores or competing plants (Figure 42-17). For instance, the chemicals released by some chaparral plants accumulate in the soil beneath the plants, blocking the germination and growth of other plants and reducing competition for scarce water and nutrients.

Sometimes, allelochemicals percolate deep into the soil and may reach high enough concentrations to kill the very plant that produced them. This phenomenon is called *autotoxicity*. Although killing oneself goes against a basic "goal" of life—survival—some biologists argue that autotoxicity has adaptive value for the species as a whole. Since it takes many years for allelochemicals to accumulate to toxic levels, only older plants with low reproductive ability die from autotoxicity, reopening space for new, reproductively vigorous individuals.

▮▮▶ Some plants defend themselves against herbivores by fatally poisoning them. Members of the crucifer family (cabbages, broccoli, brussels sprouts, mustards, radishes, and so on) produce mustard oils, chemicals that are lethal to many herbivores, and disease-causing fungi and bacteria. In response, some herbivores have evolved *counteradaptations* that detoxify poisonous allelochemicals. For example, cabbage white butterfly caterpillars (*Pieris brassicae*) have been successfully reared on cultures that contain more than ten times the concentration of mustard oils found in cabbage plants. Apparently, what began as a means of protection against herbivores has backfired; cabbage white but-

FIGURE 42-16

One very unusual habitat for a fish is the anus of a sea cucumber. A young pearl fish has poor eyesight and is barely able to swim, making this animal quite vulnerable to predators. Soon after hatching, the pearl fish locates a sea cucumber. When the sea cucumber opens its anus to draw in water, the parasitic pearl fish enters this unusual, but effective, shelter.

FIGURE 42-17
Plant versus plant. Coastal sages *(Salvia leucophylla)* emit oils that prevent the germination and growth of grasses and other herbs, helping to create bare zones that encircle each plant. The bare areas are effective in keeping away other plants that would compete with the sages for scarce water supplies in the southern California chaparral.

terflies now use the scent of mustard oil to locate plants on which they lay more eggs.

During the 1960s, investigators accidentally discovered that plants produce allelochemicals that disrupt the normal growth and development of insect herbivores. While visiting the United States on a sabbatical leave, Dr. Karel Slama, a Czechoslovakian researcher, attempted to continue his investigations on the development of *Pyrrochoris apterus*, an insect he had been studying for a number of years. After having reared thousands of bugs in his native research laboratory, Slama was unable to raise reproductive adults in the United States, even though he was using the exact procedures he had always followed in Czechoslovakia.

After painstakingly reviewing every step, Dr. Slama discovered a single variable that was different. In Czechoslovakia, he had always reared the bug on filter paper, whereas in the United States he was using common laboratory paper towels. When he extracted and analyzed the chemicals found in U.S. laboratory paper towels, Slama discovered a compound that was virtually identical to a critical hormone that triggers metamorphosis in insects. He tested the effect of the chemical by rearing insects on Czechoslovakian filter paper that had been treated with the chemical and compared the results to the development of insects reared on untreated filter paper. Slama's results were conclusive: The plant chemical disrupted the development of the insect, preventing the development of adult insects with reproductive organs.

Since Slama's experiments, other researchers have discovered similar allelochemicals in some ferns, conifers, and flowering plants. Once again, these chemicals were virtually identical to those insect hormones that coordinate development during metamorphosis. As larvae consume these plants, the allelochemicals cause premature metamorphosis or produce sterile adults. Either way, herbivore reproduction is disrupted, illustrating a very effective plant adaptation for protection against increasing numbers of herbivores. Some of these hormone-mimicking allelochemicals are being considered for use as natural pesticides because these chemicals would cause considerably less environmental damage than do synthetic insecticides.

Organisms secure needed energy and nutrients from the abiotic environment or from other members of the biotic community. Most species on earth (over 70 percent) secure food by consuming part or all of another organism, leading to a diversity of interactions that benefit one species and harm or kill the other. Through natural selection, organisms have evolved a variety of adaptations that help them secure the food they need. Organisms have also evolved adaptations that help them avoid being eaten themselves. (See CTQ #4.)

COMMENSALISM: INTERACTIONS THAT BENEFIT ONE ORGANISM AND HAVE NO AFFECT ON THE OTHER

The benefits of **commensalism** are one sided: Only one of the participants (the commensal) profits, while the other is virtually unaffected. Nature exhibits many examples of commensalism. For instance, remoras are fish that attach themselves by suckers to the undersides of sharks and gather food scraps as the sharks feed. Remoras benefit from this interaction, but their presence apparently has little or no impact on the shark.

Some commensals simply live in a habitat that is created by another organism. The burrows of large "innkeeper" sea worms, for instance, house an array of "guests" that use the burrow for shelter but do not hinder or benefit the innkeeper worm in any way (Figure 42-18). Epiphytes are commensal plants that grow on the branches of taller plants. Being higher up in the forest canopy, the epiphyte captures more light than it could if it occupied a position lower in the canopy. Barnacles encrusted on a humpback whale are also commensal who gain a habitat as well as a means of transportation to new sources of food.

Organisms inhabit a tremendous variety of habitats, including other organisms. As long as there is no disadvantage for either party, natural selection does not select against such associations. (See CTQ #5.)

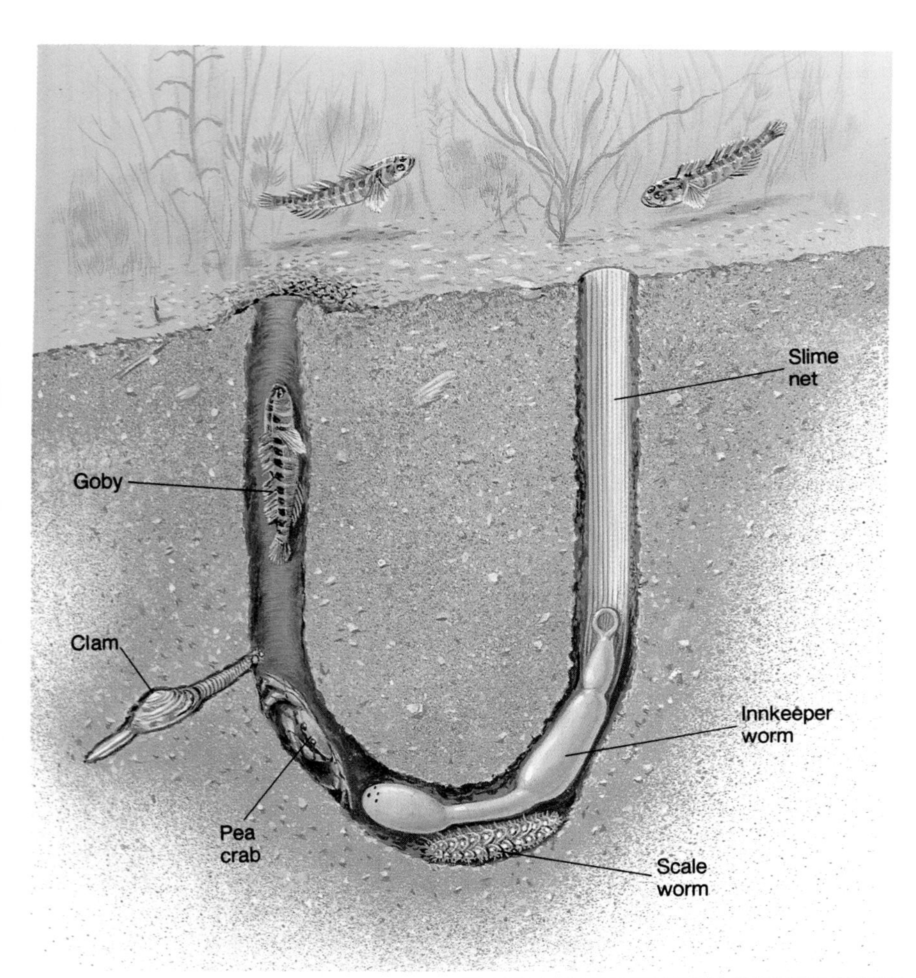

FIGURE 42-18

Commensalism: one benefits, the other is unaffected. The innkeeper worm bores a tunnel in the mud of shallow coastal waters. The worm then spins a slime net that traps minute organisms as the worm pumps water into the tunnel and through the net. When the net is full, the innkeeper gulps down the whole thing—net, trapped food, and all. But the innkeeper worm is not the only occupant of its tunnel. The goby uses the burrow for protection, while the pea crab, clam, and scale worm feast on the innkeeper's leftovers. Although the guests benefit from the association, the innkeeper worm apparently neither profits nor suffers from their presence.

INTERACTIONS THAT BENEFIT BOTH ORGANISMS

Throughout this text, we have seen how natural selection favors characteristics that improve an organism's survival and reproductive success. Interactions between organisms in an ecosystem which provide benefits to both participants are favored by natural selection because the positive interactions contribute to each organism's survival or reproduction. There are many examples of beneficial interactions, some of which are compulsory to both participants, and others that are optional.

PROTOCOOPERATION

Protocooperation interactions benefit both participants but they are noncompulsory. For example, protocooperation between a fungus and algae forms a lichen (page 821). The fungus uses some of the food produced by the algae, while the algae gain a habitat as well as some of the water and minerals absorbed by the fungus. Both organisms benefit from this form of symbiosis, yet they could live successfully on their own as well.

Another example of protocooperation is the relationship between an oxtail bird and a rhinoceros (Figure 42-19). The bird perches on the back of the rhinoceros and removes

FIGURE 42-19 **Protocooperation between oxtail birds and rhinos.** Although neither animal requires the other for its survival, both benefit from their interaction. The bird removes bloodsucking ticks from the rhino, while the rhino supplies the bird with an abundant food supply and warmth.

pests (bloodsucking ticks and flies). The oxtail benefits by receiving food, warmth, and protection from predators; the rhino benefits by receiving protection against parasites. The sharp-eyed oxtail also alerts the dim-visioned rhino of approaching intruders. Again, the relationship is facultative because both animals are capable of surviving on their own.

MUTUALISM

Mutualism is another form of interaction in which both participants benefit. Unlike protocooperation, however, the mutualistic interaction is essential to the survival or reproduction of both participants. Many mutualistic interactions are symbiotic, involving a close association between the participants. The pollination of some flowers by specific insects, birds, or bats (Chapter 21), and the interaction between ants and the *Acacia* plant found in the tropics (Chapter 40) are examples of mutualism that have been discussed earlier in this book.

Through coevolution, the adaptations of many mutualistic partners have become functionally interlocked. The partnership between many species of termites and their intestinal protozoa, for instance, goes beyond the termite's simply providing food and housing for the protozoa, and the protozoa's digesting the cellulose in wood for the termite (Figure 42-20). Coevolution has led to synchronized life cycles between these organisms. In fact, the synchrony is so precise that the internal protozoa are transmitted from one developmental stage of the termite to the next during molting. The same hormones that trigger the termite to molt also trigger the protozoa to encyst. When the termite reingests its gut lining after molting, it "reinfects" itself with its mutually beneficial partner.

Interactions between organisms in an ecosystem which enhance both participant's survival and reproduction are strongly favored by natural selection. (See CTQ #6.)

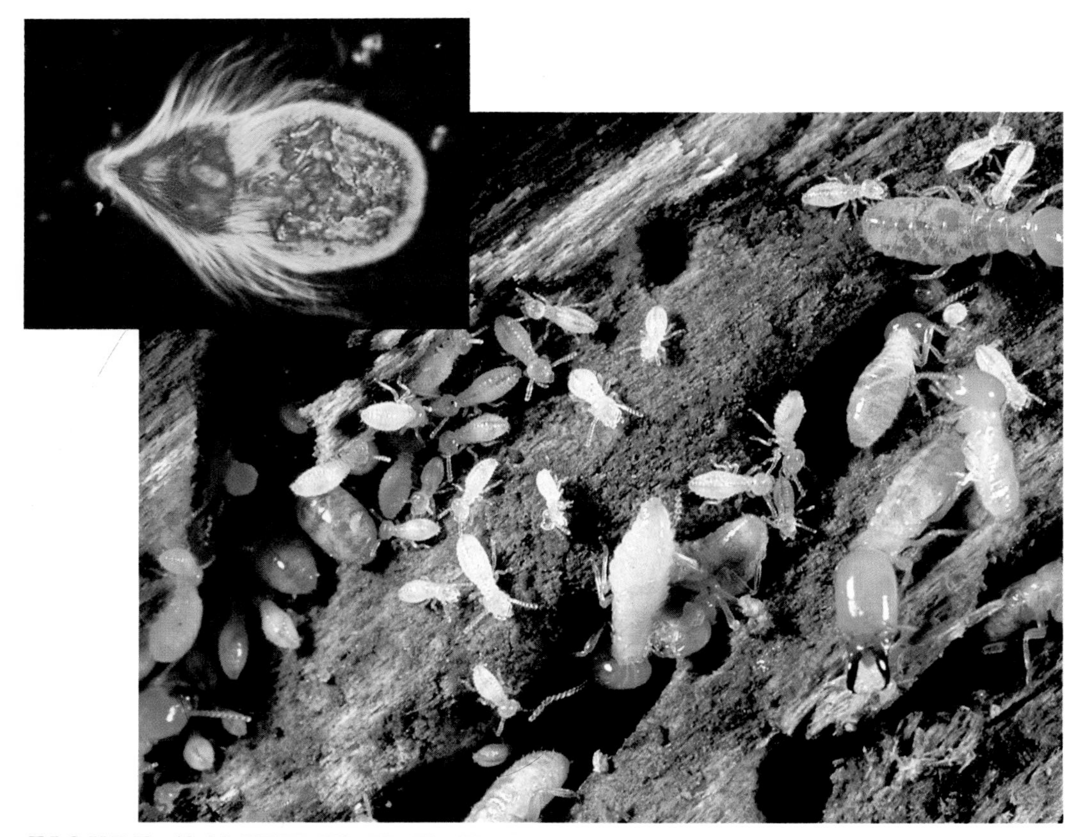

FIGURE 42-20
Mutualism: obligate partnerships with mutual benefits. Without internal protozoa (inset), termites would starve to death because they are unable to digest the wood they consume. Linked through coevolution, protozoa inhabit the gut of termites and obtain a habitat and food supply, while the termites receive a supply of usable nutrients from the digestion of wood by the protozoa.

REEXAMINING THE THEMES

Relationship between Form and Function

Organisms must have adaptations that enable them to survive and reproduce, even if the organisms live in very unusual habitats. Internal parasites have evolved a number of adaptations that enable them to live inside another organism. The hooks on a tapeworm's "head" anchors the parasite in an animal's intestine, preventing it from being flushed out in the current of passing food. Since the probability of meeting a mature individual of the opposite sex in the intestine of another animal is so remote, tapeworms are hermaphrodites, possessing both male and female reproductive structures, enabling them to reproduce alone.

Acquiring and Using Energy

Character displacement helps reduce competition for food. For example, competition is reduced if one type of bill is naturally selected in one bird species, while another type is naturally selected in another competing species. Repeated selection of different bills in competing bird species eventually reduces competition because each bird ends up with a bill that harvests food energy in different ways so the species no longer rely on the identical food resource. Resource partitioning also reduces competition, as species share the same food resource.

Unity within Diversity

Unrelated species sometimes evolve similar adaptations. Müllerian mimicry enables unrelated yet equally repugnant species to benefit from their similar appearance by deterring potential predators. Batesian mimicry allows a palatable species to benefit by resembling an unrelated, yet obnoxious species. All forms of mimicry are examples of similarity in structure and function among diverse species.

Evolution and Adaptation

Camouflage, escape responses, chemical and physical defenses against predators, allelopathy, and mimicry are all examples of adaptations that help organisms survive. Each adaptation evolved as a result of repeated selection of individuals with traits that increase survivability. For example, cryptic coloration allows individuals with coloration and patterns that harmonize best with the background to escape hungry predators more easily than can individuals that stand out. These more cryptically colored individuals will survive and produce more offspring than will less cryptically colored individuals, passing on the adaptive traits. Behaviors can also be adaptive. For example, individual and group escape behaviors are adaptations that result from natural selection.

SYNOPSIS

The organisms that make up the biotic community interact with one another in a variety of ways:

- Competition between organisms harms both participants.
- Predators gain energy and nutrients by consuming prey.
- Parasites live in or on a host organism, damaging or killing the host in the process.
- Commensalism interactions benefit one organism but do not harm the other.
- Both organisms benefit from protocooperation, yet each is able to survive independently.
- Mutualism benefits both interacting organisms, but neither can survive without the other.

When the niches of two species overlap, members of both species compete for the limited resources they require. Intense competition may lead to the sharing of different parts of a resource or of the entire resource at different times or to evolutionary changes in characteristics that reduce competition.

When the ecological niches of two species in a community are identical, competition between the two species results in one rival eliminating or excluding the other from the community. Species can coexist in the same community when they have slightly different ecological niches.

Evolution has resulted in a number of physical and behavioral adaptations that enhance organisms' predatory skills or help organisms escape predators. These adaptations include camouflage, which helps organisms blend in with their surroundings, conceal their shape, or protect vital parts; individual or group behaviors that confuse or distract attackers; anatomic features, such as shells, spines, or armor, that discourage attackers; a foul taste; and chemicals that kill or discourage predators.

Key Terms

symbiosis (p. 962)
competition (p. 963)
intraspecific competition (p. 963)
interspecific competition (p. 963)
exploitative competition (p. 964)
interference competition (p. 964)
territoriality (p. 964)
competitive exclusion (p. 964)
resource partitioning (p. 965)
character displacement (p. 965)
predation (p. 966)
predator (p. 966)
prey (p. 966)
herbivory (p. 966)
herbivore (p. 966)
carnivore (p. 966)
omnivore (p. 966)
camouflage (p. 967)
cryptic coloration (p. 968)
disruptive coloration (p. 968)
aposematic coloring (p. 970)
mimicry (p. 971)
Müllerian mimicry (p. 971)
Batesian mimicry (p. 971)
allelochemical (p. 972)
parasitism (p. 972)
parasite (p. 972)
host (p. 972)
parasitoid (p. 972)
social parasitism (p. 974)
allelopathy (p. 974)
commensalism (p. 976)
protocooperation (p. 977)
mutualism (p. 978)

Review Questions

1. Consider two seedlings of different plant species growing right next to each other in a community. Both grow at about the same rate and develop roots to the same depth. List the resources for which the seedlings will compete as they grow. What is the probable outcome of this situation? What will happen if one plant suddenly outgrows the other?
2. Is there greater opportunity for resource partitioning in a tropical rain forest, a deciduous forest, or a desert? Why? How does each of these terrestrial biomes compare to potential resource partitioning in the pelagic zone of oceans?
3. Match the example with the correct term.

   ____one bee species chases away another species from flowers.
   ____similar monkey species with different-size teeth
   ____similar species of whales visit a feeding bay in different seasons
   ____a plant releases chemicals that stop other plants from growing
   ____contrasting colors distort the shape of a fish as it swims through a reef
   ____a harmless fly looks like a stinging wasp

   a. Müllerian mimicry
   b. Batesian mimicry
   c. disruptive coloration
   d. allelopathy
   e. exploitative competition
   f. interference competition
   g. character displacement
   h. resource partitioning

4. Under what conditions would competitive exclusion not take place in an ecosystem?
5. List some of the reasons why competitive exclusion is rarely observed in natural ecosystems.
6. Use examples to distinguish between cryptic coloration and aposematic coloring. How do these adaptations help prey escape predators?
7. Monarch butterflies and their larvae do not manufacture toxic chemicals, yet both are poisonous to birds. How is this possible? If the poison kills the birds, why wouldn't it kill the larvae or butterfly as well?
8. With the exception of parasitoids, the vast majority of parasites do not kill their host. What advantage is there to killing a host? Must there be an advantage, from a natural selection/evolutionary point of view, in order for there to be any host-killing parasitoids at all?
9. Are herbivores really predators, and are the plants they eat really their prey? If so, why do you think ecologists make this distinction? If not, list the reasons why they should be considered separate.
10. For each of the following pairs of terms, state how they are similar and how they are different.
    a. predation and allelopathy
    b. mutualism and protocooperation
    c. Müllerian and Batesian mimicry

Critical Thinking Questions

1. Each of the six scientists mentioned in the Steps to Discovery (Darwin, Lotka, Boltzman, Volterra, Gause, and Tansley) was an expert in a particular, yet different, scientific discipline (save two). Specifically, what insight was gained from each scientist (discipline) that eventually led to the formulation of the Principle of Competitive Exclusion? Why did this phenomenon become known as a "principle" rather than a "theory" or "hypothesis?"
2. Many symbioses are very specific and permanent. As we discussed in Chapter 20, the pollination of the Spanish dagger *(Yucca whipplei)* by only female pronuba moths is an example of such a relationship. Neither the plant nor the moth can reproduce without the other. From an evolutionary point of view, there are advantages *and* disadvantages to such compulsory and exclusive interactions. List and explain as many advantages and disadvantages as you can. Since there are disadvantages, why would narrow and binding relationships be favored by natural selection at all?
3. In what sense is competition, which is always harmful to the organisms involved, good for the species? How does this concept connect ecology with evolution?
4. More than 70 percent of species obtain their energy and nutrients by consuming all or part of another organism, while only about 30 percent of all species on earth harvest energy and nutrients from the physical environment. These proportions are not always the same for all ecosystems, however. In fact, in some ecosystems (or biomes) the percentages may even be reversed. In which ecosystems would you expect the percentages to be the same, and in which would you expect the percentages to be reversed? Are there any ecosystems in which the biotic community is completely one or the other? With energy being so abundant in the abiotic environment of most ecosystems, explain why 70 percent of species consume other organisms for energy.
5. A biologist who has heard the phrase "nature abhors a vacuum" on many occasions wants to test whether this idea is true or not. As stated, is this a testable hypothesis? If so, design an experiment or series of experiments to test the hypothesis. If not, how could the phrase be reworded so that it could be tested? Design an experiment to test your new hypothesis.

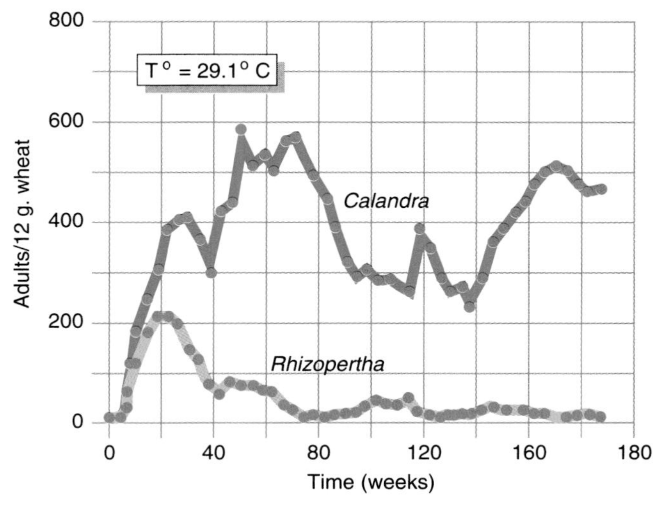

Graph A

6. Study the following graphs on competition between two grain beetles living in wheat at 29.1°C (graph A) and at 32.3°C (graph B). Is the principle of competitive exclusion supported by these data? Altering only one factor (temperature) changed the outcome of competition. Can you offer an explanation for this change?

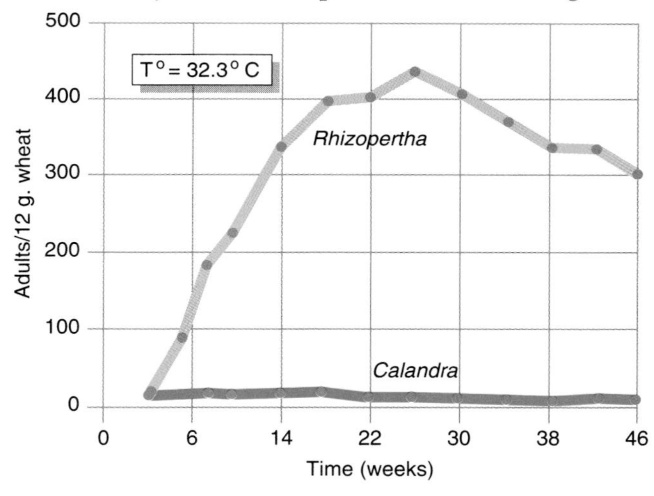

Graph B

Additional Readings

Boltzmann, L. 1905. Populare Schriften. Leipzig. (Advanced)

Barbour, M., J. Burk, and W. Pitts. 1987. *Terrestrial plant ecology.* Menlo Park, CA: Benjamin/Cummings. (Intermediate)

Brewer, R. 1979. *Principles of ecology.* Philadelphia: Saunders. (Intermediate)

Darwin, Charles, 1859. *On the origin of species.* London: John Murphy, p. 76. (Intermediate)

Fogden, M., and P. Fogden. 1974. *Animals and their colors.* New York: Crown Publishers. (Introductory)

Gause, G. F. 1934. *The struggle for existence.* Baltimore: Williams & Wilkins. (Intermediate)

Owen, D. 1980. *Survival in the wild. Camouflage and mimicry.* Chicago: University of Chicago Press. (Introductory)

Putman, R., and S. Wratten. 1984. *Principles of ecology.* Berkeley: University of California Press. (Intermediate)

Tanner, O. 1978. *Animal defenses, wild, wild world of animals.* A Time-Life Television Book. Time-Life Films. (Introductory)

CHAPTER

43

Population Ecology

STEPS TO DISCOVERY
Threatening a Giant

POPULATION STRUCTURE

Population Density

Patterns of Distribution

FACTORS AFFECTING POPULATION GROWTH

Age and Sex Ratios

Mortality and Survivorship Curves

Biotic Potential

Exponential Growth and the J-Shaped Curve

Environmental Resistance and Carrying Capacity

Logistic Growth and the Sigmoid Growth Curve

The "*r* and *K*" Continuum

FACTORS CONTROLLING POPULATION GROWTH

Density-Dependent Factors

Density-Independent Factors

HUMAN POPULATION GROWTH

Historical Overview

Growth Rates and Doubling Times

Current Age–Sex Structure

Fertility Rates and Population Growth

The Earth's Carrying Capacity and Future Population Trends

BIOLINE

Accelerating Species Extinction

THE HUMAN PERSPECTIVE

Impacts of Poisoned Air, Land, and Water

STEPS TO DISCOVERY
Threatening a Giant

Driving across Arizona, you will notice dense forests of giant saguaro cactuses *(carnegiea gigantea)* that extend for miles in all directions. It seems inconceivable that scientists argue that saguaro populations are declining and that this majestic giant may someday disappear from earth. But after conducting a study for the federal government in 1910, Forrest Shreve, a research associate with the Carnegie Institute in Washington, D.C., made just such a conclusion. Based on previous studies that measured growth rates, Shreve estimated the ages of saguaros from their height. He then calculated the age structure of a stand of 240 saguaros on Tumanmoc Hill in central Arizona and noted that the majority of individuals (64 percent) were over 60 years of age. In fact, Shreve found no saguaros younger than 15 years of age and no young seedlings anywhere in the population. Shreve concluded that the saguaro was not reproducing enough offspring to replace the older, dying individuals.

Shreve's conclusion verified what other Southwest desert scientists had also observed in other saguaro populations: For some reason, saguaros were dying out. In an effort to understand why this could be happening, and in order to provide a safe haven for saguaros while researchers determined the causes, The Saguaro National Monument was established in 1933 near Tucson, Arizona. Research within the monument also documented population decline. Since the 1930s, a great deal of research has focused on analyzing every aspect of the saguaro's life cycle and population dynamics in an effort to pinpoint the reason(s) why saguaros were failing to reproduce in adequate numbers.

The primary factors that have led to the decreased reproduction of the Saguaro cactus include ants harvesting seeds, rodents

The saguaro, aptly named the giant cactus, is a massive columnar cactus that weighs several tons and grows in the Southwest deserts in Arizona and Sonora, Mexico. An individual saguaro may live for 175 years or more and grow more than 15 meters (45 feet) tall. Although a saguaro may not bloom until it is at least 30 to 50 years old, its reproductive lifetime still stretches well over 100 years, which helps explain why saguaros have such an enormous reproductive potential. Each year, a single saguaro produces an average of 200 fruits, containing a total of 400,000 seeds. Over its reproductive lifetime, an individual saguaro produces some 50 million seeds.

It takes only *one* seed per saguaro to grow, become established, and reach reproductive maturity in order to maintain stable saguaro populations. Despite prodigious seed production, however, this is not happening in many populations. The problem is clearly not with flowering, pollination, or fruit and seed development since each individual generally produces several million seeds in its lifetime, so the explanation for the saguaro's reproductive failure must lie with seed survival, germination, and/or seedling and adult survival.

In 1969, Warren Steenbergh and Charles Lowe of the Saguaro National Monument and the University of Arizona, Tucson, documented the fact that saguaro seeds disappear at a high rate once they fall to the ground. During a 5-week period, mammals, birds, and insects (particularly harvester ants) consumed nearly all of the seeds produced in their study site. Only 4 in 1,000 seeds survived to germination, significantly reducing the saguaro's reproductive potential. To make matters even worse, saguaro seeds do not survive from one year to the next so reproduction is always limited to the current year's seed supply.

Of the seeds that did germinate, all but a few died within the first year. To study the causes of seedling death, Raymond Turner of the U.S. Geological Survey, Stanley Alcorn of the U.S. Department of Agriculture, and George Olin of the National Park Service transplanted 1,600 young saguaro seedlings in the Saguaro National Monument in 1957. The researchers enclosed some of the young saguaros with cages to protect the seedlings from grazing ground squirrels, rodents, and rabbits; other saguaros were left uncaged. All the uncaged plants were killed within just 1 year by grazing; only 1.9 percent of the caged seedlings remained alive after 10 years. The uncaged seedlings were being eaten by rodents, who eat the saguaros for their water content, a very limited resource in the desert.

Turner, Alcorn, and Olin also found that drought caused high losses of saguaro seedlings during the first few years of life. Small saguaros are usually found in the shade of desert trees or shrubs, places where water loss is reduced. Small saguaros have a small water-storage capacity and become dehydrated easily in direct sun. In another study, Turner, Alcorn, and Olin protected seedlings from rodents and studied the effects of shading on these protected seedlings. All 1,200 unshaded seedlings died within 1 year, while 35 percent of 1,200 shaded seedlings survived. The researchers concluded that the survival of young saguaro seedlings is closely tied to that of other perennials that provide the seedlings with shade. This relationship led to speculation that the rapidly growing cattle industry may also be contributing to saguaro population declines since trampling by cattle reduces tree and shrub cover. In fact, a 1965 study led by James Hastings of the Institute of Atmospheric Physics at the University of Arizona and Raymond Turner reported a general deterioration of woody perennial survival in some parts of the Southwest, assuming a combination of changing climate and cattle grazing. Young saguaro seedlings are also crushed outright by grazing cattle.

In 1976, Steenbergh and Lowe identified another important factor associated with saguaro seedling survival: freezing weather. Young saguaros freeze at −3°C (26°F) to −12°C (10°F) or when exposed to more than 19 hours of freezing temperatures. This means that during certain years, all saguaro seedlings may be killed by freezing. Steenbergh and Lowe demonstrated that this is precisely what had occurred during a severe January freeze in 1971 and likely occurred in 1894 and 1913. Such vulnerability to freezing temperatures decreases as saguaros grow older.

As you can see, many saguaro populations may be declining primarily because virtually all the seeds produced in a year are quickly eaten by mammals, birds, and ants. If weather conditions are just right, only a few of the remaining seeds germinate and grow, provided that they are in the shade of a tree or shrub and are not eaten by rodents and rabbits or trampled by cattle. Periodic freezing can also kill saguaro seedlings.

Like all organisms, saguaro reproduction and its resulting population dynamics are affected by a number of environmental factors. For the saguaro, some factors, such as shade, low temperature, and camouflage, are more critical than are others. No matter what they are, critical factors affect how quickly or slowly a population grows.

consuming seedlings, cattle trampling small individuals, and decreased protective shade.

When we consider . . . how soon some species of trees would equal in mass the earth itself, if all their seeds became full-grown trees, how soon some fishes would fill the ocean if all their ova became full-grown fishes, we are tempted to say that every organism, whether animal or vegetable, is contending for the possession of the planet. Nature opposes to this many obstacles, as climate, myriads of brute and also human foes, and of competitors . . . Each suggests an immense and wonderful greediness and tenacity of life. . . .

Henry David Thoreau, journal entry, March 22, 1861.

Elephants are among the slowest reproducers on earth. Over its lifetime, a female elephant can give birth to a maximum of only six babies. Even so, the number of possible descendants from just one pair of mating elephants could total 5 billion (5×10^9) after just 1,000 years. After 100,000 years, the number of potential descendants from one mating pair would theoretically pack the visible universe with elephants.

If such outlandish growth is possible for a slow reproducer like the elephant, imagine what could happen with organisms that have faster reproduction rates, such as house flies. In less than *1 year*, the number of possible descendants from a single pair of house flies would exceed 5.5 trillion (5.5×10^{12})! Consider the magnitude of this number this way: 1 trillion seconds amounts to about 31,700 years. Humans were in the Stone Age only 1 trillion seconds ago.

Clearly, animals have a tremendous capacity to reproduce, and the reproductive potential of plants is often even greater. Yet, the world is not tightly packed with elephants, flies, or any other kind of organism. Disease, parasitism, predation, and limited food and space curb the potential number of individuals, often leading to a balance between the number of individuals living in an area and the availability of resources to support them.

Predator and prey relationships illustrate such a balance. The number of prey in an area is controlled by both the availability of food and the number of predators: More predators eat more prey. In turn, the number of prey determines how many predators can survive: More prey means more predators; fewer prey means fewer predators. As a result, a dynamic balance is often produced between the number of predators and the number of prey in an area (see Figure 42-4).

▼ ▼ ▼

(a)

FIGURE 43-1

Distribution patterns. ***(a)*** *Clumped:* A grove of clumped palms. A school of fish or a herd of elephants are other examples of clumped distributions. ***(b)*** *Uniform:* Oaks secrete chemicals that prevent growth of nearby oaks, creating more-or-less equal distances between trees. Similarly, when animals defend their territories, the individuals remain separated, producing a uniform distribution. ***(c)*** *Random:* Joshua trees may have random distributions in some locations. Random distributions result when the location of one individual has no affect on another individual of the same species.

POPULATION STRUCTURE

Most communities contain many **populations,** each of which consists of the individuals of the same species that live in the same area at the same time. For example, a mountain forest not only contains a population of yellow pine trees but also populations of sugar pine trees, white fir trees, brown bears, Anna's hummingbirds, and more.

To understand the structure and dynamics of each of the populations that make up a community, ecologists examine three fundamental properties of populations:

- *population density* (the size of the population, expressed as the number of individuals in a given area at a particular time);
- *distribution* of individuals throughout the habitat; and
- *growth rate* (increases or decreases in population density per unit of time).

(b)

(c)

POPULATION DENSITY

Population density equals the number of individuals of a species that live in a particular area at the same time. The population density of people in Manhattan is 100,000 per square mile; that of sugar maple trees in Michigan is 300 per hectare (741 acres); and that of dinoflagellates in a red tide is 8 million per liter of ocean water.

Population density can be determined simply by counting every individual in an area. However, ecologists often estimate population density by counting the number of individuals in small, representative areas and then extrapolate that figure to the total area being studied. This technique is called *sampling*. To estimate the number of creosote bushes in California's Mojave Desert, for example, the number of individuals in ten randomly placed 100-square-meter plots were counted and averaged, revealing 30 bushes per sample. Projecting to an acre, the population density is estimated to be about 1,200 bushes per acre.

PATTERNS OF DISTRIBUTION

Although population density reveals the number of individuals in an area, it provides no information about how the individuals are arranged in space. The distribution of individuals is typically categorized into one of three patterns: clumped, uniform, or random (Figure 43-1).

Interactions among individuals often determine how the individuals are distributed within ecosystems. **Clumped patterns** and **uniform patterns** are nonrandom distributions that result when members of a population have some effect on one another (which is almost always the case) or when environmental conditions favor growth in suitable patches. Uniform spacing results when members of a species repel one another, such as when the roots of some plants release chemicals that inhibit the growth of other members of its species, or when animals establish and defend territories. Both chemical inhibition and territoriality maintain an even, maximum distance between members of a species, creating a more or less uniform pattern of individuals.

The most common distribution pattern is clumped, whereby individuals aggregate into groups, forming groves, schools, flocks, herds, and so on. Some animals have a clumped distribution because they cooperate in societies or gain protection from predators by remaining in herds (page 1018). At least three factors can contribute to clumping in plants: (1) favorable conditions for germination and survival occur only in suitable patches; (2) plant seeds are dispersed in groups; and (3) asexual reproduction from runners (stolons), bulbs, branches, or rhizomes concentrates offspring near the parent plant.

Random distribution is the least common pattern in natural populations. For individuals to be randomly distributed, two requirements must be met: (1) the presence of one individual can in no way affect the location of another; and (2) environmental conditions must be more or less the same throughout an area. Both of these prerequisites are very rare.

Not all distribution patterns remain permanent over time. In some animals, for instance, seasonal changes trigger migrations, causing cyclic changes in distribution. Falling temperatures at night and strong winds at the end of summer initiate migratory behavior in several alpine tundra animals (marmots, mountain goats, mountain sheep, pikas, and rosy finches), causing entire populations to move to warmer regions. At the end of winter, when conditions become less severe, the tundra animals migrate back to their previous homes (see Animal Behavior: Migration, Chapter 44).

The distinctiveness of an ecosystem depends on the physical features of the environment and on the nature of the biotic community, including the number of individuals in each population, their distribution, and the rate of growth of all the populations. (See CTQ #2.)

FACTORS AFFECTING POPULATION GROWTH

Like ecosystems, populations inevitably change. Four events trigger increases or decreases in the density of a group of organisms:

1. **Natality** *increases* density, as new individuals are born into a population.
2. **Immigration** *increases* density, as new individuals permanently move into the area.
3. **Mortality** *decreases* density, as individuals in the population die.
4. **Emigration** *decreases* density, as individuals permanently move out of the area.

If the combination of natality and immigration exceeds that of mortality and emigration, the population grows and density increases. Conversely, when the combination of mortality and emigration exceeds that of natality and immigration, population density decreases. The growth rate of a population equals the rate at which the population size changes. **Zero population growth** occurs when the combined additions and losses to a population are equal; when this happens, population density remains the same.

AGE AND SEX RATIOS

The age and sex of individuals in a population affect population growth. The age of the members of a population is a predictor of both natality and mortality. For example, red alder trees live to be about 100 years old. If the majority of red alders are older than 95 years, the mortality rate will likely be high and the population of trees will likely decline over the next 5 years. In addition, since each individual reproduces only during part of its lifetime, the ages of the members of a population can also be used to predict natality. In general, the saguaro reproduces between the ages of 40 and 150 years; female humans reproduce between ages 15 and 44.

Population biologists plot the number of individuals of a certain age and sex to determine the **age–sex structure** for a population, which helps them predict future population changes. When a large proportion of individuals in a population are at reproductive age (or younger), the age–sex structure tends to be shaped like a pyramid:

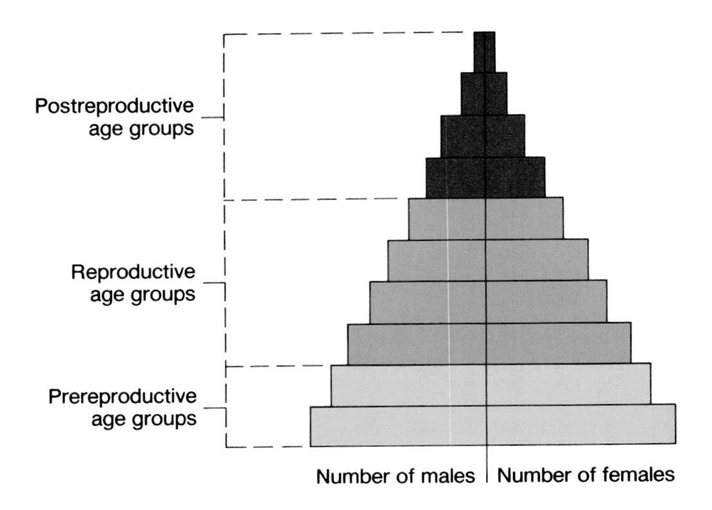

A broad-based population like this one will increase in size; in general, the broader the base, the more rapidly the population will increase. In contrast, a population with an inverted pyramid will decline because most individuals are past reproductive age.

A pyramid for a population with zero population growth would look like this:

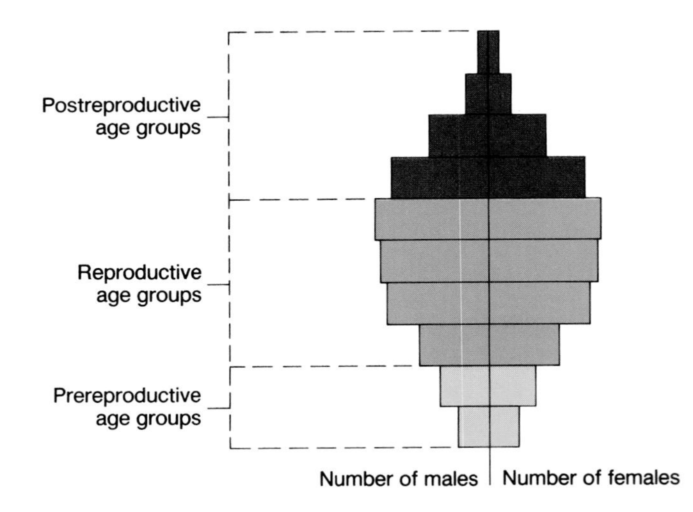

In this stable population, the number of prereproductive individuals balances the number of older individuals, so natality equals mortality. In this stable population, there is either no migration at all, or emigration and immigration are equal.

MORTALITY AND SURVIVORSHIP CURVES

To forecast population changes accurately, ecologists must consider the individual's life expectancy as well as the age–sex structure of the population. When life expectancy is plotted on a graph, a **survivorship curve** is produced. Ecologists identify three general types of survivorship curves: Type I, Type II, and Type III (Figure 43-2).

In a *Type I curve,* mortality remains low for much of the organisms' lifetimes and then increases sharply as individuals reach old age. Humans and other animals that provide long-term care for their young often exhibit a Type I survivorship curve. A *Type II curve* is a straight, diagonal line, indicating that the chances of survival remain about the same throughout an individual's lifetime. Many birds and small aquatic animals exhibit a Type II curve. A *Type III curve* is exactly the opposite of a Type I curve: Most individuals die when they are very young, and only a few adults survive to old age. Species that produce enormous numbers of offspring and provide no parental care, such as insects, frogs, and many plants, exhibit a Type III curve.

BIOTIC POTENTIAL

As we mentioned in the beginning of this chapter, all species have the capacity to produce tremendously large numbers of descendants eventually, as long as there are no

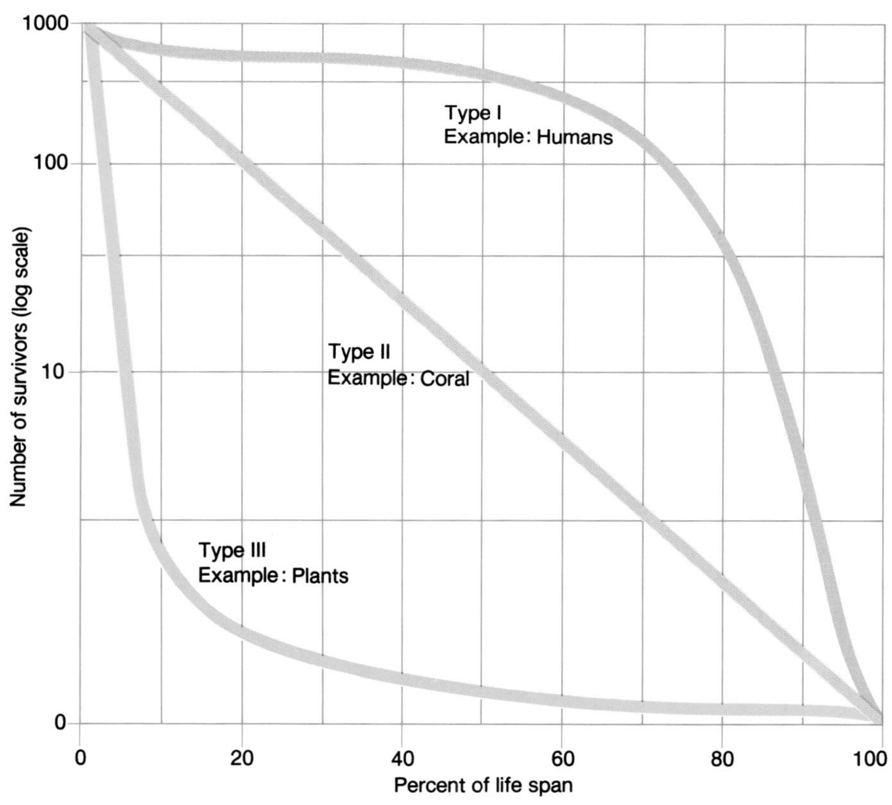

FIGURE 43-2

Three types of survivorship curves are produced by plotting the number of survivors (on a log scale) against the percent of the life span of a species. In the Type I curve, the mortality rate is low in the first years of life and then becomes higher at old age. Humans exhibit a Type I curve. In the Type II curve, the chances of surviving or dying are virtually the same throughout an organism's entire life. In a Type III curve, nearly all of the young die quickly, but the mortality rate is quite low for the few survivors, until they reach old age. Oysters, some insects, and weedy plants exhibit a Type III curve.

FIGURE 43-3
Although it may have seemed like a great idea at the time, importing plants into new habitats without natural population controls has produced some unexpected overpopulation disasters. For example, water hyacinth plants, with their orchid-like flowers and attractive stems and leaves, were brought to the United States from Venezuela to be displayed at the 1884 New Orleans Cotton Exposition. Many of the exposition visitors were given clippings of the charming plant to place in ponds and streams near their homes. With no competitors, few predators, and plenty of available space and nutrients, the growth of the water hyacinth spread rapidly. As this photo dramatically shows, today, water hyacinths are choking streams, rivers, irrigation systems, and hydroelectric installations.

restrictions to curb population growth. This innate capacity to increase in number under ideal conditions is called the **biotic potential** of a population.

▶ When conditions are optimal and no limitations exist, the population grows at its **intrinsic rate of increase,** or $\mathbf{r_o}$ (r = rate, o = optimum), the maximum increase in numbers of individuals per unit of time under optimal growth conditions. Organisms find themselves in such conditions only on rare occasion, however, such as when organisms colonize a new favorable habitat, when environmental conditions suddenly change for the better, or when organisms are introduced into new habitats in which there are no natural competitors or predators (Figure 43-3).

EXPONENTIAL GROWTH AND THE J-SHAPED GROWTH CURVE

When populations grow at their intrinsic rate of increase, the number of individuals increases exponentially. That is, the number increases by a fixed proportion, such as when a population *doubles* in size with each new generation (2, 4, 8, 16, 32, 64, 128, and so on). Such **exponential growth** can be demonstrated by placing a single *E. coli* bacterium into a nutrient culture (Figure 43-4). The bacterium divides into two cells after 20 minutes; the two cells divide into four cells in another 20 minutes; and the four cells divide into eight cells 20 minutes later. The population continues to double every 20 minutes. After 5 hours, 32,768 bacteria have been produced from the original bacterium. After 7 hours, there are more than 2 million bacteria. And after 36 hours, there would be enough bacteria to blanket the entire earth's surface with 28.8 centimeters (1 foot) of bacteria.

The growth of this bacterial culture illustrates how the number of individuals in a population affects the number of offspring produced; that is, the more reproducing individuals there are in a population, the greater the number of progeny. For example, during a 20-minute period, a population of *E. coli* increases to two when the population contains only the original bacterium, but during the same 20-minute period, the population can jump to 2 million

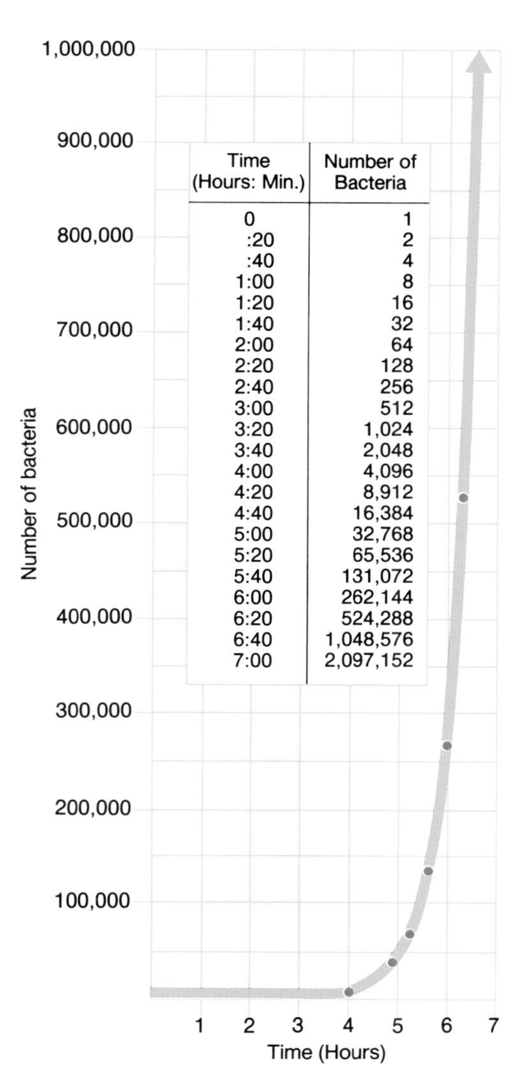

Time (Hours: Min.)	Number of Bacteria
0	1
:20	2
:40	4
1:00	8
1:20	16
1:40	32
2:00	64
2:20	128
2:40	256
3:00	512
3:20	1,024
3:40	2,048
4:00	4,096
4:20	8,912
4:40	16,384
5:00	32,768
5:20	65,536
5:40	131,072
6:00	262,144
6:20	524,288
6:40	1,048,576
7:00	2,097,152

FIGURE 43-4
Exponential growth of an *E. coli* bacterial culture produces a J-shaped growth curve. After rounding the "bend" of the J, the curve gets steeper and steeper, until some limitation, such as depletion of food or the buildup of contaminating waste products, curbs further growth.

when the population starts with 1 million bacteria. Expressed mathematically, the rate of exponential growth ($\Delta N/T$, where Δ = change, N = number of individuals, and T = time) equals the intrinsic rate of increase (r_o) multiplied by the number of individuals in the population (N):

$$\Delta N/T = r_o N$$

The pattern of exponential growth is always the same: The size of a population increases gradually at first and then grows larger and larger in progressively shorter periods of time, as more and more individuals are added to the population. Plotting exponential growth on a graph produces a curve that resembles the letter J; not surprisingly, it is called a **J-shaped curve.** As Figure 43-4 illustrates, once a population "rounds the bend" of a J-shaped curve, the number of individuals added to a population begins to skyrocket.

ENVIRONMENTAL RESISTANCE AND CARRYING CAPACITY

No natural ecosystem can support continuous exponential growth for any species; that is, no ecosystem has unlimited resources, and environmental conditions never remain constantly favorable for limitless growth. Eventually, some environmental limitation imposes a restriction on continued population growth. The factor(s) that eventually limit the size of a population create an **environmental resistance** to population growth. Environmental resistances include competition, predation, hostile weather, limited food or water supplies, restricted space, depleted soil nutrients, and the buildup of toxic byproducts from the organisms themselves.

The combined limitations imposed by the environment establish a ceiling for the number of individuals that can be

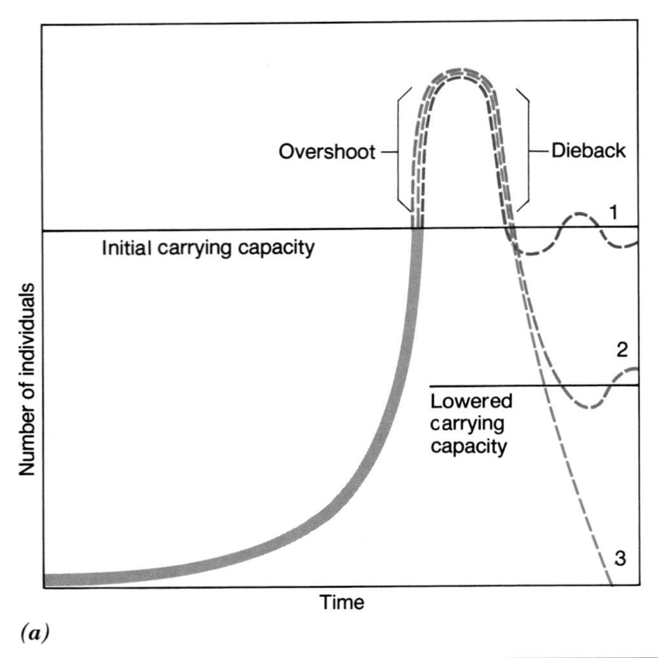

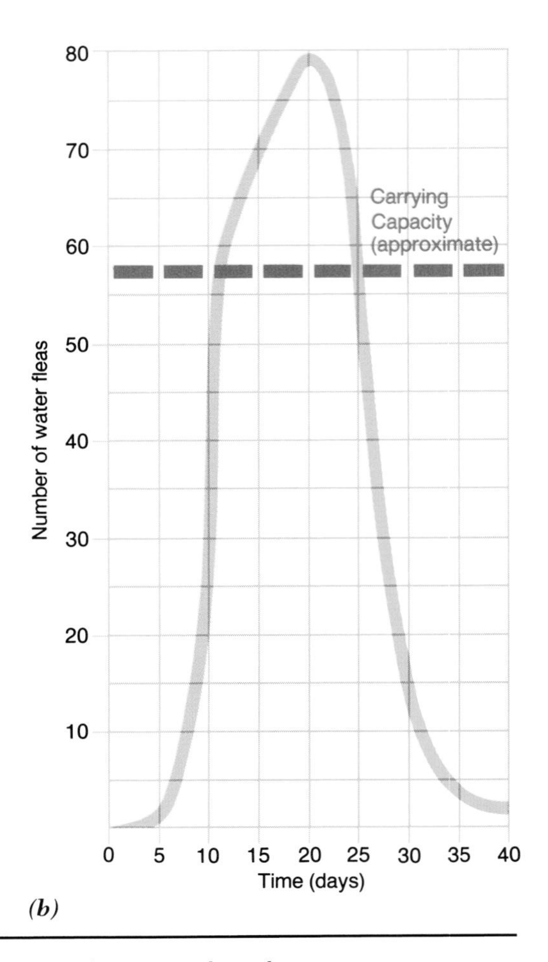

FIGURE 43-5
The risks of overpopulation. ***(a)*** A population undergoing exponential growth may overshoot the carrying capacity of the environment. When this happens, the population experiences a dieback to one of three levels: (1) to the original carrying capacity (the least common outcome), (2) to a population density at a lower carrying capacity, or (3) to extinction or near extinction levels. ***(b)*** A natural population of water fleas *(Daphnia)* illustrates the third fate: extinction.

supported in an area. The size of a population that can be supported indefinitely is called the **carrying capacity,** or ***K***, of the environment. Populations undergoing exponential growth sometimes overshoot their carrying capacity before environmental resistances are able to curb their growth. The greater the reproductive momentum, the more likely it is that the population will exceed the environment's carrying capacity for some period of time.

Laboratory experiments and field observations identify three possible fates for populations that overshoot the carrying capacity of their environment (Figure 43-5). All the repercussions begin with a *dieback,* the death of a portion of the population. The least dramatic of the three fates is also the least common: The population simply dies back to the level of the original carrying capacity (curve 1 on the graph). In most cases, however, the excess population damages the environment in some way, which in turn, *reduces* the carrying capacity. In other words, the organisms themselves impact the carrying capacity. For example, an excessively large population of caterpillars could consume all of the leaves on a tree, weakening or possibly killing the plant. When the carrying capacity is lowered, the population plunges. If the damage caused by overpopulation is not too severe, the population eventually comes into balance with a lower carrying capacity (curve 2). When damage is extensive, however, or when a vital resource is drastically depleted (as in the case of the caterpillars, killing the plant they live on), the population crashes to a very low level or suffers the third fate: It disappears altogether (curve 3).

Occasionally, a population exceeds the carrying capacity because a limiting factor does not take effect until a threshold level is reached. The population continues to grow exponentially until it reaches the threshold level, at which point the factor causes a sharp decrease in population size. For example, the gradual buildup of toxic waste products may have no effect on a population until these chemicals reach a critical level, at which point large numbers of individuals will die. Some biologists warn that this might be the case for the human population, as our air, soil, and water become more and more polluted (see The Human Perspective: Impacts of Poisoned Air, Land, and Water).

◁ THE HUMAN PERSPECTIVE ▷

Impacts of Poisoned Air, Land, and Water

On December 3, 1984, in Bhopal, India, a huge cloud of methyl isocyanate gas leaked out from a storage tank at the Union Carbide chemical plant, killing 3,000 people. It is estimated that an additional 2,000 people will die from side effects by 1995, and 17,000 of the 200,000 people injured from the gas leak have been permanently disabled as a result of lung ailments.

This event dramatically underscores the hazards of toxic substances that surround us in modern society. We are being exposed to more and more toxic substances in the air, in our water, and on the land—chemicals that adversely affect living organisms. In the United States alone, 60,000 chemicals are added to our food or are used to make cosmetics or to combat pests. Hundreds of these chemicals are known to be hazardous. Each year, 700 to 1,000 new chemicals enter the marketplace; fewer than 10 percent are tested to assess their health effects. Over 170 million metric tons (378 billion pounds) of potentially hazardous chemicals are manufactured each year in the United States alone, exposing people to hazardous chemicals in their homes, at schools, at work, and even while playing outdoors.

Toxic chemicals can affect virtually every cell in an organism and can cause cancer, mutations, birth defects, or reproductive impairment. These chemicals can affect cells in several ways: (1) by disturbing enzyme activities that regulate critical chemical reactions (mercury and arsenic inactivate enzymes); (2) by binding directly to cells or to essential molecules in the cell (carbon monoxide binds with hemoglobin in the blood, preventing it from carrying oxygen to cells); or (3) by releasing chemicals that have an adverse effect (in addition to its immediate destructive effects, carbon tetrachloride triggers nerve cells to release large amounts of epinephrine, which is believed to cause long-term liver damage).

Some chemicals gradually accumulate in the bodies of organisms, eventually building to toxic levels. The buildup of chemicals in the organisms in a food chain exposes organisms toward the top of a food chain to potentially dangerous levels of chemicals. The use of DDT as a pesticide is a good example of this phenomenon. Through bioconcentration and biological magnification, DDT concentrations are several million times greater in the fish-eating birds than they are in the water (page 946 and Figure 41-11). Large doses of DDT severely reduced the amount of calcium deposited in eggshells, causing bird eggs to break easily.

Like air pollutants, water pollutants can cause a physical or chemical change that adversely affects life. In the United States, the water in 40 states is already hazardously polluted, and more than half of Poland's water supply is so polluted that it cannot be used even by industry. Water pollutants include poisonous *toxic chemicals,* such as mercury, nitrates, and chlorine; microscopic *pathogens* that cause disease; various *physical agents,* such as soil sediments; and *excess nutrients and organic matter,* such as the remains of plants and animals, feces, debris from food-processing plants, and runoff from feedlots, sewage treatment plants, and fertilized agricultural land.

Pollutants are either dumped into or seep into all the earth's water sources, including surface waters (lakes, ponds, streams, and rivers), ground-water aquifers (which, in the United States, supply more than one-quarter of the annual water demands), and oceans, especially biologically rich coastlines, coastal wetlands (bays, swamps, marshes, and lagoons), and estuaries. Not only are coastal zones rich with myriad organisms, they are also the most vulnerable of the ocean's regions to numerous sources of pollution, including wastes from sewage plants and factories, sediment from erosion, and oil spills. Combine these sources with the fact that many cities may draw huge quantities of fresh water from streams during droughts, and you can see how water flow into important regions has diminished, increasing pollutant concentrations. As a result, coastal zones are being destroyed by pollution, water loss, sedimentation, dredging, and filling, at alarming rates. In the United States alone, more than 40 percent of estuaries have been destroyed, despite the implementation of state and federal laws to protect them.

LOGISTIC GROWTH AND THE SIGMOID GROWTH CURVE

An alternative to unbridled exponential growth occurs when environmental resistance increases as a population approaches the carrying capacity of the environment, slowing growth. When this happens, a **sigmoid growth curve** is produced (Figure 43-6). The "lazy S"-shaped sigmoid curve begins the same as does the J-shaped curve; that is, numbers of individuals increase slowly at first, then, as the reproductive base builds, the population goes into a period of exponential growth.

The rate of growth gradually slows down as the population approaches the carrying capacity, forming the top of the S-shaped curve. Population density eventually fluctuates around the carrying capacity: When the population

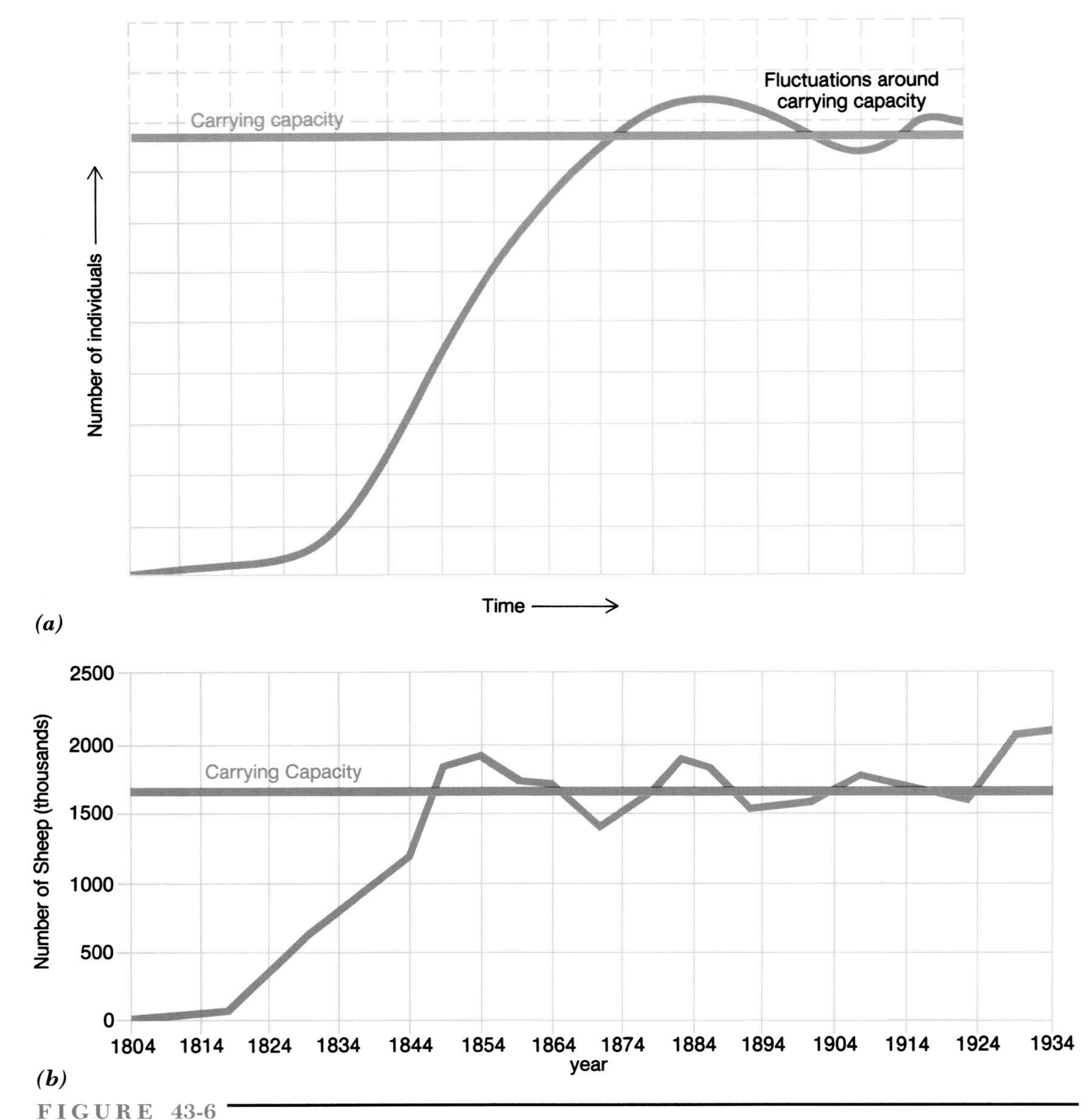

FIGURE 43-6

The sigmoid growth curve. ***(a)*** In a new habitat, a population increases slowly at first and then undergoes exponential growth, until environmental resistances begin to reduce growth rates and stabilize population density around the carrying capacity. This common logistic growth pattern produces a sigmoid growth curve. ***(b)*** Sheep were introduced on Tasmania in the early 1800s. The population growth pattern produced a sigmoid growth curve, with the population oscillating between 1.5 million and 2.0 million individuals.

slightly exceeds the carrying capacity, environmental resistance intensifies and causes a slight dieback; when the population falls below the carrying capacity, environmental resistance relaxes, and the population gradually increases. In this way, the size of a population remains fairly stable, and the growth curve flattens out at the carrying capacity level. Populations that produce a sigmoid growth curve exhibit **logistic growth.**

Oscillations are not always small, however. In 1957, for example, A. J. Nicholson raised laboratory colonies of Australian sheep-blowflies *(Lucilia cuprina)* by introducing a small number of blowflies into a cage. All the conditions were kept constant, including food and water supplies. Blowfly populations grew rapidly because of their high biotic potential. Nicholson documented that blowfly populations underwent "violent oscillations," changes that could not have been caused by environmental fluctuations.

Nicholson suggested that since sheep-blowflies "scramble" for their food, overcrowding triggered excessive mortality; as a result, density dropped far below that which would be governed by normal environmental factors, such as seasonal changes in weather. As the population recovered in size, the blowflies' extremely high reproductive potential caused the population to overshoot the carrying capacity once again, triggering overcrowding and excessive mortality again.

THE "r AND K" CONTINUUM

Just as natural selection favors physical traits that improve survival (such as camouflage), traits that increase reproduction are also naturally selected. Over the course of evolution, species have evolved a range of reproductive strategies that offer advantages in different types of habitats. For example, species with the ability to reproduce large numbers of offspring will dominate a disturbed or uninhabited area before slower reproducing species can do so. In stable climax communities, however, where competition for resources is intense, species that provide long-term care to a few offspring (koalas, for example) often have a greater chance of success than do species that produce many offspring and provide no care (such as turtles).

Species that produce many offspring at one time are said to be ***r*-selected species,** *r* referring to a high intrinsic rate of increase (r_m; r = rate of increase; m = maximum). In other words, *r*-selected species have adaptations that maximize r_m. In contrast, species that produce one or a few well-cared-for individuals at a time are said to be ***K*-selected species,** *K* referring to those strategies that are more favorable for populations near the carrying capacity of the environment. These opposite strategies are the extremes in a continuum of reproductive strategies found in organisms. The main components of reproductive strategies and their relation to *r*-selected and *K*-selected species are presented in Table 43-1.

Differences between *r*- and *K*-selection strategies become very important when we consider the fate of rare, threatened, or endangered species, the most probable candidates for extinction. A rare species is one with only a few individuals remaining. Although a rare species is in no particular danger of extinction, because of its small numbers a rare species could be quickly wiped out. Endangered and threatened species are near extinction as a direct result of human activities. The passenger pigeon is an example of how quickly a species can become extinct when the combination of reproductive strategy and human activities works against the survival of a species. It took less than 100 years for the passenger pigeon to become extinct, even though many millions of pigeons existed in the early 1800s. Since

TABLE 43-1
RANGES OF REPRODUCTIVE STRATEGIES[a]

	r-*Selected*				K-*Selected*
Number of offspring	many				few
Number of times an individual reproduces	once				many times
Size of young	small				large
Rate of development	fast				slow
Parental care	minimal				intensive
Life span	short				long
Survivorship curve	Type III		Type II		Type I
Energy to reproduction	high				low
Energy to increasing body size	low				high
Examples					
Animals	oysters	insects	birds	elephants	humans
Plants	weeds	saguaro	oaks	pears	mangroves

[a] *r*-selected and *K*-selected strategies are at the extremes of each continuum.

the passenger pigeon was *K*-selected—females laid only one egg each year—the production of offspring could not keep up with the enormous number of pigeons that were killed by sportsmen. Thus, the species died out.

With an innate ability to reproduce quickly, *r*-selected species may not become extinct as quickly as *K*-selected species do. However, no amount of reproduction can save a species if its environment is destroyed or severely contaminated (see Bioline: Accelerating Species Extinction).

The size of a population and the rate at which the population size increases or decreases depend on a balance between the pace at which new individuals are added to and the pace at which individuals are removed from a population. (See CTQ #3.)

FACTORS CONTROLLING POPULATION GROWTH

A change in the physical environment can have an impact on population growth, leading to a change in the size of a population. For example, global warming and its influence on rainfall patterns is expected to trigger reductions of redwood tree populations along the Pacific coast of North America. In addition to environmental impacts, all interactions within an ecosystem—between organisms and between organisms and their physical environment—can have an impact on population growth. The impact of some population-controlling factors increases or decreases in intensity as the size of the population changes, while the intensity of other population controls remains the same, regardless of the population size.

DENSITY-DEPENDENT FACTORS

As the size of a population increases, the increasing density of the number of individuals within the population sometimes limits the population growth rate. For example, when population density is low, young locusts develop normal-length wings. When locust density is high, however, hormonal changes trigger the development of longer wings in offspring. Long wings increase emigration, which, in turn, reduces the population density in that area. Such factors that are influenced by the number of individuals in the population and ultimately affect population density are called **density-dependent factors.**

There is a direct relationship between the intensity of density-dependent factors and population size: As population density rises, the intensity of density-dependent regulatory mechanisms increases, dampening population growth by reducing natality or by boosting mortality or emigration (as in locust populations). Conversely, when population density falls, density-dependent mechanisms decrease in intensity, allowing population growth to accelerate. Density-dependent mechanisms explain why there are small fluctuations around the carrying capacity of the environment during logistic growth.

In addition to locust wing length, examples of density-dependent factors include the following:

- *Disease:* As density increases, the number of contacts between members of the population multiplies, intensifying the likelihood of the spread of disease-causing pathogens or parasites.
- *Competition:* Since members of the same species have very similar needs, intraspecific competition intensifies as density increases.
- *Predation:* The number of predators increases as the size of a prey population increases.
- *Stress:* Crowing increases stress in animal populations, which, in turn, increases aggression, infertility, and other growth-limiting factors.

Laboratory studies on overcrowding in rats show that crowding increases aggressive behavior, delays sexual maturation, reduces sperm production in male rats, and causes irregular menstrual cycles in females. Overcrowding also reduces sexual contacts between males and females, and increases homosexual contact between males. Such density-dependent factors quickly curb population growth by reducing natality. Whether such dramatic effects occur in populations outside the laboratory or in other species is not yet known. However, many demographers (scientists who study human population dynamics) believe that increased crime, drug abuse, and suicide may be partly the result of overcrowding, a situation that worsens as the world's human population swells by more than 250,000 people each day (an increase of 3 persons per second).

DENSITY-INDEPENDENT FACTORS

Not all regulatory factors are affected by the density of a population. Factors that are not influenced by population size are called **density-independent factors.** A killer earthquake, such as the earthquake that jolted the city of Erzincan, Turkey, on March 20, 1992, measuring 6.2 and killing over 500 people in less than 1 minute, is an example of a density-independent factor: The population density neither caused the earthquake nor affected the magnitude of the quake or the percentage of individuals killed. Many catastrophic events, including fires, floods, hurricanes, tornadoes, volcanic eruptions, and avalanches, are density-independent factors.

Density-dependent and density-independent factors often combine to regulate population size. For example, the number of aphids feeding on a sycamore leaf is affected not only by competition (a density-dependent factor that determines how many aphids may feed on the same leaf vein) but also by wind velocity (a density-independent factor; as

◁ BIOLINE ▷

Accelerating Species Extinction

One alarming consequence of human activity is the diminished diversity of life on earth. One of every five species that was thriving when you were born is now extinct, the direct result of human impact on the environment. The biosphere is an extraordinarily complex structure, much more complicated than the space shuttle. Yet, in both the biosphere and space shuttle, the failure of vital parts will likely lead to the failure of the whole; all occupants will perish. Furthermore, the vital parts may be as small as the rivets that hold the craft together.

The species that comprise the living component of the biosphere can be considered the "rivets" that hold together spaceship Earth. Loss of these species is tantamount to "popping the rivets." If enough rivets are removed, the entire structure will inevitably fall apart. Many biologists predict that we are rapidly approaching the point of popping some critical rivets. If certain key species disappear, the ripple effect could spread to most of the planet's life forms. For example, extinction of the microscopic phytoplankton that supports virtually all life in the oceans and inland waterways would lead to starvation of every fish and aquatic mammal, animals that are an important source of food for humans. The extinction of phytoplankton would also reduce the amount of breathable oxygen in the atmosphere to less than half its current content, which could lead to the deaths of all air-breathing animals.

What would you do if you were an astronaut about to fly on the space shuttle and you noticed a person popping out rivets with a crowbar? You ask him why he is doing such a crazy thing, and he responds that rivets bring a good price and it's a way to make a bigger profit. "But the ship will crash if you remove its rivets," you protest. The profiteer responds, "I have removed 200 rivets on past flights and the shuttle hasn't crashed. What are you so worried about?" Any rational astronaut would refuse to be a passenger on such a compromised space shuttle.

Unfortunately, we see rivets being removed from spaceship Earth in unprecedented numbers, but we don't have the option to take another flight or to decline the trip altogether. All we can do is try to stop the practice, by slowing down species extinction and by giving the biosphere the time and opportunity it needs to repair the damage. Yet, a few countries continue to pop out the rivets in the face of danger, as was evident in 1992, at The Earth Summit held in Rio de Janeiro, Brazil. At the Summit, some countries refused to sign the Biodiversity Treaty, which would help protect the vast variety of life that exists on earth; that is, the earth's **biodiversity.** These countries claimed that scientists don't know for sure what the consequences of reduced biodiversity will be, so why should they endure economic hardship based on unproven predictions? Such arguments often delay scientific investigations into the possible ramifications of continued reductions in biodiversity. These arguments also ignore clear warnings from respected ecologists who continue to remind us of the popping rivets analogy, even though we don't know exactly how many more loosened rivets it will take before serious damage is done to the earth's ability to support life.

winds increase, sycamore leaves brush up against one another, scraping off the aphids).

The rate of population growth and the resulting population size is governed by limitations of the physical environment as well as by factors within the biotic community. The intensity of some of these factors changes with the size of the population, while the intensity of others do not. (See CTQ #4.)

HUMAN POPULATION GROWTH

In 1987, the world human population reached *5.0 billion people.* By the end of 1992, there were more than 5.4 billion people, and the number continues to climb quickly. More than 380,000 people are born each day (over four babies every second), and nearly 130,000 people die each day. This means that the world's human population is growing by some 250,000 people every single day. At this rate, nearly 100 million people are added to the human population every year. What will the future be like for those babies born into an exploding human population?

The answer has a great deal to do with geography. If these children are born in the United States in 1993, many will begin kindergarten in 1998, enter college around 2011, and be eligible to retire in the year 2058. By the time their children enter college, the world's human population may have swelled to more than 8 billion. By the time they retire, at the current growth rate, the world's human population will have climbed to 17 billion people, more than three times the number of people living on earth today, unless human endeavors or density-dependent controls can change the human population growth rate.

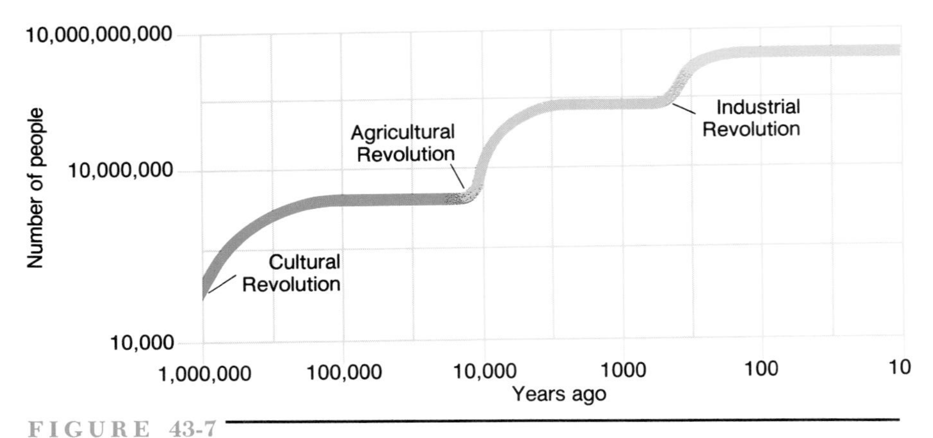

FIGURE 43-7

Surges of human population growth become clear when the size of the human population is plotted against time on a log–log scale (In log–log graphs, both population size and time are plotted on log scales, concentrating huge numbers of individuals and enormous time spans on a single sheet of graph paper.) Bursts of exponential growth occurred on three occasions, corresponding to the cultural, agricultural, and industrial revolutions.

HISTORICAL OVERVIEW

Modern humans appeared on earth less than 1 million years ago. The size of early human populations remained more or less in balance with the available food supply, increasing slowly as our early ancestors spread out and discovered new lands.

When the size of the human population is plotted against time on a log scale (Figure 43-7), three surges of population growth become evident. Each surge is the result of a major technological invention that improved food supply and/or human health.

➡ The use of tools and the movement from nomadic tribes to a stationary society were responsible for the first population surge, about 600,000 years ago. The development of agriculture (cultivating food in a village-farming society) produced the second population surge, about 10,000 years ago. The most recent growth spurt occurred about 200 years ago, as humans entered the Industrial Revolution. At the same time, humans continued to make advances in agriculture, medicine, and hygiene, decreasing the death rate and increasing the average life span. These advances lessened some of the former controls on human population growth, mainly food shortages and disease.

The overall growth curve for the world's human population is clearly a J-shaped curve (Figure 43-8). It took hundreds of thousands of years for the world human population to reach 1 billion people. Once a reproductive base of 1 billion people was established in 1800, however, it took progressively less time to add another 1 billion people to the human population: 130 years to reach 2 billion people; 30 years to reach 3 billion; 15 years to reach 4 billion; and finally only 13 years to reach 5 billion people in 1987.

GROWTH RATES AND DOUBLING TIMES

The current human birth rate is 27.7 babies per 1,000 people per year. The death rate is 9.5 people per 1,000 per year, making the current human population growth rate equal to 18.2 humans per 1,000 people per year:

$$\underset{\text{(Birth rate)}}{27.7} - \underset{\text{(Death rate)}}{9.5} = \underset{\text{(Growth rate)}}{18.2}$$

Population increases are often expressed as the number of people added to the population per 100 individuals, giving the **percent annual increase** in population. The average annual increase for all nations is currently 1.8 percent (see (Table 43-2). This means that it will take less than 56 years for the world's human population to double. That is, if you add 1.8 people to a population of 100 each year, it will take 55.6 years for the population to reach 200 ($55.6 \times 1.8 = 100$). As we discussed earlier, however, population size affects the rate of increase; thus, a growth rate of 1.8 actually has a doubling time of only 39 years instead of 55.6 years because of the large reproductive base. Doubling times for the world human population are given in Table 43-3.

TABLE 43-2

PERCENT GROWTH RATE FOR VARIOUS COUNTRIES

Country	*Percent Annual Increase*
Developing countries	
India	2.3
Brazil	2.9
Uganda	3.3
Mexico	3.5
Kenya	4.1
Developed countries	
United Kingdom	0.0
United States	0.6
Japan	1.1

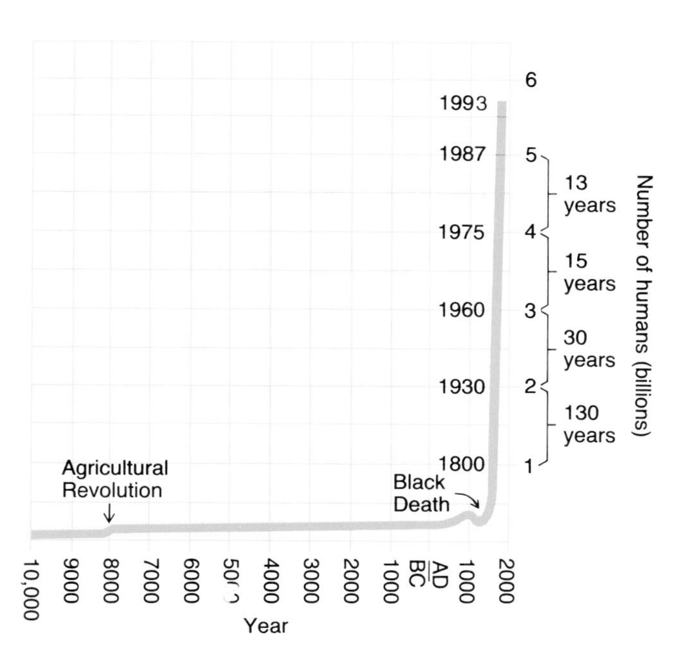

FIGURE 43-8

The human population growth curve. The world's human population grew slowly for 2 million years until the 1800s, at which time population growth began to soar. By the year 2000, there will be an estimated 7 billion to 14 billion people on earth, if our population continues to grow at its present rate.

CURRENT AGE–SEX STRUCTURE

There is a big difference between the age-sex structure of developing nations (many countries in Asia, Africa, and South America) and that of developed countries (Figure 43-9). The populations of developing nations are growing rapidly because large numbers of individuals in these populations are moving into their reproductive years. Since developing countries are heavily populated, they have a large effect on the world's human population growth. As a result, the world's human population will continue to increase at a rapid rate.

FERTILITY RATES AND POPULATION GROWTH

Although age–sex structure diagrams help predict population growth, they do not take into consideration another factor that affects human population growth: **fertility rates.** A fertility rate is the average number of children born to each woman between 15 and 44 years of age (the reproductive years). The average fertility rate for all nations in the world is now slightly below 2.1 births per woman. In developed countries, however, fertility rates average 1.9 births, compared to 4.5 births in developing nations.

Given the current mortality rates, fertility rates of between 2.1 and 2.5 are required to maintain zero population growth. With an average fertility rate of 1.9 in developed countries, populations in these countries will decline. In contrast, with an average fertility rate of 4.5, populations in developing countries will rapidly increase. As a result, the world's human population will continue to increase by greater and greater numbers each year. Some scientists believe that the world's human population will begin leveling off in 30 to 40 years, to between 8 billion and 14 billion people, while others believe that the human population will not reach these levels because we are very close to, or have already exceeded, the earth's carrying capacity for humans.

TABLE 43-3

HUMAN POPULATION SIZE ESTIMATES, DOUBLING TIMES, AND PERCENT GROWTH RATES

Date	*Estimated Human Population (bil.)*	*Doubling Time (yrs.)*	*Percent Annual Increase*
Pre 8000 B.C.			0.0007
800 B.C.	0.005	1500 million	0.0015
1650 A.D.	0.500	200	0.1
1850 B.C.	1.000	80	0.8
1930 A.D.	2.000	45	1.9
1975 A.D.	4.000	35	2.0
1987 A.D.	5.000	39	1.8

THE EARTH'S CARRYING CAPACITY AND FUTURE POPULATION TRENDS

Density-dependent controls are already beginning to curb human population growth, as the every-burgeoning human population is reducing the ability of the earth to support life (Figure 43-10). Lowering the earth's carrying capacity affects not only millions of other organisms but ultimately humans as well. We have learned now important photosynthetic plants are to life on earth; they supply chemical energy and nutrients to virtually all heterotrophs, including humans. With only 15 staple plant species standing between humankind and starvation, it is imperative that we minimize the damage to our environment. If just one of these staple crops were to become extinct, millions, or possibly billions, of people might perish.

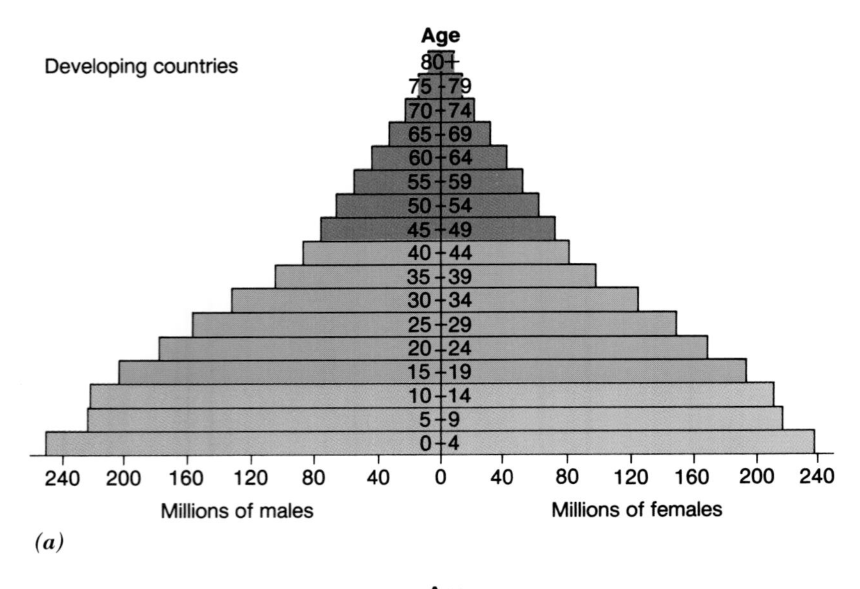

(a)

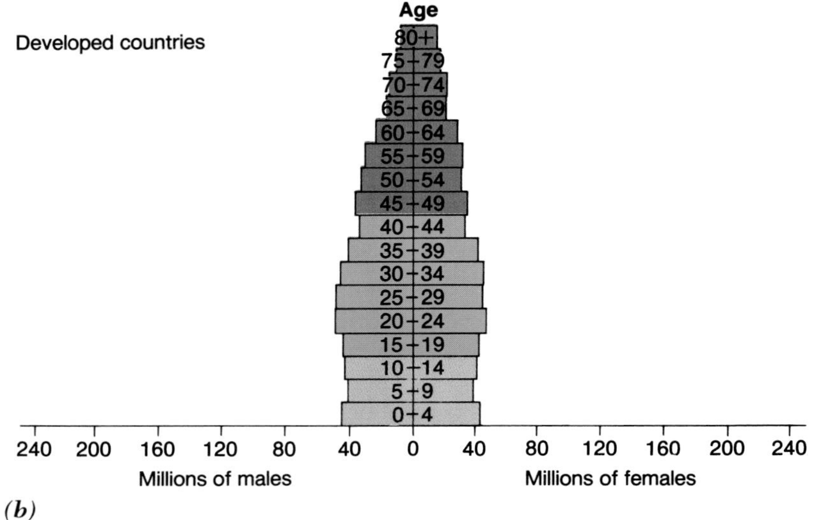

(b)

FIGURE 43-9

Age–sex structure diagrams for *(a)* developing and *(b)* developed countries differ radically. With a tremendous reproductive base, developing countries will continue to grow rapidly, while developed countries will decline in population. Since 75 percent of humans live in developing countries, the world's human population is expected to continue its perilous increase.

Even simple, microscopic organisms, like bacteria, fungi, and nematodes, are important to human life (and life in general) for they contribute to the recycling of nutrients. Many organisms provide medical and agricultural remedies. The extinction of species not only reduces ecological diversity but also cuts into the genetic bank, a bank of genes we have tapped a number of times to solve medical, industrial, and agricultural problems.

Between 10 million and 20 million people—mostly children—die each year from starvation or malnutrition-related disease. Even in affluent countries, overpopulation has accelerated the rate of environmental deterioration, lowering the quality of life and most likely reducing the environment's carrying capacity. Is it possible that the world's human population is already in overshoot? If so, what lies ahead? A dieback to the original carrying capacity,

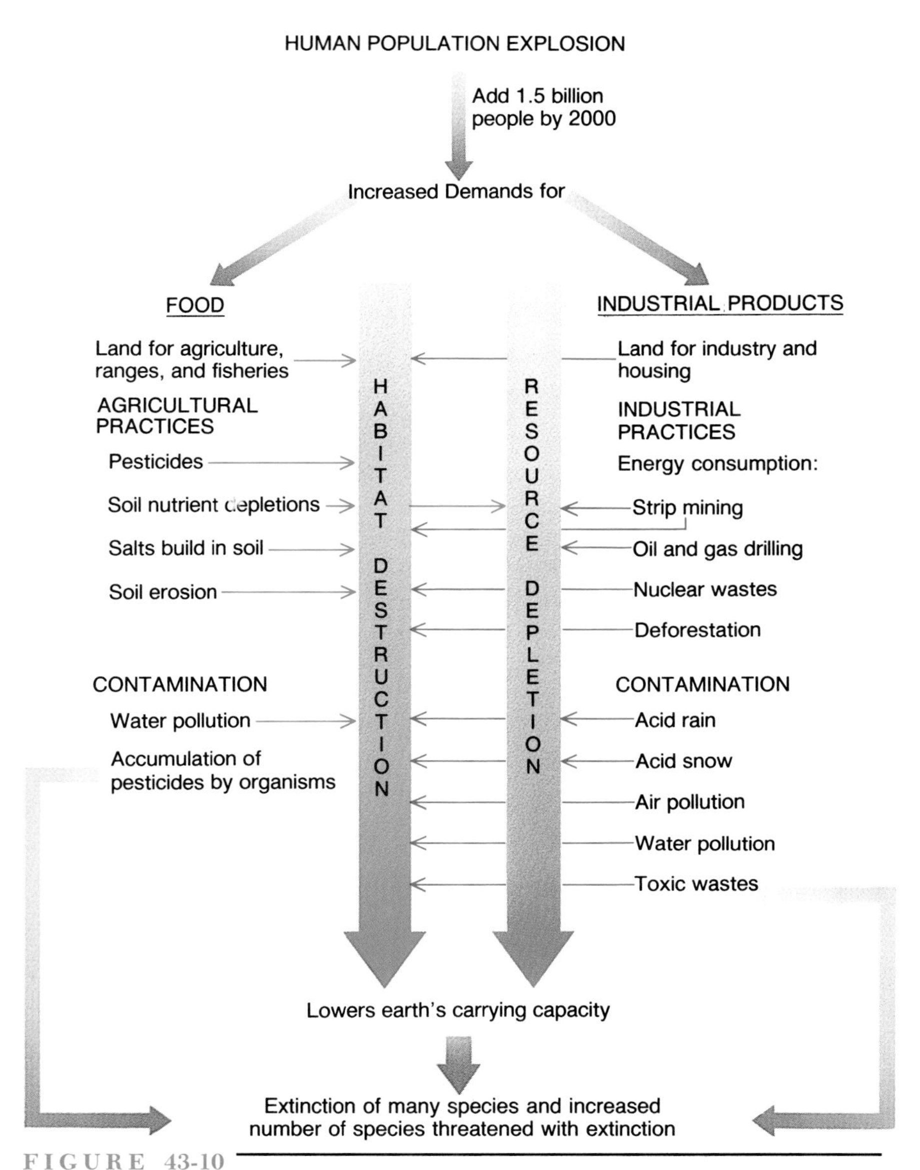

FIGURE 43-10
Impacts of the human population explosion. Increased numbers of people mean increased demands for food and products. These demands deplete resources and destroy habitats, which, in turn, lower the earth's carrying capacity and increase the number of species threatened with extinction.

a dieback to a lower carrying capacity, or a dieback to extinction?

Many biologists believe that we can still avoid these undesirable consequences through efforts to curb population growth and to protect the environment; these are the challenges facing us today. Biotechnological advances for increasing food production, methods for disposing of wastes that otherwise contaminate the biosphere, and other advances that would help control growth rates may someday help us avert the disasters of overpopulation.

The world's human population is currently growing exponentially. As with all species, a population ceiling for humans will ultimately be reached. Humans are unique in their ability to analyze and anticipate future trends as well as to modify their environment to their advantage. All of these skills will be necessary in order to solve the problem of the human population explosion. (See CTQ #5.)

REEXAMINING THE THEMES

Biological Order, Regulation, and Homeostasis

The populations that make up an ecosystem interact with one another in many ways. Prey populations provide energy and nutrients for predator populations, just as plant populations provide energy and nutrients for herbivores. A change in the size of one of these interacting populations eventually changes the size of the other. Interacting populations also affect each other's distribution pattern in an ecosystem. For example, when an animal population disperses seeds in groups, the result is a clumped distribution for the plant. When plants produce chemicals that inhibit the growth of other plants, or when animals establish territories, uniform distribution patterns are formed.

Unity within Diversity

All organisms, from the slowest reproducer to the fastest, have the potential eventually to produce huge numbers of descendants. This capacity for reproduction enables species to colonize new habitats, to recover from disasters that kill a large proportion of the population, and to provide the variability upon which natural selection operates.

Evolution and Adaptation

Reproduction is a basic characteristic of life. Organisms have evolved a range of reproductive strategies, from *r*-selected species that (1) grow fast, (2) reach reproductive maturity early, and (3) invest the greatest amount of energy and nutrients they acquire into a single, large reproductive event to *K*-selected species that (1) grow slowly, (2) delay reproduction, (3) invest the largest proportion of energy and nutrients into growth, increasing competitive abilities, and (4) have small, multiple reproductive events. Between these extremes lies a full range of species with intermediate reproductive strategies, each an adaptation that increases survival in particular ecosystems.

SYNOPSIS

Communities are made up of populations. Each population contains all of the individuals of the same species that occupy an area at the same time.

Every population has the potential to reproduce large numbers of offspring. A population growing at its maximum rate increases in size exponentially and produces a J-shaped growth curve. Eventually, some factor or combination of factors (limited food or the buildup of toxic wastes, for instance) halts continued exponential growth, causing a population dieback, sometimes to extinction.

For most populations, growth starts off slowly, increases rapidly, and then slows down again and levels off. The population levels off at the carrying capacity of the environment, where growth remains in dynamic balance with available nutrients and appropriate conditions.

Populations grow when natality and immigration exceed mortality and emigration. In populations where there is no immigration or emigration, the rate of growth equals the difference between the birth rate and the death rate.

Organisms have evolved a broad range of reproductive strategies. Some species produce enormous numbers of offspring at one time, while others reproduce just one or a few offspring, which they provide with extended care. Each strategy has its advantages in certain environments.

The world's human population exceeded 5.5 million people in 1993. The human population is continuing to grow exponentially, mainly as a result of growth in developing countries.

Key Terms

population (p. 986)
population density (p. 987)
clumped pattern (p. 987)
uniform pattern (p. 987)
random distribution (p. 987)
natality (p. 988)

immigration (p. 988)
mortality (p. 988)
emigration (p. 988)
zero population growth (p. 988)
age–sex structure (p. 988)
survivorship curve (p. 989)
biotic potential (p. 990)
intrinsic rate of increase (r_o) (p. 990)
exponential growth (p. 990)
J-shaped curve (p. 991)
environmental resistance (p. 991)
carrying capacity (K) (p. 992)
sigmoid growth curve (p. 993)
logistic growth (p. 994)
r-selected species (p. 995)
K-selected species (p. 995)
biodiversity (p. 997)
density-dependent factor (p. 996)
density-independent factor (p. 996)
percent annual increase (p. 998)
fertility rate (p. 999)

Review Questions

1. What is the relationship between interspecific competition and density-dependent factors that limit population growth?
2. Match the process (numbered column) with the resulting outcome (lettered column).

____1. logistic growth	a. J-shaped growth curve
____2. intrinsic rate of increase	b. sigmoid growth curve
	c. a population dieback
____3. exponential growth	d. increased density
____4. immigration	

3. Check those conditions that would cause a population to *decrease* in density.
   ____1. natality that greatly exceeds mortality, immigration, and emigration
   ____2. pyramid-shaped age–sex structure
   ____3. a Type III survivorship curve, high emigration, and high mortality
   ____4. a population that has greatly exceeded the carrying capacity of the environment
4. In the following habitats, which reproductive strategy —*K*-selected or *r*-selected—would be of greater advantage? Give a brief explanation for your answer.

Environment	*r or K Strategy*	*Explanation*
a. cool lava		
b. recent burn area		
c. climax rainforest		
d. abandoned farm		
e. desert		
f. tundra		
g. streamside		
h. recent flood plain		

5. List two density-dependent and two density-independent factors that would affect human population growth. For example, a collision between two planes as a result of crowded airways would be a density-dependent factor, whereas a plane crash into a mountain as a result of adverse weather would be a density-independent factor.
6. What is the relationship between age–sex ratios and survivorship curves to biotic potential?
7. Describe the reproductive characteristics of an organism that is precisely in the middle of the *r* and *K* continuum. In what types of ecosystems would such an organism be at a clear advantage over organisms at either extreme of the continuum?
8. Carrying capacity of the environment, biotic potential, and environmental resistance all affect the rate of population growth. Consider two distinctly different types of organisms in two distinctly different habitats: elephants in an African savanna, and phytoplankton in the open ocean. Describe the similarities and differences between population growth factors for these two organisms. Be sure to consider all aspects.
9. List the conditions that determine the carrying capacity of a prey population. If the prey population exceeded its carrying capacity, would the predator population also exceed its carrying capacity? Explain your answer.
10. Name four factors or events that have blocked or postponed traditional limits on human population growth. Given the current rate of growth, name two ways humans can continue to block traditional limits.

Critical Thinking Questions

1. As in all plants, the critical phases in the life cycle of the saguaro cactus include seed germination, seedling growth and development, growth to reproductive maturity, flowering, pollination, fertilization, fruit and seed development, and seed dispersal. Which critical phases were tested by the research presented in the chapter opening vignette and which phases were not? Of those phases that were tested, are there any other elements that should be looked at to be sure that nothing was missed? Of those phases that were not tested,

choose one that you believe may explain why saguaros are not reproducing adequate numbers of offspring to maintain stable populations, and devise an experiment to test whether you are correct.

2. Each ecosystem supports many populations of species, each growing at different rates. Since all populations share the same abiotic environment, explain how the growth rates of each population can be different. Could any populations have the same growth rate? How?

3. The growth rates of populations can be calculated from birth rates and death rates, assuming there is no emigration or immigration. By convention, birth and death rates are given as the number per 1,000 people, whereas growth rate is given as a percent (i.e., per 100). Thus, the calculation for growth rate is

   growth rate (%) = birth rate − death rate/10.

   You can estimate the time in years that it will take a population to double by dividing 70 by the annual growth rate.

   Calculate the growth rate and doubling time of the countries in the table below:

Country	*Birth Rate*	*Death Rate*	*Growth Rate*	*Doubling Time*
United States	17	9		
India	31	10		
China	21	7		
Somalia	49	19		
Poland	15	10		
France	14	9		

4. Wildebeests and birds migrate from one location to another in search of food and other resources. Barnacles and mussels are sessile and cannot travel to locate richer sources of needed materials. Yet, the populations of both migratory and sessile species are subject to limitations to growth. List some of the differences in the ways limiting factors would affect the size of migratory populations, compared to sessile ones. Are there common features to the limiting factors in each category that may lead you to draw a general conclusion about how environmental resistances differ between migratory and sessile organisms?

5. Describe your opinion concerning the future trends in human population growth. Listen to the news and read a major newspaper over the next week and document which stories support your opinion and which stories do not. Based on these stories, should your projection of future population trends be modified? In what way? Choose one of the following statements that you feel best supports your revised opinion, and devise an experiment for testing whether the statement is correct or not:

   a. Humans are fundamentally different from all other organisms (behaviorally, intellectually, and physiologically) and are not subject to the same population controls as are other species.

   b. Humans are governed by the same population controls as are all other species. Although humans cannot change the types of natural population controls, their intellect and ingenuity empower them to modify levels of natural controls.

Additional Readings

Ayensu, E., V. Heywood, G. Lucas, and R. Defilipps. 1984. *Our green and living world. The wisdom to save it.* Cambridge, London: Cambridge University Press. (Introductory)

Council on Environmental Quality, 1981. *Global future: Time to act. Report to the President on Global Resources, Environment and Population.* Washington, D.C.: U.S. Government Printing Office. (Introductory)

Ehrlich, P., and A. Ehrlich 1972. *Population, resources, environment. Issues in human ecology.* San Francisco: W.H. Freeman. (Introductory)

Ehrlich, P., and A. Ehrlich. 1981. *Extinction.* New York: Random House. (Intermediate)

Population Reference Bureau, Inc. 1972. *The world population dilemma.* Washington, D.C.: Columbia Books. (Introductory)

World Resources. 1990–91. *A report by the World Resource Institute, in collaboration with The United Nations Environment and Development Programs.* New York: Oxford University Press. (Intermediate)

CHAPTER

44

Animal Behavior

STEPS TO DISCOVERY

Mechanisms and Functions of Territorial Behavior

MECHANISMS OF BEHAVIOR

Primarily Innate Behavior

Genes and Behavior

LEARNING

Habituation

Classical Conditioning

Operant Conditioning

Insight Learning

Social Learning

Play

Learning as an Adaptation

DEVELOPMENT OF BEHAVIOR

EVOLUTION AND FUNCTION OF BEHAVIOR

Optimality and Territoriality

SOCIAL BEHAVIOR

Costs and Benefits of Group Living

Animal Communication

ALTRUISM

Alarm Calls

Helping

Cooperation in Mate Acquisition

Food Sharing

Eusociality

BIOLINE

Animal Cognition

STEPS TO DISCOVERY

Mechanisms and Functions of Territorial Behavior

Susan Reichert has explored both the mechanisms and function of territorial behavior of the funnel-web building spider *Agelenopsis aperta.* These spiders compete for web sites and defend an area around the web as a territory. The territory must be large enough to provide sufficient food for survival and reproduction. This species occupies a wide variety of habitats, ranging from northern Wyoming to southern Mexico. Spiders living in the relatively lush vegetation along the rivers and lakes in Arizona (riparian populations) allow neighbors to build webs closer to their own than do spiders in desert grassland populations of New Mexico. Furthermore, the intensity of territorial disputes between desert grassland spiders is greater than is that between spiders that live near water. Threat displays of grassland spiders are more likely to escalate into battles, and the fighting more often results in physical injury.

A series of interesting experiments has demonstrated the genetic basis of territorial behavior in these spiders. Spiders were collected from a desert grassland environment in New Mexico and from a riparian environment in Arizona. Pure-bred lines were established by allowing individuals from a particular habitat to mate only with one another. After the spiderlings emerged from the eggs, each was raised separately on a mixed diet of all they could eat. When they were mature, sixteen females from each population line were placed into experimental enclosures where they could build webs. Just as in field populations, the average distance between laboratory raised females from riparian populations was less than that between females from desert grassland populations. Thus, territory size is an inherited characteristic and is not determined by recent feeding history or learned from previous territorial disputes.

With limited resources, competition is greater among funnel spiders living in the desert as opposed to funnel spiders living in

Working with John Maynard Smith, Reichert continued to explore the genetic mechanisms underlying territoriality and aggression in these spiders. Pure-bred lines of grassland spiders were mated with pure-bred lines of riparian spiders. Smith and Reichart found that fights between the hybrid offspring were even more likely to end in injury or death than were those between pure grassland spiders. Why should hybrid spiders be more combative than either of their parents? The simplest explanation for these results is that the behavior of a spider is determined by two conflicting tendencies: "aggression" and "fear." Each tendency is controlled by a gene or, more likely, a gene complex. The allele(s) for high aggression (A) is dominant to that for low aggression (a), and the allele(s) for low fear (B) is dominant to that for high fear (b). Researchers proposed that aggression and fear are low in riparian spiders and high in grassland spiders. Thus, grassland spiders would be homozygous for high aggression (AA) and for high fear (bb). In contrast, riparian spiders would be homozygous for low aggression (aa) and for low fear (BB). As a result, the hybrid offspring would have high aggression and low fear, a situation that would be expected to lead to costly fights. In additional crosses in which the hybrid offspring were mated with one another to create an F_2 generation, and in which they were mated with spiders thought to be homozygous recessive for these traits (backcrosses), the results were consistent with this model. The results of these later crosses also indicated that aggression was inherited on sex chromosomes and that fear was inherited on autosomal chromosomes.

Reichert also examined the degree of genetic diversity between the two populations of spiders. To do this, she used electrophoresis, a technique that reveals the number of alleles that exist for a given gene. Electrophoresis also can be used to estimate the degree of genetic variability among individuals in a population (page 328).

Those genetic differences and the differences in territorial behavior presumably arose as natural selection favored traits in the populations that suited the local environments. In relatively lush riparian areas found along rivers and lakes, for instance, prey are more abundant and there are more suitable web sites than in the desert grassland habitat. As a result, spiders living along a river are able to capture adequate prey in a smaller area. In contrast, the desert grassland environment is severe. Prey are scarce, and the scorching sun makes it difficult for spiders to forage during much of the day. Thus, larger territories are suited to stringent environmental conditions. Evidence suggests that territory size is genetically set. One reason for the intensity of territorial disputes among grassland spiders may be that the scarcity of web sites increases the value of a web. So, differences in genetic diversity between species of spiders is increased as a response to its environment.

Territoriality in the funnel-web spider seems to be an adaptation to the existing ecological conditions. However, there are no data to prove that owners of large territories in desert grassland populations leave more offspring than do individuals with small territories. In another locality, a recent lava bed in central New Mexico, there are data to support the hypothesis that territory quality may influence an animal's reproductive success. Here, spiders with quality web sites have thirteen times the reproductive potential of their neighbors in poorer quality areas.

rich riparian habitats.

"Why is that animal doing that?" This is the fundamental question of ethology, the study of animal behavior. This seemingly simple question has been interpreted in several ways. For example, the Dutch biologist Niko Tinbergen, a corecipient of the Nobel Prize in medicine and psychology in 1973, identified four related questions: (1) What are the mechanisms that cause the behavior? (2) How does it develop? (3) What is its survival value? (4) How did it evolve? Tinbergen believed that the biological study of behavior should "give equal attention to each and to their integration."

To better appreciate the types of questions we may ask about animal behavior, consider those that may be raised regarding a massive herd of caribou in their migratory march across the frozen tundra (Fig. 44-1). (1) How do the caribou "know" when it is time to migrate? How do they find their way along a predictable migratory path? Such questions focus on the mechanisms underlying a behavior. (2) Do those making this journey for the first time learn the route from experienced travelers, or do they inherit a directional tendency from their parents? Questions such as these concern development. (3) Why do they migrate? How do the advantages they gain outweigh the risks and demands of such a journey? These are questions regarding the survival value, or adaptiveness, of migration. (4) Finally, how did caribou migration begin? This question centers on the evolution of the behavior.

▼ ▼ ▼

MECHANISMS OF BEHAVIOR

When we observe an animal in nature, we generally find that its behavior is adaptive; that is, the behavior enhances the animal's chances of surviving and reproducing. Adaptation includes not only traits with known genetic causes but also the inherited potential for learning and even the learned behaviors themselves. The relative importance of genes and experience may vary tremendously, but neither is ever equal to zero. Genes generally code for a range of potential phenotypes. Sometimes genes specify a precise behavior, leaving little room for modification by learning. Behaviors that are precisely specified by genes are often those that must be expressed in nearly perfect form, even on the very first trial. For example, if an animal fails to respond

FIGURE 44-1
Migrating caribou. Many questions may come to mind when watching the behavior of animals. Some deal with the details of the behavior, others concern its advantages and evolution.

appropriately the first time it encounters a predator, it may not get a second chance to refine its escape response. Genes also play an important role in determining the actions of animals that have little opportunity to learn. For example, fruit fly parents are generally not present when their offspring emerge from the pupal cases. Nonetheless, a male fruit fly who has been isolated from the larval stage until adulthood still exhibits specific courtship behavior. In other cases, the behavioral blueprint is more general so that the behavior is almost entirely shaped by experience. A predator that fails to capture food on the first attempt may be hungry, but it learns from its experience, increasing its chances of success the next time.

PRIMARILY INNATE BEHAVIOR

Innate behaviors are those that are under fairly precise genetic control. Innate behaviors are often species-specific and highly stereotyped.

The Fixed Action Pattern

Among the primarily innate behaviors are **fixed action patterns (FAPs).** These are motor responses that are triggered by some environmental stimulus. Once started, FAPs continue to completion without the help of external stimuli. For example, a brooding female greylag goose will retrieve an egg that has rolled just outside her nest by reaching beyond it with her bill and rolling it toward her with the underside of the bill. Once the rolling behavior has begun, if the egg is experimentally removed, the goose will continue the retrieval response until the now imaginary egg is safely returned to the nest. The egg retrieval response of the female greylag goose illustrates other characteristics that are generally true of most FAPs. An FAP is performed by all appropriate members of a species. Furthermore, in the case of the greylag goose, each time an egg is retrieved, the sequence of actions is virtually identical, modified very little by experience. As evidence, an FAP will be exhibited even in inappropriate circumstances. For example, a brooding female will retrieve a beer bottle or any small object outside the nest, as if it were her egg.

Stimuli and Triggers

A fixed action pattern is produced in response to something in the environment. Ethologists called such a stimulus a **sign stimulus.** If the sign stimulus is given by a member of the same species, it is termed a **releaser.** Releasers are important in communication among animals.

Sign stimuli may be only a small part of any environmental situation. For example, a male European robin will attack another male robin that enters its territory. Experiments have shown, however, that a tuft of red feathers is attacked as vigorously as is an intruding male. Of course, in the world of male robins, red feathers usually appear on the breast of a competitor.

Any of the traits possessed by an animal or an object may serve as a sign stimulus. It may have a certain color, a special shape, or a particular pattern of movement, or it may produce a sound or have an odor. How do we know which character serves as the sign stimulus?

One way ethologists can identify the sign stimuli from the barrage of information reaching an animal is with models in which only one trait is presented at a time. The model is presented to an individual in the appropriate physiological state to see whether it will respond as it would to the normal stimulus. For example, a male stickleback in reproductive condition defends his territory from any intruding males. By constructing dummies of sticklebacks of varying degrees of likeness to the real male and painting them red, pale silver, or green. Niko Tinbergen and his co-workers demonstrated that in male sticklebacks, a red tint on the undersurface of the trespasser releases an aggressive territorial response. A very realistic replica lacking the red color was not attacked, but a model barely resembling a fish, on which the underside was painted red, provoked an assault.

Chain of Reactions

So far, we have considered only relatively simple behaviors; more complex behaviors can also be built from sequences of FAPs. The final product is an intricate pattern called a *chain of reactions*, whereby each component FAP brings the animal into the situation that triggers the next FAP.

An early analysis of a chain of reactions was conducted on the courtship ritual of the three-spined stickleback. This sequence of behaviors culminates in synchronized gamete release, an event of obvious adaptive value in an aquatic environment. Each female behavior is triggered by the preceding male behavior which, in turn, was triggered by the preceding feminine behavior (Fig. 44-2).

FIGURE 44-2
The courtship ritual of sticklebacks is built from a sequence of fixed action patterns.

Sometimes a male stickleback attacks a female entering his territory. If the female displays the appropriate "head-up" posture, in which she hangs in the water, exposing her egg-swollen abdomen, the male will begin his courtship with a zig-zag dance. He repeatedly alternates a quick movement toward her with a sideways turn away. This dance releases approach behavior of the female. Her movement induces the male to turn and swim rapidly toward the nest, an action that entices the female to follow. Once at the nest, the male stickleback lies on his side and makes a series of rapid thrusts with his snout into the entrance to the nest, while raising his dorsal spines toward his mate. This action is the releaser for the female to enter the nest. The presence of the female in the nest, in turn, is the releaser for the male to begin to prod the base of the female's tail with his snout, causing the female to release her eggs. The female then swims out of the nest, making room for the male to enter and fertilize the eggs. We can see that this complex sequence is largely a chain of FAPs, each triggered by its own sign stimulus or releaser.

GENES AND BEHAVIOR

We have seen that some behaviors have a strong genetic basis. Genes do, in fact, influence all behavior to some extent. We might wonder then what it means to say that a behavior has a genetic basis. It simply means that an animal with a certain gene will be able to perform the behavior. If the animal lacks that gene, it may be unable to perform the behavior, or the behavior may be expressed in a different manner.

As you already know, genes direct the synthesis of proteins, proteins that may affect some of the connections within the nervous system, or may act as regulators, such as enzymes that have a regulatory function. In a few cases, we know the link between the protein product of a specific gene and a behavior. For example, the gene behind egg-laying behavior of the sea hare *Aplysia* codes for a long chain of amino acids that is later cleaved into many proteins, three of which are known to be important in egg-laying behavior (Fig. 44-3). One of the three proteins is egg-laying

FIGURE 44-3
The sea hare Aplysia lays eggs in a stereotyped sequence of actions that is controlled by a single gene that codes for a long chain of amino acids. This chain is cleaved into three shorter proteins, three of which are important in orchestrating the behavior.

hormone (ELH), which increases the firing rate of the abdominal ganglion, increases heart and respiration rates, and stimulates the ducts of the reproductive system to contract and expel the egg string. The two other proteins function as neurotransmitters that increase or decrease the activity of neurons involved in egg-laying behavior.

Many animals perform complex behaviors that are either entirely or largely directed by the animal's genes. Such innate behaviors are particularly adaptive in species whose offspring have little or no contact with their parents. Innate behavior often reveals itself by its rigid, stereotypic pattern and by the fact that it can be evoked by highly specific environmental stimuli, such as a red feather or a specific aroma. (See CTQ #1.)

LEARNING

Learning is a process in which the animal benefits from experience so that its behavior is better suited to environmental conditions. As we look at learning in various situations, it appears that the process can occur in several fundamentally distinct ways. Although not everyone agrees on how it should be done, it is often useful to group types of learning into categories, including habituation, classical conditioning, operant conditioning, insight learning, social learning, and play.

HABITUATION

In **habituation,** the simplest form of learning, the animal learns *not* to show a characteristic response to a particular stimulus because the stimulus was shown to be unimportant during repeated encounters.

For an example that illustrates the essential characteristics of habituation we turn to the clamworm, a marine polychaete that lives in underwater burrows which it constructs out of mud. A clamworm partially emerges from the burrow while feeding. However, certain sudden stimuli, such as a shadow that could herald the approach of a predator, cause the clamworm to withdraw quickly for protection. If the stimulus is repeated and there are no adverse consequences, the withdrawal response gradually wanes. If the stimulus is changed, however, say from a shadow to a touch, the clamworm will again respond. Thus, the loss of responsiveness to the shadow is not due to muscular fatigue (page 483).

Habituation is beneficial in that it eliminates responses to frequently occurring stimuli that have no bearing on the animal's welfare, without diminishing reactions to significant stimuli. Obviously, it is important for the clamworm to withdraw to the safety of its burrow when a shadow belongs to an approaching predator. If the shadow is encountered often without a predator's attack, however, it is probably caused by something harmless, perhaps a patch of algae repeatedly blocking the sun. In this case, responding to the shadow each time it appears would waste energy and leave the worm little time for other essential activities, such as feeding or reproducing. Thus, habituation is one of the mechanisms that focuses attention and energy on the important aspects of the environment.

CLASSICAL CONDITIONING

In **classical conditioning,** an animal learns a new association between a stimulus and a response. Because the new stimulus repeatedly occurs before the usual one, it gradually begins to serve as a signal that the usual stimulus will occur. Eventually the new stimulus alone is sufficient to cause the response.

The most familiar example of classical conditioning is that of Pavlov's dogs, who learned to associate the sound of a bell with the presence of food (Fig. 44-4). During training, Pavlov rang a bell immediately before feeding a hungry dog. When the dog saw the food, it began to salivate. The procedure was then repeated many times. Eventually, the bell became a signal that food would be delivered soon, and the dog began to salivate at the mere sound of the bell. This response—salivating at the sound of the bell—is an example of a **conditioned reflex.**

In more general terms, an animal has a particular inborn response to a certain stimulus. This stimulus (e.g., food) is called the **unconditioned stimulus (US)** because the animal did not have to learn the response to it. A second

FIGURE 44-4
Ivan Pavlov, a Russian physiologist, demonstrated classical conditioning in the salivary reflex of a dog. The dog is prepared so that the saliva it produces can be collected and measured. In this experiment, food serves as the unconditioned stimulus since it triggers the desired response (salivation) without prior training. A tone is sounded immediately before food is offered to the dog. Because of the repeated pairing of the tone and food, the dog eventually begins to salivate in response to the tone alone.

FIGURE 44-5
A male blue gourami that is classically conditioned to cues signaling the approach of a rival has a competitive edge in territorial disputes.

stimulus—one that does not initially elicit the response—is repeatedly presented immediately before the US. After several pairings, the second stimulus is able to elicit the response. The new stimulus (e.g., a bell) is now called the **conditioned stimulus (CS)** since the animal's response has become conditional upon its presentation.

Karen Hollis has experimentally tested the hypothesis that the adaptive function of classical conditioning is to prepare animals for important events. Her studies have centered on territorial and reproductive behaviors in blue gouramis (Fig. 44-5). Successful territorial defense is important because female blue gouramis rarely mate with a male without a territory. A conditioned response to signals indicating the approach of a rival gourami might prepare a male for battle and give him a competitive edge. In nature, as the rival approached, he would inadvertently send visual, chemical, or mechanical signals that territorial invasion was imminent. In the laboratory, male blue gouramis that are classically conditioned to a signal (a brief light) that predicts an encounter with a rival are more successful in aggressive contests. It may be that conditioned males are the winners because the light (CS) increased the level of androgens, male sex hormones known to heighten aggressiveness in many species. A more aggressive male has a better chance of winning the battle and defending his territory, thereby increasing his chances of mating.

OPERANT CONDITIONING

When a behavior has favorable consequences, the probability that the act will be repeated is increased. This relationship may result because the animal learns to perform the behavior in order to be rewarded. This type of learning has been named **operant conditioning** to emphasize the fact that the animal operates on the environment to produce consequences. The behavior must be spontaneously emitted, not elicited by a stimulus, as it is in classical conditioning, and the favorable result, or reinforcement, must follow the behavior closely. The timing of events is critical. In a sense, a cause-and-effect relationship develops between the performance of the act and the delivery of the reinforcer.

B. F. Skinner devised an apparatus used to study operant conditioning in the laboratory. Typically, a hungry animal is placed into a "Skinner box" and must learn to manipulate a mechanism that yields food. For example, a hungry rat placed in a Skinner box will move about randomly, investigating each nook and cranny. Eventually, the rat will put its weight on a lever provided in the box. When the lever is pressed, a bit of food drops into a tray. The rat will usually press the lever again within a few minutes. In other words, the rat first presses the lever as a random act; then, when the action is rewarded, the probability of its being repeated increases.

INSIGHT LEARNING

Insight learning is a sudden solution to a problem without obvious trial-and-error procedures. For example, captive chimpanzees have been known to stack boxes in order to climb up and reach a banana hanging from the ceiling of their cage. One interpretation of the chimps' problem-solving abilities was that they saw new relationships among events—relationships that were not specifically learned in the past—and that they were able to consider the problem as a whole, not just a stimulus–response association between certain elements of the problem. It has been suggested, for instance, that the chimp forms a mental representation of the problem and then mentally applies trial-and-error patterns to the problem.

Other researchers explain sudden problem solving as the result of associations among previously learned components. It has been argued, for instance, that chimps that moved boxes and then climbed on them to reach a banana had previously acquired two separate behaviors: moving boxes toward targets, and climbing on an object to reach another object.

SOCIAL LEARNING

Some organisms are able to learn from others. The possibility for such **social learning** is much greater in social species since they spend more time close to others.

Although each member of a population may have the capacity to learn appropriate responses for himself or herself, it is often more efficient, and perhaps less dangerous, to learn about the world from others. Individuals of some species may learn to avoid dangerous situations by watching their fellow members. For example, rhesus monkeys can learn to fear and avoid snakes by watching other monkeys show fear of snakes. In other species, interaction with adults is *critical* to learning appropriate behaviors by the young. For example, a juvenile wren must perfect its crude rendition of the song used for territorial defense by countersinging with a neighboring adult.

When a tradition spreads through a population, it is not always because animals have learned the trick simply by seeing it done. For instance, a tradition of washing sweet potatoes in the sea was begun by a young Japanese snow monkey, and it spread rapidly to other members of the troop (Fig. 44-6). Although the habit clearly spread throughout the population, we don't know whether the snow monkeys learned to wash food by imitating others or whether they were trained to do so by the caretaker. Because food washing amuses tourists, the caretakers may have given more sweet potatoes to those members of the troop that were known to wash them. The habit may have spread because the monkeys near those who washed their sweet potatoes (who, by the way, were likely to be relatives) were also close to the caretaker and the source of reinforcement.

FIGURE 44-6
Snow monkeys washing food. The tradition of washing sweet potatoes was begun by a young Japanese snow monkey and spread rapidly throughout the troop.

Some traditions are due to social learning but not because the individual observed another performing the activity and then imitated it. For example, rats can learn what to eat, not by watching others, but by smelling the breath of others. In one experiment, a "demonstrator" rat ate food flavored with cocoa or cinnamon. The "demonstrator" was then anesthetized and placed 2 inches away from the wire cage of an awake "observer" rat. Although the "demonstrator" slept through the demonstration, the observer later showed a preference for the food the demonstrator had eaten.

PLAY

Although it is easy to spot "play," it is a difficult term to define. One reason that play eludes a simple definition is that there is no specific behavior pattern or series of activities that exclusively characterize play. Play borrows pieces of other behavior patterns, usually incomplete sequences and often in an exaggerated form.

What function does frolicking serve? One hypothesis is that it is physical training for strength, endurance, and muscular coordination. Indeed, it is thought that the sensory and motor stimulation of play causes the formation of a network of synapses in the cerebellum, a part of the brain responsible for sensory-motor coordination. Another hypothesis maintains that play allows individuals to practice social skills, such as grooming and sexual behavior, that are important in establishing and maintaining social bonds. Hatchling sea turtles, for example, take turns vibrating a foot in front of another hatchling's face, a gesture that will later be part of a male's courtship display. Finally, play may be a mechanism for learning specific skills or improving overall perceptual abilities.

Filial Imprinting in Birds

Young chicks, ducklings, and goslings generally follow their mother wherever she goes. How does such following behavior develop? Konrad Lorenz, an Austrian biologist and corecipient of the 1973 Nobel prize, was the first to systematically study this behavior, working with newly hatched goslings. In one experiment, Lorenz divided a clutch of eggs laid by a greylag goose into two groups. One group was hatched by its mother; as expected, these goslings trailed behind her. The second group was hatched in an incubator. The first moving object these goslings encountered was Lorenz, and the goslings responded to him as they normally would to their mother. Lorenz marked the goslings so that he could determine in which group they belonged and placed them all under a box. When the box was lifted, the goslings streamed toward their respective "parents," normally reared goslings toward their mother, and incubator-reared youngsters toward Lorenz. The goslings had developed a preference for characteristics associated with their "mother" and expressed this preference through their following behavior. The attachment was unfailing, and from that point on Lorenz had goslings following in his footsteps.

Today, the process by which young birds develop a preference for following their mother is called **filial imprinting.** Imprinting is distinguished from other types of learning by several characteristics: (1) It is relatively quick; (2) it occurs only during a limited time (called the **critical period**); and (3) it occurs without any obvious reward. Presumably, the biological function of filial imprinting is to allow young birds to recognize close relatives and thereby distinguish their parents from other adults that might attack them.

Early experience also has important consequences for development of mate preferences in birds. In many species, experience with parents and siblings early in life influences the sexual preferences that form in adulthood. The learning process in this case is called **sexual imprinting.** Sexual imprinting is typically exhibited in the preferences of sexually mature birds for individuals of the opposite sex. One dramatic demonstration of the importance of early experience to subsequent mate preference came from cross-fostering experiments with finches. Eggs of zebra finches were placed in clutches belonging to Bengalese finches. The Bengalese foster parents raised the entire brood until the young were old enough to feed themselves. From then on, young zebra finch males were reared in isolation until they were sexually mature. When later given a choice between a zebra finch female and a Bengalese finch female, zebra finch males courted Bengalese females almost exclusively.

LEARNING AS AN ADAPTATION

Many scientists now accept that evolution shapes the learning "styles" of different species to suit their ecological demands. That is, individuals may learn certain things more easily because, in the natural ecological setting, those that do learn have a better chance of surviving and leaving offspring than those that do not.

Species-specific differences in spatial learning and memory among Clark's nutcrackers, pinyon jays, and scrub jays illustrate how ecology and evolution may influence a species' learning skills. These birds are among those that store seeds and recover them later, when food is more difficult to find. To ensure that their seeds won't be stolen by other animals, the birds hide them in small holes they dig in the ground and then cover over. The seeds are cached in this way in the autumn and are used, as needed, throughout the winter and spring. This means that the birds must be able to return to cache sites months after the seeds are buried. Recovery of the seeds is quite an impressive feat for all three species but particularly for Clark's nutcrackers, which may have as many as 9,000 caches, covering many square kilometers of ground.

If natural selection shapes the learning ability of species, we might predict that those species that depend more heavily on cached food for survival would be better at recovering caches than would other species that are less reliant on cached food. Clark's nutcracker, pinyon jays, and scrub jays are related species (members of the same family) that differ in their dependence on cached food. The nutcrackers live at high altitudes, where they have little else to eat during the winter and spring but their stored seeds. Their winter diet consists almost entirely of cached pine seeds. Nutcrackers prepare for winter by storing as many as 33,000 seeds. Pinyon jays live at slightly lower altitudes, where food is a little easier to find during the winter. Nonetheless, 70 to 90 percent of the pinyon jays' winter diet consists of some of the 20,000 seeds they cached in preparation for winter. In contrast, winter stresses are not as severe for a scrub jay. Scrub jays are smaller than the other two species, so they require less energy to maintain themselves. Furthermore, scrub jays live at much lower altitudes, where food is somewhat easier to find in the winter. Thus, scrub jays store only about 6,000 seeds a year, and these account for less than 60 percent of the winter diet.

A study comparing cache recovery by Clark's nutcrackers, pinyon jays, and scrub jays found species differences that are correlated with their relative dependence on stored seeds. In these experiments, the birds were permitted to store seeds in sand-filled holes in the floor of an indoor aviary. Each bird's ability to remember where it hid its seeds was tested 1 week later. Pinyon jays and nutcrackers, the species that depend most heavily on finding their stored seeds to survive the winter, remembered where they had hidden their seeds more accurately than did the scrub jays.

Unlike innate behaviors, learned behaviors are the result of prior experience. Learned behaviors range from simple reflex responses that change over time to highly complex behaviors, such as those exhibited by chimpanzees who stack boxes to reach bananas. The types of behavior learned by a species is suited to the ecological demands it faces. For example, those birds that depend on stored seeds to survive the winter are better able to remember seed-hiding places than are birds that live in areas where food remains available. (See CTQ #2.)

DEVELOPMENT OF BEHAVIOR

During the development of a behavior, there is an intimate relationship between the behavior's genetic component and its learned component. Although genes and experience work together throughout the organism's development to

FIGURE 44-7
Song development in this species depends on both genes and experience. A young male must learn to sing its song by hearing the song of adult males. However, it inherits a "template" of its own species' song.

produce most behaviors, as the organism changes over time, the nature of the interaction will vary. For example, as a result of its genetic makeup, an animal may be more or less responsive to certain environmental stimuli at different times of its life. Conversely, the presence or absence of certain stimuli at critical times during development may alter the expression of the genes.

The interaction of genes and learning is seen in the development of the song of a male white crowned sparrow (Fig. 44-7). Young males must learn to sing by hearing adult males sing. However, if a young male white-crowned sparrow is isolated from other members of its species and is allowed to hear recordings of bird songs, including one of its own species, it will learn to sing correctly. How does the sparrow know which song is correct? It must have inherited a template, or image, of its species' song. The template allows the male to recognize his own song, but the experience of hearing the song is also needed before the male can sing properly.

Most complex behaviors cannot be divided neatly into either the innate or learned categories but, rather, are influenced by both genetics and prior experience. (See CTQ #3.)

EVOLUTION AND FUNCTION OF BEHAVIOR

Ecological conditions will determine which traits are favored during evolution. Therefore, as a result of natural selection, we would expect traits of individuals to become better fitted to the environment from generation to generation. The traits that allow individuals to survive and reproduce better than their competitors can are called adaptations. When we ask questions about the adaptiveness of behavior, we are asking about its survival value. The aim in answering such a question is to understand why those animals that behave in a certain way survive and reproduce better than do those that behave in some other way.

Studies of parental behavior among species of gulls provide an example of how behavior may be shaped by natural selection to suit the environmental conditions. Kittiwakes, which have low predation rates, leave eggshell pieces in the nest, while ground nesting gulls, which have high predation rates, generally remove broken eggshells. Niko Tinbergen and his colleagues experimentally tested the hypothesis that eggshell removal reduced predation on chicks. It was noted that the eggs, chicks, and nest were camouflaged and might be difficult for a predator to spot. However, the bright white inner surface of a piece of eggshell might catch a predator's eye and reveal the nest site. The experimenters painted some blackheaded gull eggs white to test the idea that white eggs might be more vulnerable to predators than would the naturally camouflaged eggs. The difference in predation rates supported the hypothesis that the white inner surface of egg shell pieces might endanger nearby eggs or chicks. Tinbergen then painted hen's eggs to resemble those of a gull and placed white pieces of shell at various distances from some of the nests. The broken egg shell bits did attract predators. Furthermore, the risk of predation decreased with increasing distance of eggshell pieces from the nest. Thus, fastidious parents leave more offspring to perpetuate their genes. Eggshell removal is clearly adaptive.

The timing of eggshell removal is also adaptive. Whereas oystercatchers remove eggshell pieces almost immediately after a chick hatches, a blackheaded gull stays near its chick an hour or two after hatching before removing the eggshell pieces. The difference in the timing of eggshell removal between these species is related to their nesting habits. Oystercatchers generally nest alone, so neighbors do not pose a threat. In contrast, blackheaded gulls live in colonies, and their chicks are commonly eaten by neighboring gulls. Furthermore, a newly hatched chick is wet and easier to swallow than is one that has dried and become fluffy. Thus, although removing the shells reduces predation by other species of birds, delaying removal until the chicks are dry decreases the likelihood of the chick's being cannibalized by neighboring gulls while its parents are away from the nest.

By focusing on the survival value of adaptation, we have considered only the benefits gained through certain behaviors. But most actions also have costs. *Optimality theory* views natural selection as "weighing" the costs and benefits of each available alternative. First, natural selection translates all costs and benefits into common units: fitness. Then it chooses the behavioral alternative that maximizes the difference between costs and benefits. The choice that maximizes fitness is the alternative that would continue into the next generation.

OPTIMALITY AND TERRITORIALITY

A **territory** is an area that is defended against intruders, generally in the protection of some resource. Optimality theory predicts that an individual should be territorial if the benefits from enhanced access to the resource are greater than is the cost of defending the resource. The benefits of territoriality include increased mate attraction, decreased predation, protection of young and/or mates, reduced transmission of disease, and a guaranteed food source. The costs of defending the territory include energy expenditure, risk of injury, and increased visibility to predators.

Optimality theory predicts that territoriality will evolve only when the benefits exceed the costs. On the one hand, if

FIGURE 44-8
Marine iguanas on the Galapagos Islands. Optimality theory predicts that territoriality will evolve only when its benefits exceed its costs. Female iguanas defend nest sites on Hood Island, where suitable sites are rare, but not on other Galapagos Islands, where nest sites are abundant.

the resource is scarce, an individual may not gain enough to pay the defense bill; it may be economically wiser to look for greener pastures. Accordingly, the golden-winged sunbird will abandon a territory when it no longer contains enough food to meet the energy costs of daily activities as well as defense. On the other hand, if there is more than enough of the resource to go around, it would be energetically wasteful and economically unsound to defend it. Water striders are among the species that will cease to defend territories if supplied with abundant food. Likewise, we find that female marine iguanas don't bother defending territories with nest sites on most of the Galapagos Islands (Fig. 44-8); they defend territories only on Hood Island, the only Galapagos island where nest sites are in short supply.

We have been discussing optimality as if an individual's reproductive success depends solely on its own behavior. There are situations, however, in which the individual is not the sole master of its fate; instead, the reproductive success gained by behaving a certain way depends on what the other members of the population are doing. Such an optimal course of action is called an **evolutionary stable strategy (ESS).** By definition, an ESS is a strategy that cannot be bettered and, therefore, cannot be replaced by any other strategy when most of the members of the population have adopted it. An ESS cannot be bettered because when it is adopted by most of the members of the population, it results in maximum reproductive success for the individuals employing it. As a result, an ESS is both unbeatable and uncheatable.

ESS theory has improved our understanding of the logic of animal combat, a situation in which the best strategy depends on what competitors are doing. Consider, for instance, aggressive games in which there are only two strategies that might be employed. The "hawk" strategy is to fight to win. A hawk will continue to fight either until it is seriously injured or its opponent retreats. The "dove" strategy is to display but never engage in a serious battle. If a dove is attacked, it retreats before it can be injured seriously. Many factors will determine a strategy's success. As the cost of fighting increases, the dove strategy should be favored because the potential cost of injury is greater in species that possess weapons, such as horns or sharp teeth. Therefore, the hawk–dove model predicts that contests between members of well-armed species should rarely escalate from display to battle (Fig. 44-9). Conversely, the hawk strategy should be more common when the risk of injury is low. Accordingly, animals without weapons, such as toads, fight fiercely. In this case, although some individuals are injured, the risk of injury is much less than it would be if the individuals had the prowess of lions.

It seems reasonable to assume that the intensity of fighting should increase with the value of the resource. Offspring—a direct measure of fitness—are obviously of great value. It is not surprising, then, that a mother will fight fiercely to defend her young. Similarly, male elephant seals, for instance, fight brutal and often bloody battles for the right to mate. Battles are titanic because a male's entire reproductive success is at stake. All matings are performed by a few dominant males who defend harems of females. The duels between males are so strenuous that a male can usually be harem master for only a year or two before he dies.

Behaviors are shaped by natural selection to suit the environmental conditions in which the species lives. Such behaviors are adaptive because they increase the chances of survival and reproduction of its members. Even though a particular behavior may have costs, such as an increased risk of predation or the expenditure of energy, the benefits from that behavior are presumed to exceed the costs; otherwise, the behavior would not have evolved. (See CTQ #4.)

FIGURE 44-9
Optimality theory predicts that fights are not likely to escalate when the cost of injury is great. These males could kill one another, but they generally do not bite.

SOCIAL BEHAVIOR

Not all animal species live in groups. Tigers, for example, may live and hunt alone in large territories. Furthermore, not all groups are social (as anyone who has experienced a rush hour traffic jam will testify). Sometimes animals that live alone come together only to share a vital resource, such as a water hole in the desert. But some groups are social. We may wonder, then, what makes the difference?

COSTS AND BENEFITS OF GROUP LIVING

Soon after hatching, a *Holocnemus pluchei* spiderling is faced with a momentous "decision"—build a web of its own or join large individuals of its kind on an existing web. The small spiderlings are generally outcompeted for food by larger individuals whose web they share. Indeed, they catch fewer prey items than do solitary spiderlings. It is energetically expensive to build a web, however, and spiderlings add little or no silk to an existing web. The spiderlings must weigh the value of food lost against that of silk saved.

Living in groups, then, has both costs and benefits. Among the costs may be increased competition for mates, nest sites, or food. Other costs include increased exposure to parasites or disease, increased conspicuousness to predators or prey, and an increased risk of wasted energy in raising offspring that are not one's own.

Group living also has many benefits, particularly those relating to eating and not being eaten. Some predators, for example, engage in cooperative hunting. As a result, they have a higher capture success rate and average energy intake per individual than do solitary hunters. For example, groups of predatory jackfish may work together in catching small prey. Some members of the group form a circle around swarms of prey while other fish make strikes into the midst of the swarm.

In addition, groups are usually more successful than are individuals at detecting, confusing, and repelling predators. With all the members of the group on alert, a predator is likely to be spotted. Many species call to warn others in the group of a predator's approach. Vervet monkeys have a particularly sophisticated warning system. Different calls warn of the approach of an eagle, a leopard, or a poisonous snake. This system makes sense because the best manner of escape would differ depending on whether the predator were striking from the air or land.

During an attack, an individual within a group has a smaller chance of becoming the next meal. In addition, since predators find it easier to attack a single individual, they often swoop at a group, causing the individuals to scatter. Then one member may be singled out. It isn't surprising then that many prey species, such as flocks of starlings or schools of fish, counter this "divide and conquer" predator strategy by forming even tighter clusters when a predator is detected.

Many prey species will engage in mobbing behavior, whereby adults harry a predator that is frequently far superior to any individuals of the group. For example, small birds will frequently mob a predator, such as an owl. Baboons and chimpanzees use a similar strategy against leopards. Screaming, the baboons or chimpanzees charge and retreat; they may even throw sticks at the leopard. Finally, living in groups may aid in defending young, food, or space against other members of one's species.

An increase in the density of reproducing animals may also facilitate reproduction by providing mutual stimulation. The sights, sounds, and smells of other courting individuals appears to enhance and synchronize breeding. It is thought that this may have been a factor in the evolution of group courtship displays. Such group "advertising" is observed in frogs and insects. Although it is relatively rare in higher vertebrates, the *lek* displays of some bird species, such as the sage grouse, may be thought of in this way as well. In a lek display, as many as 50 or 60 male sage grouse may congregate at dawn. As they congregate for a communal display, the males may stimulate the production of reproductive hormones in one another as well as in the females about to choose a mate from among the madly displaying cocks.

ANIMAL COMMUNICATION

Communication is essential to social life. It underlies cooperative as well as aggressive behaviors.

Reasons For Communication

There are numerous functions for communication. Species recognition is important to ensure that reproductive efforts are not wasted on members of the wrong species and that aggression is directed toward those individuals who are competing for the same resources.

Sexual reproduction is often dependent on communication. First a mate must be attracted and then courted. During courtship, the male often advertises himself, and the female assesses his qualities. Aggression is reduced, and the behavior and physiology of the mates is coordinated. Following mating, the offspring may communicate its desire to be fed, or the parent may have to indicate its willingness to feed the young.

Communication is also important in aggressive interactions. Most animal disputes are settled without fighting; rather, the animals threaten one another with stereotyped displays. Threat displays may allow competitors to assess one another's abilities. In this way, the weaker individual may accept its loss and leave without risking injury. The display is often tied to a physical characteristic, such as size or strength, that cannot be faked. Thus, rivals are kept honest. Male red deer stags, for example, challenge each

FIGURE 44-10
Male red deer roaring. Male red deer challenge one another in a roaring contest. Because roaring is strenuous, it provides an honest cue for assessing a rival's fighting ability.

other in a vocal duel in which each male takes a turn roaring at the other (Fig. 44-10). As the pace of the bellowing increases, one contestant usually gives up. Because roaring is strenuous, it is a reliable indicator of the rival's fighting ability.

Other signals allow animals to keep in touch with one another. Sometimes, it is important to remain in contact so that future association is possible. At other times, signals bring animals together immediately. *Recruitment* occurs when individuals are brought together to perform a specific duty, as occurs in social insects. For example, fire ants leave an odor trial that will guide recruits to a food source.

Honeybees have a remarkable system of recruitment, which was discovered by the Austrian biologist Karl von Frisch, the third corecipient of the 1973 Nobel prize. When a scout bee finds food, she recruits nestmates to help in the harvest by dancing on the vertical surface of the comb. When the food is close to the hive, the bee performs a *round dance,* in which she walks in circles, first to the right and then to the left (Fig.44-11a). By feeling the dancer with their antennae, other bees follow the dance. The dancer informs recruits that food is within a certain distance of the hive. The *waggle dance* is performed when food is farther from the hive (Fig. 44-11b). A path similar to a figure eight is traced on the comb. During the straight part of her run, the dancer vigorously shakes her abdomen. The speed of the dance, the number of waggles on the straightaway, and the duration of the bee's buzzing sound correlate well with the distance to the food source. The orientation of the straight part of the run indicates the direction of the food source. The angle a forager leaving the hive must assume with the sun as it flies to the food source is indicated by the angle of the waggle run relative to gravity. If the scout bee dances straight up on the comb, recruits should fly directly toward the sun. A dance oriented straight downward indicates a food source directly away from the sun. To indicate other directions, the dancer changes the angle of the waggle run relative to gravity so that it matches the angle between the sun and the food source.

Types of Communication Signals

Any sensory modality may be used for communication. The particular sensory modality that has been naturally selected during evolution to be used to communicate in a given species will have been influenced by the nature of the message as well as by the ecological situation of the organism. What factors may influence the nature of the communication? One factor is how well the animal sending the

(a)

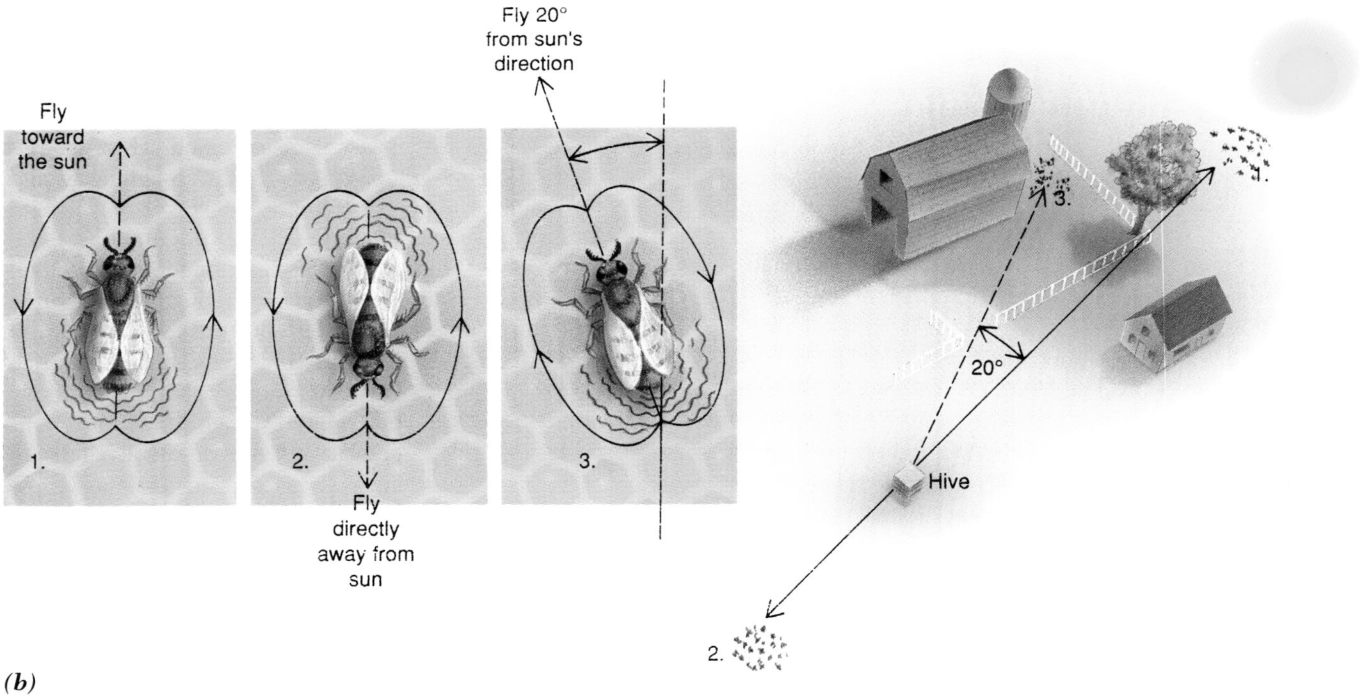

Fly toward the sun

1.

2.

Fly directly away from sun

Fly 20° from sun's direction

3.

Hive

20°

1.

2.

3.

(b)

◀ FIGURE 44-11
Round dance of honeybee scout. ***(a)*** After finding food close to the hive, a scout returns and does a round dance on the vertical surface of the comb. The dance consists of circling alternately to the left and right. The dance informs recruits that food can be found within a certain distance of the hive. ***(b)*** Waggle dance of honeybee is performed when a scout finds food at some distance from the hive. The dancer traces the pattern of a figure eight and waggles her abdomen during the central straight part of the dance. Aspects of the dance correlate with the distance and direction to the food source. The dancer indicates direction by the orientation of her waggle run relative to gravity. The direction in which recruits should fly is equal to the angle of the waggle run relative to vertical. When the food is in the direction of the sun, the dancer waggles straight up. When the recruits should fly directly way from the sun, the waggle run is oriented straight down. A food source located 20 degrees to the right of the sun would be indicated by a dance oriented so that the waggle run were 20 degrees to the right of vertical.

message can be localized. Visual signals that may be enriched with color and brightness are easiest to localize. Therefore, visual displays are frequently used by animals that are active during the day for short-range communication in open environments. Sound can also be localized and is generally employed by animals that are active at night or that live in dense vegetation, where vision is limited. As an example, consider the melodies of insect and amphibian calls that are so obvious on summer nights in the country.

Another important consideration is the distance over which the signal is effective. Pheromones—chemical signals secreted by animals which influence the behavior of other members of the same species—may be the best signal when long distance communication is desired. Perhaps the best known pheromones are sex attractants, such as those produced by moths. For instance, the gypsy moth sex attractant is carried by the wind and may attract males more than a mile away from the female. Sound is not usually as good as are chemicals for long-distance communication, particularly when the sound must be transmitted through the air. Sound travels much farther in water, however. The songs of whales can be heard hundreds of miles away, for example. Light is rapidly attenuated in water, so if a signal must be sent over a distance, visual displays are generally not used by aquatic animals. Likewise, in any environment where long-distance vision is limited, such as in a dense forest, auditory signals are often employed.

A third important factor is the duration of the signal. Chemical signals are generally among the most durable, which is why chemicals are often used to mark territories. Not all chemicals are long-lived, however; some, like the alarm signals of some ants, fade within 30 seconds.

Bodily Contact

Social bonds may be cemented by physical contact. Some species of animals reduce tensions by touching. Greeting ceremonies are common in social animals and often involve touching and sometimes even embracing. It is not uncommon to observe even nonhuman primates sitting with their arms around one another. Some animals, such as black-tailed prairie dogs, may kiss to establish community membership. Sea lions and chimps may kiss one another as a greeting. The members of a wolf pack may surround the dominant male and lick his face and poke his mouth with their muscles. This ceremony occurs at those times when it is useful to reinforce social ties, such as when the wolves awake in the morning, when they have been separated for a period of time, and when they are ready to go out on a hunt (Fig. 44-12).

FIGURE 44-12
The wolf greeting ceremony illustrates the importance of bodily contact in cementing social bonds.

FIGURE 44-13
Grooming in primates. Although grooming originally functioned only for skin care, its social functions have now become more important. Primates spend a major portion of their day grooming; it helps form and maintain social bonds.

Social grooming (Fig. 44-13) is a form of social behavior that is found in a variety of animals, but it is especially prominent among higher primates. One effect of such behavior is to rid the animal being groomed of parasites, hardened skin secretions, and debris on the skin. Equally important, grooming allows animals to cement social bonds.

Living in a group may provide significant benefits for an individual over a solitary lifestyle. Group living provides safety in numbers, available mates, and an increased opportunity to obtain food. Social life depends on communication between its members, allowing essential activities to be coordinated. Different types of signals—visual, auditory, tactile, or chemical—may be suited for communicating different types of information in different environments. (See CTQ #5.)

ALTRUISM

Altruism is the performance of a service that benefits a *conspecific* (another member of the species) at some cost to the *altruist* (the one who does the deed). Strictly speaking, the benefits and costs of altruism are measured in units of fitness (the reproductive success of a gene, organism, or behavior). Since changes in fitness are nearly impossible to ascertain, however, the gains and losses are usually arbitrarily defined by researchers as certain goods or services that seem to influence the participants' chances of survival.

On the surface, the existence of altruism seems to contradict evolutionary theory. Evolution involves a change in the frequency of certain alleles in the gene pool of a population. If aiding a conspecific costs the altruist, then the altruist should be less successful in leaving offspring that bear copies of its alleles than are the recipients of its services. As a result, the alleles for altruism would be expected to decrease in the population.

The hypotheses for the evolution of altruism can be arbitrarily classified into several overlapping classes:

1. *Individual selection.* The general thrust of these hypotheses is that, when the interaction is examined closely enough, the altruist will be found to be gaining, rather than losing, as a result of its actions. The benefit may not be immediate; sometimes the gain is in the individual's *future* reproductive potential.

2. *Kin selection.* One of the most cogent explanations for altruism is kin selection. Kin selection is based on the theoretical work of the British geneticist W. D. Hamilton. Central to kin selection is the idea of **inclusive fitness** which considers all the adult offspring of an individual as well as those of its relatives that are alive because of the actions of that individual. If family members are assisted in a way that increases their reproductive success, the alleles that the altruist has in common with them are also duplicated, just as they would be if the altruist reproduced personally. In other words, since common descent makes it likely that a certain percentage of the alleles of family members are identical, assisting kin is another way to perpetuate one's own alleles. Since an individual shares more alleles with certain relatives by common descent, the possibility of genetic gain increases with the closeness of the relationship. For example, aiding a cousin, who shares an average of only one-eighth of the same alleles, is less productive than is assisting a brother or sister, who is likely to share half of its alleles with the altruist. Thus, the fitness gained through family members must be devalued in proportion to their genetic distance (diminished relatedness).

How can relatives be identified? There are several possibilities. One way might be to use location as a cue: The individuals who share one's home are likely to be kin. Alternatively, individuals might be identified as kin because they are recognized from prior social contact or as a result of their association with a known relative. Another possible way to recognize a relative is by certain traits that characterize family members. In other words, an image of a family member may be developed that is matched or compared to the appearance of a stranger. Finally, recognition may be genetically programmed. Perhaps there are alleles that, in addition to labeling relatives with a noticable characteristic, may cause the altruist to assist others who bear the label.

3. *Reciprocal Altruism.* Altruism may evolve—despite the initial cost to the alturist—if the service is repaid with interest by other members of the population. In other words, altruism will be favored if the final gain to the altruist exceeds the initial cost. In order for reciprocal altruism to work, however, individuals who fail to make restitution must be discriminated against. Because of this latter requirement, certain conditions make reciprocal altruism more likely in particular species: First, there should be a good chance that an opportunity for future repayment will arise; second, the individuals must be able to recognize one another.

There is no one explanation for the evolution of altruism that applies to every example. Different life histories and ecological conditions may alter the relative importance of a particular evolutionary mechanism. Thus, similar behaviors may evolve by different mechanisms in different species. Furthermore, as we will see, the mechanisms suggested are not mutually exclusive and may be working simultaneously. Let's look at several examples of altruism.

ALARM CALLS

Belding's ground squirrels are often victims of aerial predators, such as hawks, or terrestrial predators, such as coyotes, long-tailed weasels, badgers, and pine martens. The alarm calls warning of these classes of predators are different, and the selective forces behind the evolution of these two types of alarm calls appear to be different. Whereas the alarm calls warning of aerial predators appear to promote *self*-preservation, those warning of terrestrial predators do not. When a hawk is spotted overhead or when an alarm whistle is heard, near pandemonium breaks out in a Belding's ground squirrel colony. Following the first warning, other squirrels whistle a similar alarm, and all scurry to shelter. As a result, a hawk is rarely successful in attracting a group of Belding's ground squirrels. In those cases where the hawk is successful, the victim is most likely to be a noncaller. Thus, it seems that the alarm whistles given at the sight of a predatory bird directly benefit the caller by increasing its chances of escaping predation. This behavior would appear to be an example of individual selection.

In contrast, kin selection seems to be behind the evolution of ground squirrel alarm trills, which are issued in response to terrestrial predators. In this case, the caller is truly assuming risk; significantly more callers than noncallers are attacked. However, those saved by the warning are likely to be the caller's relatives. Because daughters tend to settle and breed near their birthplace, the females within any small area are usually genetically related to one another. In contrast, the sons set off independently before the first winter hibernation. When a terrestrial predator appears, females are more likely to sound an alarm than are males. This is consistent with kinship theory since females are more likely to have nearby relatives who would benefit from the warning. Reproductive females are even more likely to call than are nonreproductive females. Furthermore, reproductive females with living relatives call more frequently than do reproductive females with no living family!

HELPING

Another form of altruism is helping. A helper is an individual who assists in the rearing of offspring that are not its own, usually by providing food or by protecting the young. In most species, helpers are offspring who are helping their parents raise their siblings. Thus, kin selection seems to be a reasonable explanation for these occurrences of helping. The helpers are not always relatives, however. In these cases, helping may be a means of maximizing individual fitness in the future. Helping commonly accompanies ecological conditions that make reproduction difficult or costly. Under such conditions, helping may be a means of obtaining permission to remain in a high-quality territory, of maintaining group or territory cohesiveness, of earning the future assistance of those helped, of obtaining a mate, or of protecting young from predators.

COOPERATION IN MATE ACQUISITION

Another apparently altruistic behavior is helping another individual acquire a mate. Both individual selection and kin selection seem to have played a role in the evolution of this behavior among lions. The males of a pride are usually related. The group of males, called a coalition, generally consists of brothers, half-brothers, and cousins who left their natal pride as a group. These lions challenge the males of other prides (Fig. 44-14). In such contests, the larger coalition usually wins, and the reward is a harem of lionesses. Females of a harem often come into reproductive condition simultaneously. During the 2 to 3 days of estrus (page 668), any of the males in the coalition may be the first to mate with a female and thereby gain fitness directly. When another male takes over, it is likely to be a relative. At that point, the first male may still gain fitness indirectly.

BIOLINE
Animal Cognition

Many people have wondered what it is like to be an animal. Do nonhuman animals have thoughts or subjective feelings? Such musings have led some investigators to consider whether or not nonhuman animals are cognitive, conscious, aware beings.

Donald Griffin has suggested that tapping animal communication lines is a way that we might find out whether animals have conscious thoughts or feelings. After all, the only way we know about the thoughts or feelings of other people is when they *tell* us what they are thinking or feeling, through either verbal or nonverbal communication. So, if nonhuman animals have thoughts and feelings, they probably communicate these to others via their communication signals as well. If we could learn to speak their language, we could "eavesdrop" and thereby glimpse into the animal mind.

Most people agree that one sign of cognition is the ability to form mental representations of objects or events that are out of sight. We might ask, then, whether animal signals are symbolic; that is, whether they refer to things that are not present. Certain apes can learn a language that uses symbols. Kanzi, the pygmy chimpanzee, for instance, can communicate by using a computer keyboard that has over 250 symbols, called lexigrams. In addition, Alex, an African gray parrot, is able to request more than 80 different items vocally, even if they are out of sight. In addition, Alex can quantify and categorize those objects. He has shown an understanding of the concepts of color, shape, and same versus different with both familiar and novel objects. But these animals have been taught to use language.

Are the communication signals that animals use in nature symbolic? The apparent simplicity of this question is deceptive because observations are often open to alternative interpretations. For example, the waggle dance of the honeybee that was described earlier is symbolic in that it contains information regarding the distance and direction to a distant food source. However, most scientists agree that this behavior is not evidence of thought since the dances are genetically preprogrammed; bees can perform and understand dances without previous experience.

Another sign of cognition might be whether or not animals adjust signals according to conditions at the time. One way we might see such an adjustment in signaling is if an individual determined whether or not to signal on the basis of the composition of its audience. In other words, if an individual sees a predator, does it always sound an alarm, or does this action depend on the individual's present company? The company does seem to affect the likelihood of alarm calling among domestic chickens. A cock is more likely to call in the presence of an unaltered companion. This has been interpreted by some as an indication that the cock chooses whether or not to call. The choice of calling or withholding the call would be taken as evidence of cognition.

Learning studies may also shed some light on the issue of animal cognition. Some scientists believe that insight learning shows that the animal is *thinking*, an animal that thinks about objects or events can be said to experience a simple level of consciousness. An animal that thinks must also form mental representations of objects or events. Therefore, insight has been used as evidence of animal awareness or cognition. But not everyone agrees that animals, or even *some* animals, might be aware. Some might be willing to accept the idea of awareness in a chimp but not in a pigeon that shows similar behavior.

Roughly half of all male coalitions contain at least one unrelated male. Unrelated males may be accepted because the larger the coalition, the greater a male's reproductive success. Larger coalitions have a better chance of ousting the current coalition in a pride, of maintaining control of that pride, and perhaps even of gaining residence in a succession of prides. A solitary male has little chance of reproducing and, therefore, much to gain by joining another coalition. A small coalition may also gain by accepting an unrelated male because the extra member may help it take over prides.

FOOD SHARING

Vampire bats share food with needy familiar roostmates even if they are not related. This generosity may mean the difference between life and death for the recipient. If a vampire bat fails to find food on two successive nights, it will starve to death, unless a bat that has successfully fed regurgitates part of its blood meal. The hungry bat begs for food first by grooming, which involves licking the roostmate under the wings, and then by licking the donor's lips. A receptive donor will then regurgitate blood. The regurgitated food must be plentiful enough to sustain the bat until the next night, when it may find its own meal. Although the benefit to the recipient is great, the cost to the donor is small. Since a bat's body weight decays exponentially following a meal, the recipient may gain 12 hours of life and, therefore, another chance to find food. But the donor loses less than 12 hours of time until starvation and usually has about 36 hours—another 2 nights of hunting—before it would starve.

Generally, only individuals who have had a prior association share food. In one experiment, a group of bats was formed from two natural clusters in different areas and maintained in the laboratory. Aside from a grandmother and granddaughter, all the bats were unrelated. The bats were fed nightly from plastic measuring bottles in order to determine the amount of blood consumed by each bat. Each night, one bat was chosen at random, removed from the cage, and deprived of food. When it was reunited with its cagemates the following morning, the hungry bat would beg for food. In almost every instance, blood was shared by a bat that came from the starving bat's population in nature. Furthermore, there seemed to be pairs of unrelated bats that regurgitated almost exclusively to one another, suggesting a system of reciprocal exchange.

EUSOCIALITY

Eusocial species are those that have sterile workers, engage in cooperative care of the young, and have an overlap of generations so that the colony labor is a family affair. The eusocial insects, (e.g., ants, bees, and wasps) behave altruistically in several ways: Food is shared; those colony members specialized for defense often die performing their duty; and some members of the colony are sterile but care for the young of the colony's royalty.

Eusociality is also found in African naked mole-rats, burrowing rodents that look like rats but act like moles. Naked mole-rat societies are similar to honeybee societies in several ways. First, in both societies, breeding is restricted to a single female, the queen. Second, colonies contain overlapping generations of offspring. Third, there is differentiation of labor among individuals within the colony. During the first 12 or so days after a honeybee emerges from her pupal case, she specializes almost entirely on cleaning nest cells. For the next week, she performs a variety of tasks associated with brood and queen care, nest maintenance, and food storage. When she is about 3 weeks old, she begins to gather pollen and nectar. The duties assumed by the nonbreeding members of mole-rat colonies seem to depend on both their size and age. The duties of the smaller members generally include gathering food and transporting nest material. As they grow, they begin to clear the elaborate tunnel system of obstruction and debris. Larger members dig tunnels and defend the colony.

Both honeybees and naked mole-rats may have been predisposed to a eusocial lifestyle by a close genetic relationship among colony members along with ecological factors that maximize the benefits of group life or the costs of dispersal. The close genetic relationship among honeybees results from their system of sex determination, called **haplodiploidy,** in which fertilized eggs develop into females

FIGURE 44-14

Male lions from different prides. Coalitions of male lions, most of them relatives, fight other coalitions for control of a harem.

and nonfertilized eggs develop into males. Haplodiploidy results in a closer relationship between sister workers and their siblings than between females and their own offspring. The female workers are likely to share 75 percent of their alleles with the reproductively capable siblings they helped to raise, but they share only 50 percent of their alleles with the offspring they produced. As a result, a female worker makes greater gains in inclusive fitness by raising siblings than she would by producing offspring.

There is also an unusually high degree of genetic relatedness among the members of any single mole-rat colony. This similarity is thought to be a consequence of the extreme inbreeding within a colony. Naked mole-rats live for about 15 years and are prolific breeders. Generally, only two or three males mate with the queen mole-rat, who may give birth to a litter consisting of up to 12 young every 70 to 80 days. Furthermore, there is little mixing of genes between colonies. In fact, members of different colonies are quite aggressive toward one another; intruders may even be killed.

The high degree of genetic relatedness cannot, by itself, explain the evolution of eusociality in mole-rats. Another predisposing factor may be the mole-rats subterranean lifestyle. Within their underground tunnels, naked mole-rats are fairly safe from predators. Furthermore, the dry regions they inhabit have many plants with subterranean roots, tubers, and bulbs. Since naked mole-rats feed primarily on these parts of the plants, they need not leave the safety of the tunnel system to forage. The tunnel system can also be expanded easily as the colony grows. Finally, dispersal is risky because the tubers and bulbs the mole-rats eat are distributed unevenly throughout the habitat. The natal colony may have access to a patch of food, but a group that sets off on its own may have to burrow extensively before encountering another rich area. Burrowing uses quite a bit of energy so members of a small group might die of starvation before locating a new food resource.

The discovery that, in certain species, individuals will exhibit behaviors that help other members of the species, even if that behavior increases the risk to themselves, at first seemed to contradict accepted evolutionary theory. Several explanations for the evolution of altruistic behavior were formulated and are now generally accepted. Altruistic behavior reaches its greatest expression among eusocial species, such as bees, wasps, and naked mole-rats, where many or all of the members of a colony are closely related. (See CTQ #6.)

REEXAMINING THE THEMES

Relationship between Form and Function

The form of an object—the shape of an egg or the color of a part of the body—may play a key role in triggering a particular innate behavior. For example, a tuft of red feathers introduced into the territory of a male red robin provides a sufficient sign stimulus to evoke a "full-blown"attack by the territory's occupant. Similarly, the sight of a female stickleback in a "head-up" posture with an exposed, egg-swollen abdomen, or a facsimile thereof, provides the sign stimulus for triggering the courtship dance of a male member of the species. Similarly, the form of a communication signal—whether visual, vocal, chemical, or tactile—can be correlated with the function of the signal and the ecology of the organism.

Acquiring and Using Energy

Many diverse behaviors exhibited by animals can be explained on the basis of conservation of energy. For example, a clamworm that stops withdrawing into its burrow with every passing shadow saves energy that would otherwise be wasted. Similarly, a male bird or mammal that defends a territory against outside invaders is able to utilize the resources available on that territory to feed itself and any mate that he might attract. Some types of social behavior may be explained, at least in part, in terms of energy conservation. A small spiderling, for example, saves energy if it shares a web with other individuals of its species rather than bearing the considerable expense involved in constructing a web of its own.

Evolution and Adaptation

As in the case of anatomic traits or physiologic activities, specific behaviors evolve because they are adaptive; that is, they increase the likelihood of an individual's surviving and reproducing. The adaptive behavior of simpler behaviors, such as habituation, which prevents an animal from repeatedly continuing an inappropriate physiologic response, or filial imprinting, which increases the likelihood that a newly hatched bird will remain in the company of a parent, is usually evident. While the adaptive quality of altruistic behaviors is less evident, these behaviors are also thought to provide selective advantage, if not always for the individual that displays the behavior, then for close relatives that share the same genes.

SYNOPSIS

Tinbergen's four questions about animal behavior ask about the mechanisms that underlay the behavior, its development, its survival value and its evolution.

Behavior is adaptive, whether it is primarily innate or learned. Innate behaviors are largely controlled by genes and are generally stereotyped species-specific actions. An FAP is an innate behavior that, once triggered, will continue to completion without further stimulation. The stimulus for an FAP, called a sign stimulus or a releaser, is usually only a small part of the total environmental situation. One FAP may bring an animal into a stimulus situation that triggers another FAP. The sequence of FAPs that results is called a chain of reactions.

Learning is a process through which an animal benefits from its experience. Habituation is a simple form of learning, whereby the animal learns not to show a characteristic response to a stimulus because the stimulus was shown to be unimportant during repeated encounters. Thus, habituation is a mechanism that focuses attention on important aspects of the environment. In classical conditioning, the animal learns a new association between a stimulus and response. Classical conditioning begins with an unlearned association between an unconditioned stimulus (US) and a response. A new stimulus repeatedly occurs immediately before the US. Eventually, the new stimulus alone is sufficient to cause the response. The new stimulus is then called a conditioned stimulus (CS), and the response is called a conditioned reflex. In operant conditioning, the probability of an animal repeating an action increases because the action met with favorable consequences. Insight learning is the sudden solution to a problem without obvious trial-and-error procedures. Social learning occurs when an animal learns from others. A proposed function of play is to serve as a mechanism for learning specific skills or improving overall perceptual abilities. Imprinting is a form of learning that occurs quickly during a restricted interval of time, called a critical period, and without any obvious reinforcement.

Many scientists accept that the inherited potential for learning is an adaptation. Studies comparing the spatial memory of three species of seed-caching birds support this hypothesis. Pinyon jays and Clark's nutcrackers rely more heavily on cached seeds to survive the winter than do scrub jays; the former species also remember where they hide their seeds more accurately than do scrub jays.

Genes and experience interact during development to produce each behavior. Song development in male white-crowned sparrows provides a good example of this interaction. Young male white-crowned sparrows must learn the correct song by hearing adult males of its species sing. The young male inherits a template, or image, of its species' song that allows him to recognize the correct song, but the experience of hearing the song is needed for its development.

Comparisons of parental behavior among certain gull species have supported the hypothesis that behavior is adaptive. The white inner surface of eggshell pieces may catch a predator's attention. Species, such as the blackheaded gull, that experience high predation rates remove eggshell pieces from the nest after the chicks hatch, but those that experience low predation rates, such as kittiwakes, leave eggshell pieces in the nest. Furthermore, oystercatchers, which generally nest alone, remove the eggshell pieces immediately. However, blackheaded gulls, which nest in colonies, and may cannibalize the chicks of absent neighbors, delay removing the eggshell pieces until the chicks have dried and are more difficult to swallow.

Optimality theory views natural selection as "weighing" the costs and benefits of each available behavioral alternative. For example, optimality theory predicts that an animal would defend a territory when the benefits from enhanced access to the resource within the territory are greater than the costs of defending the territory.

When an individual's reproductive success depends on what others in the population are doing, the optimal course of action is called an evolutionarily stable strategy (ESS). An ESS is a course of action that cannot be bettered and, therefore, cannot be replaced by any other strategy when most of the members of a population have adopted it.

Communication is essential to social life. The functions of communication include: species recognition, mate attraction and courtship, aggressive interactions, and recruitment. Some displays allow individuals to assess the qualities of a potential mate or competitor. Any sensory channel can be used for communication. The evolutionary choice of a sensory channel for a particular communication signal will be influenced by both the nature of the message and the ecological situation of the organism.

Altruism is the performance of a service that benefits a conspecific at some cost to the altruist. Hypotheses for the evolution of altruism can be grouped into several overlapping classes: (1) individual selection, in which there is a net gain directly to the altruist; (2) kin selection, in which the altruist gains indirectly through increased reproductive success of the altruist's relatives; and (3) reciprocal altruism, in which the altruist is later repaid for its service by other members of the population. These hypothetical mechanisms for the evolution of altruism may work simultaneously, and none applies to every example of altruism. There are several actions that are often considered to be examples of altruism. Many species alert their companions to danger. Another form of altruism is helping, whereby an individual assists in rearing offspring that are not its own. In some species, some individuals may assist others in acquiring a mate. Eusociality is the highest form of altruism. Eusocial species are those that have sterile workers, engage in cooperative care of the young, and have an overlap of generations. The close genetic relationship among eusocial insects is thought be a result of their form of sex determinism, called haplodiploidy, in which fertilized eggs develop into females and unfertilized eggs develop into males. As a result of haplodiploidy, sister workers are more closely related to their siblings than they would be to their own offspring.

Key Terms

fixed action patterns (FAPs) (p. 1009)
sign stimulus (p. 1009)
releaser (p. 1009)
habituation (p. 1011)
classical conditioning (p. 1011)
conditioned reflex (p. 1011)
unconditioned stimulus (p. 1011)
conditioned stimulus (p. 1012)
operant conditioning (p. 1012)
insight learning (p. 1012)
social learning (p. 1013)
filial imprinting (p. 1014)
critical period (p. 1014)
sexual imprinting (p.1014)
territory (p. 1016)
evolutionarily stable strategy (ESS) (p. 1017)
inclusive fitness (p. 1022)
eusocial species (p. 1025)
haplodiploidy (p.1025)

Review Questions

1. Define and give an example of each of the following: fixed action pattern, releaser, and chain of reactions. Explain how the three are related.
2. What is habituation? Explain how habituation may be adaptive.
3. What is the difference between an unconditioned stimulus (US) and a conditioned stimulus (CS)? Describe a situation in which a conditioned reflex might be beneficial to an animal.
4. What characteristics of imprinting are often used to distinguish it from other types of learning?
5. Explain how genes and experience interact during the development of song in male white-crowned sparrows.
6. What are some costs and benefits of defending a territory? What conditions would favor territorial defense? What conditions would make territorial defense economically unsound?
7. Define (1) evolutionarily stable strategy (ESS); (2) "hawk" strategy; and (3) "dove" strategy. What factors might favor the "dove" strategy? What factors might favor the "hawk" strategy?
8. List three factors that will influence the evolutionary choice of the nature of a communication signal.
9. Explain the three hypotheses for the evolution of altruism. Explain why each of the following may be considered examples of altruism: alarm calling by ground squirrels; helping; male lions cooperating in acquiring a mate; food sharing among vampire bats; and eusociality.
10. Explain how the close genetic relationship among honeybees within a colony develops. How is this different from the way that a close genetic relationship among naked mole rats within a colony develops?

Critical Thinking Questions

1. Herring gull chicks peck at the adult's beak, which is yellow with a red spot near the tip, until the adult regurgitates food into the chick's beak. Describe how you would determine which characteristic(s) of an adult herring gull's head serve(s) to release pecking behavior in the chicks. Would you expect this behavior to be innate or learned?
2. In *The Life and Times of Archie and Mehitabel* by Don Marquis, Archie the cockroach says, "as a representative of the insect world I have often wondered on what man bases his claims to superiority. Everything he knows he has had to learn whereas insects are born knowing everything we need to know." On the basis of what you have learned in this chapter, write a reaction to this statement.
3. Chimpanzees in the wild can be observed stripping leaves from a stem and poking the stem into termite or ant hills. They then withdraw the stem and lick off any insects clinging to it. Describe how you would determine whether this behavior is genetically determined or learned.
4. Discuss the probable survival value of each of the following behaviors: (a) Butterfly courtship involves a series of signals in a particular order between male and female. Failure to produce the right signal at the right time interrupts the courtship. (b) Toads exhibit a striking and swallowing reflex to any elongated shape moving lengthwise. (c) Male bower birds, which lack colorful feathers, decorate nests with brightly colored objects. (d) Tawny owls, which are long-lived, territorial predators, can lay up to four eggs a year. Typically, however, not all pairs in an area breed every year, and some that do breed fail to incubate the eggs.
5. Ostriches practice cooperative breeding; that is, most nests contain eggs from more than one female. During the breeding season, males scrape out many depressions in the ground, only some of which will be used as nests. The eggs—the largest of any bird (weighing about 1,600 grams)—are laid every 2 days for about 24 days. The female who lays the first egg in a nest becomes the one to incubate all the subsequent eggs, including those laid by other females. The incubating female contributes 8 to 16 eggs, while other females contribute 3 to 20 eggs to a nest. Outsiders tend to lay eggs in a nest on the "off" days of the incubating female. Nocturnal carnivores, such as jackals and hyenas, destroy 40 percent of the nests, while daytime predators, such as vultures, destroy about 10 percent. Discuss the reproductive advantages and disadvantages of this pattern of breeding and nesting behavior.
6. Incubating eggs laid by other females appear to be a case of altruism. However, scientists have found that the degree of relatedness between a chick and the incubating female among ostriches is very low. Furthermore, females are able to recognize their own eggs, and those that are pushed out of the nest are more likely to belong to a female other than the one incubating the nest. How does this evidence argue against altruism as an explanation of communal nesting in ostriches?

Additional Readings

Caro, T. M. 1988. "Adaptive significance of play. Are we getting closer?" *Trends Ecol. and Evol.* 3:50–54.

Carter, C. S. and Getz, L. L. 1993. Monogamy and the prairie vole. *Sci. Amer.* June:100–106. (Intermediate)

Goodenough, J. E. 1984. "Animal Communication." *Carolina Biological Readers.* Carolina Biological Supply Co. Burlington, N.C.

Goodenough, J. E., B. McGuire, and R. A. Wallace. 1992. *Perspectives on Animal Behavior.* John Wiley & Sons, NY.

Hailman, J. P. 1969. "How an instinct is learned." *Sci. Am.* Dec:98–106.

Heinsohn, R. G., A. Cockburn, and R. A. Mulder. 1990. "Avian cooperative breeding: Old hypotheses and new directions." *Trends Ecol. and Evol.* 5:403–407.

Hoage, R. J., and L. Goldman. 1986. *Animal Intelligence. Insights into the Animal Mind.* Smithsonian Institution Press. Washington, D.C.

Honeycutt, R. L. 1992. "Naked mole rats." *Am. Sci.* 80:43–53.

Linden, E. 1992. "A curious kinship: Apes and humans." *National Geographic* 181(3):2–45.

Linden, E. 1993. Can animals think? *Time* March 22:55–61. (Introductory)

Ostfeld, R. S. 1990. "The ecology of territoriality in small mammals." *Trends Ecol. and Evol.* 5:411–415.

Reichert, S. E. 1986. "Spider fights as a test of evolutionary game theory." *Am. Sci.* 74:604–610.

Scheller, R. H., and R. Axel. 1984. "How genes control an innate behavior." *Sci. Am.* Mar:54–62.

Schneider, D. 1974. "The sex-attractant receptor of moths." *Sci. Am.* Jul:28–35.

The Marvels of Animal Behavior. National Geographic Press, Washington, D.C.

Tinbergen, N. 1952. "The curious behavior of the stickleback." *Sci. Am.* Dec:22–26.

Tinbergen, N. 1960. *The Herring Gull's World.* Doubleday, Garden City, NY.

Wilkinson, G. S. 1990. "Food sharing in vampire bats." *Sci. Am.* Feb:76–82.

Wilson, E. O. 1963. "Pheromones." *Sci. Am.* May:100–114.

Wilson, E.O. 1972. "Animal Communication." *Sci. Am.* Sept:52–60.

Winston, M. L. and K. N. Slessnor. 1992. "The essence of royalty: Honey bee queen pheromone." *Am. Sci.* 80:374–385.

APPENDIX A

Metric and Temperature Conversion Charts

Metric Unit (symbol)		*Metric to English*	*English to Metric*
Length			
kilometer (km)	= 1,000 (10^3) meters	1 km = 0.62 mile	1 mile = 1.609 km
meter (m)	= 100 centimeters	1 m = 1.09 yards	1 yard = 0.914 m
		= 3.28 feet	1 foot = 0.305 m
centimeter (cm)	= 0.01 (10^{-2}) meter	1 cm = 0.394 inch	1 inch = 2.54 cm
millimeter (mm)	= 0.001 (10^{-3}) meter	1 mm = 0.039 inch	1 inch = 25.4 mm
micrometer (μm)	= 0.000001 (10^{-6}) meter		
nanometer (nm)	= 0.000000001 (10^{-9}) meter		
angstrom (Å)	= 0.0000000001 (10^{-10}) meter		
Area			
square kilometer (km^2)	= 100 hectares	1 km^2 = 0.386 square mile	1 square mile = 2.590 km^2
hectare (ha)	= 10,000 square meters	1 ha = 2.471 acres	1 acre = 0.405 ha
square meter (m^2)	= 10,000 square centimeters	1 m^2 = 1.196 square yards	1 square yard = 0.836 m^2
		= 10.764 square feet	1 square foot = 0.093 m^2
square centimeter (cm^2)	= 100 square millimeters	1 cm^2 = 0.155 square inch	1 square inch = 6.452 cm^2
Mass			
metric ton (t)	= 1,000 kilograms = 1,000,000 grams	1 t = 1.103 tons	1 ton = 0.907 t
kilogram (kg)	= 1,000 grams	1 kg = 2.205 pounds	1 pound = 0.454 kg
gram (g)	= 1,000 milligrams	1 g = 0.035 ounce	1 ounce = 28.35 g
milligram (mg)	= 0.001 gram		
microgram (μg)	= 0.000001 gram		
Volume Solids			
1 cubic meter (m^3)	= 1,000,000 cubic centimeters	1 m^3 = 1.308 cubic yards = 35.315 cubic feet	1 cubic yard = 0.765 m^3 1 cubic foot = 0.028 m^3
1 cubic centimeter (cm^3)	= 1,000 cubic millimeters	1 cm^3 = 0.061 cubic inch	1 cubic inch = 16.387 cm^3
Volume Liquids			
kiloliter (kl)	= 1,000 liters	1 kl = 264.17 gallons	
liter (l)	= 1,000 milliliters	1 l = 1.06 quarts	1 gal = 3.785 l 1 qt = 0.94 l 1 pt = 0.47 l
milliliter (ml)	= 0.001 liter	1 ml = 0.034 fluid ounce	1 fluid ounce = 29.57 ml
microliter (μl)	= 0.000001 liter		

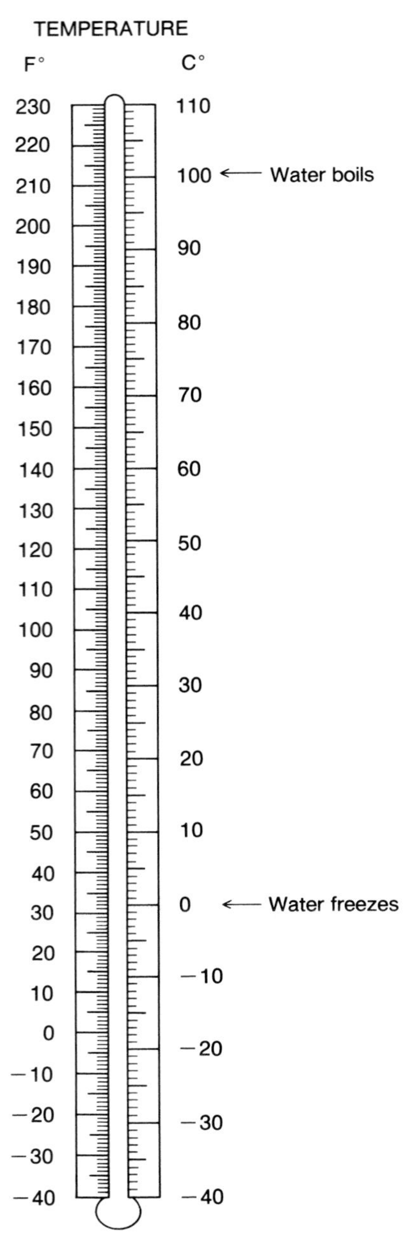

Fahrenheit to Centigrade: $°C = \frac{5}{9}(°F - 32)$

Centigrade to Fahrenheit: $°F = \frac{9}{5}(°C + 32)$

APPENDIX B

Microscopes: Exploring the Details of Life

Microscopes are the instruments that have allowed biologists to visualize objects that are vastly smaller than anything visible with the naked eye. There are broadly two types of specimens viewed in a microscope: whole mounts which consist of an intact subject, such as a hair, a living cell, or even a DNA molecule, and thin sections of a specimen, such as a cell or piece of tissue.

THE LIGHT MICROSCOPE

A light microscope consists of a series of glass lenses that bend (refract) the light coming from an illuminated specimen so as to form a visual image of the specimen that is larger than the specimen itself (a). The specimen is often stained with a colored dye to increase its visibility. A special phase contrast light microscope is best suited for observing unstained, living cells because it converts differences in the density of cell organelles, which are normally invisible to the eye, into differences in light intensity which can be seen.

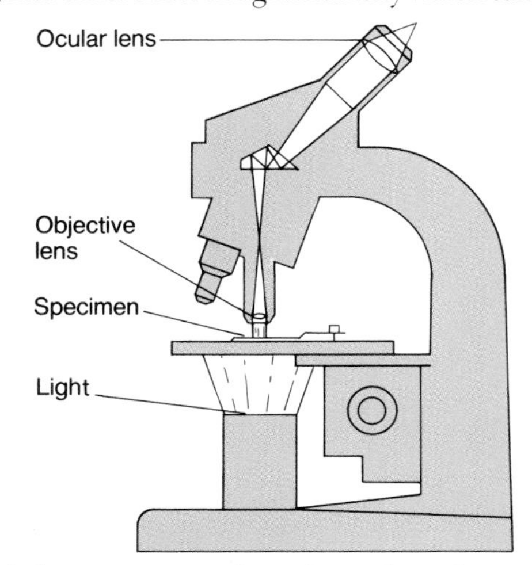

(a)

All light microscopes have limited *resolving power*—the ability to distinguish two very close objects as being separate from each other. The resolving power of the light microscope is about 0.2 μm (about 1,000 times that of the naked eye), a property determined by the wave length of visible light. Consequently, objects closer to each other than 0.2 μm, which includes many of the smaller cell organelles, will be seen as a single, blurred object through a light microscope.

THE TRANSMISSION ELECTRON MICROSCOPE

Appreciation of the wondrous complexity of cellular organization awaited the development of the transmission electron microscope (or TEM), which can deliver resolving powers 1000 times greater than the light microscope. Suddenly, biologists could see strange new structures, whose function was totally unknown—a breakthrough that has kept cell biologists busy for the past 50 years. The TEM (b) works by shooting a beam of electrons through very thinly sliced specimens that have been stained with heavy metals,

(b)

such as uranium, capable of deflecting electrons in the beam. The electrons that pass through the specimen undeflected are focused by powerful electromagnets (the lenses of a TEM) onto either a phosphorescent screen or high-contrast photographic film. The resolution of the TEM is so great—sufficient to allow us to see individual DNA molecules—because the wavelength of an electron beam is so small (about 0.0005 μm).

THE SCANNING ELECTRON MICROSCOPE

Specimens examined in the scanning electron microscope (SEM) are whole mounts whose surfaces have been coated with a thin layer of heavy metals. In the SEM, a fine beam of electrons scans back and forth across the specimen and the image is formed from electrons bouncing off the hills and valleys of its surface. The SEM produces a three-dimensional image of the surface of the specimen—which can range in size from a virus to an insect head (c,d)—with remarkable depth and clarity. The SEM produces black and white images; the colors seen in many of the micrographs in the text have been added to enhance their visual quality. Note that the insect head (d) is that of an antennapedia mutant as described on p 687.

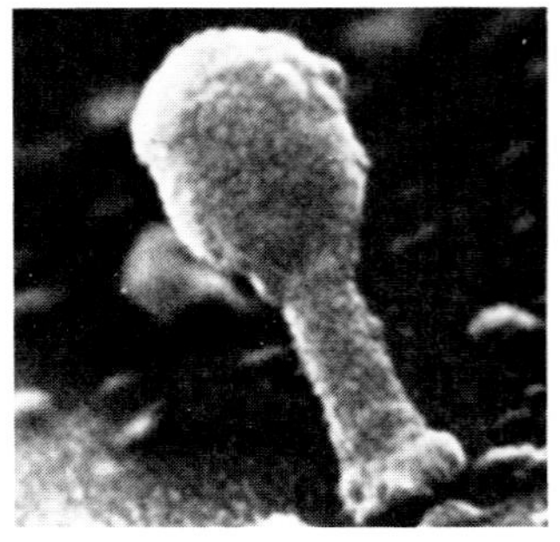

(c)

(d)

APPENDIX C

The Hardy-Weinberg Principle

If the allele for brown hair is dominant over that for blond hair, and curly hair is dominant over straight hair, then why don't all people by now have brown, curly hair? The **Hardy-Weinberg Principle** (developed independently by English mathematician G. H. Hardy and German physician W. Weinberg) demonstrates that the frequency of alleles remains the same from generation to generation unless influenced by outside factors. The outside factors that would cause allele frequencies to change are mutation, immigration and emigration (movement of individuals into and out of a breeding population, respectively), natural selection of particular traits, and breeding between members of a small population. In other words, unless one or more of these forces influence hair color and hair curl, the relative number of people with brown and curly hair will not increase over those with blond and straight hair.

To illustrate the Hardy-Weinberg Principle, consider a single gene locus with two alleles, *A* and *a*, in a breeding population. (If you wish, consider *A* to be the allele for brown hair and *a* to be the allele for blond hair.) Because there are only two alleles for the gene, the sum of the frequencies of A and *a* will equal 1.0. (By convention, allele frequencies are given in decimals instead of percentages.) Translating this into mathematical terms, if

p = the frequency of allele A, and
q = the frequency of allele *a*,

then $p + q = 1$.

If A represented 80 percent of the alleles in the breeding population ($p = 0.8$), then according to this formula the frequency of *a* must be 0.2 ($p + q = 0.8 + 0.2 = 1.0$).

After determining the allele frequency in a starting population, the predicted frequencies of alleles and genotypes in the next generation can be calculated. Setting up a Punnett square with starting allele frequencies of $p = 0.8$ and $q = 0.2$:

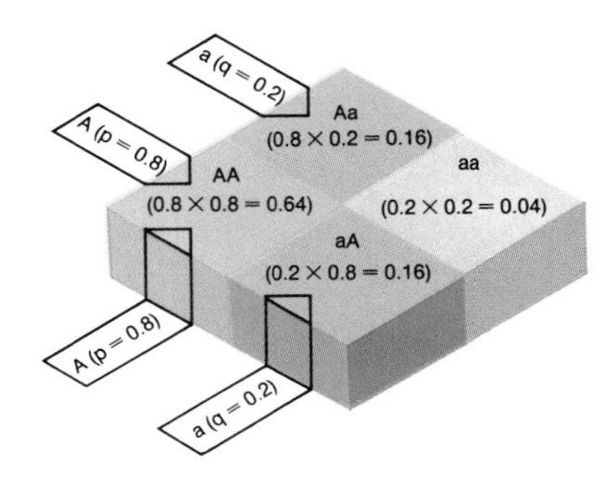

The chances of each offspring receiving any combination of the two alleles is the product of the probability of receiving one of the two alleles alone. In this example, the chances of an offspring receiving two A alleles is $p \times p = p^2$, or $0.8 \times 0.8 = 0.64$. A frequency of 0.64 means that 64 percent of the next generation will be homozygous dominant (*AA*). The chances of an offspring receiving two *a* alleles is $q^2 = 0.2 \times 0.2 = 0.04$, meaning 4 percent of the next generation is predicted to be *aa*. The predicted frequency of heterozygotes (*Aa* or *aA*) is 0.32 or $2pq$, the sum of the probability of an individual being $Aa(p \times q = 0.8 \times 0.2 = 0.16)$ plus the probability of an individual being $aA(q \times p = 0.2 \times 0.8 = 0.16)$. Just as all of the allele frequencies for a particular gene must add up to 1, so must all of the possible genotypes for a particular gene locus add up to 1. Thus, the Hardy-Weinberg Principle is

$$p^2 + 2pq + q^2 = 1$$
$$(0.64 + 0.32 + 0.04 = 1)$$

So after one generation, the frequency of possible genotypes is

$$AA = p^2 = 0.64$$
$$Aa = 2pq = 0.32$$
$$aa = q^2 = 0.04$$

Now let's determine the actual allele frequencies for A and *a* in the new generation. (Remember the original allele frequencies were 0.8 for allele A and 0.2 for allele *a*. If the Hardy-Weinberg Principle is right, there will be no change in the frequency of either allele.) To do this we sum the frequencies for each genotype containing the allele. Since heterozygotes carry both alleles, the genotype frequency must be divided in half to determine the frequency of each allele. (In our example, heterozygote *Aa* has a frequency of 0.32, 0.16 for allele A, plus 0.16 for allele *a*.) Summarizing then:

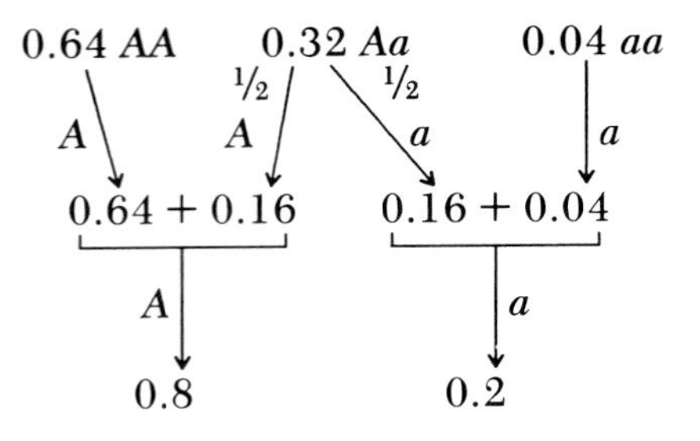

As predicted by the Hardy-Weinberg Principle, the frequency of allele A remained 0.8 and the frequency of allele *a* remained 0.2 in the new generation. Future generations can be calculated in exactly the same way, over and over again. As long as there are no mutations, no gene flow between populations, completely random mating, no natural selection, and no genetic drift, there will be no change in allele frequency, and therefore no evolution.

Population geneticists use the Hardy-Weinberg Principle to calculate a starting point allele frequency, a reference that can be compared to frequencies measured at some future time. The amount of deviation between observed allele frequencies and those predicted by the Hardy-Weinberg Principle indicates the degree of evolutionary change. Thus, this principle enables population geneticists to measure the rate of evolutionary change and identify the forces that cause changes in allele frequency.

APPENDIX D

Careers in Biology

Although many of you are enrolled in biology as a requirement for another major, some of you will become interested enough to investigate the career opportunities in life sciences. This interest in biology can grow into a satisfying livelihood. Here are some facts to consider:

- Biology is a field that offers a very wide range of possible science careers
- Biology offers high job security since many aspects of it deal with the most vital human needs: health and food
- Each year in the United States, nearly 40,000 people obtain bachelor's degrees in biology. But the number of newly created and vacated positions for biologists is increasing at a rate that exceeds the number of new graduates. Many of these jobs will be in the newer areas of biotechnology and bioservices.

Biologists not only enjoy job satisfaction, their work often changes the future for the better. Careers in medical biology help combat diseases and promote health. Biologists have been instrumental in preserving the earth's life-supporting capacity. Biotechnologists are engineering organisms that promise dramatic breakthroughs in medicine, food production, pest management, and environmental protection. Even the economic vitality of modern society will be increasingly linked to biology.

Biology also combines well with other fields of expertise. There is an increasing demand for people with backgrounds or majors in biology complexed with such areas as business, art, law, or engineering. Such a distinct blend of expertise gives a person a special advantage.

The average starting salary for all biologists with a Bachelor's degree is $22,000. A recent survey of California State University graduates in biology revealed that most were earning salaries between $20,000 and $50,000. But as important as salary is, most biologists stress job satisfaction, job security, work with sophisticated tools and scientific equipment, travel opportunities (either to the field or to scientific conferences), and opportunities to be creative in their job as the reasons they are happy in their career.

Here is a list of just a few of the careers for people with degrees in biology. For more resources, such as lists of current openings, career guides, and job banks, write to Biology Career Information, John Wiley and Sons, 605 Third Avenue, New York, NY 10158.

A SAMPLER OF JOBS THAT GRADUATES HAVE SECURED IN THE FIELD OF BIOLOGY*

Agricultural Biologist
Agricultural Economist
Agricultural Extension Officer
Agronomist
Amino-acid Analyst
Analytical Biochemist
Anatomist
Animal Behavior Specialist
Anticancer Drug Research Technician
Antiviral Therapist
Arid Soils Technician
Audio-neurobiologist
Author, Magazines & Books
Behavioral Biologist
Bioanalyst
Bioanalytical Chemist
Biochemical/Endocrine Toxicologist
Biochemical Engineer
Pharmacology Distributor
Pharmacology Technician
Biochemist
Biogeochemist
Biogeographer
Biological Engineer
Biologist
Biomedical Communication Biologist
Biometerologist
Biophysicist
Biotechnologist
Blood Analyst
Botanist
Brain Function Researcher
Cancer Biologist
Cardiovascular Biologist
Cardiovascular/Computer Specialist
Chemical Ecologist
Chromatographer
Clinical Pharmacologist
Coagulation Biochemist
Cognitive Neuroscientist
Computer Scientist
Dental Assistant
Ecological Biochemist
Electrophysiology/ Cardiovascular Technician
Energy Regulation Officer
Environmental Biochemist
Environmental Center Director
Environmental Engineer
Environmental Geographer
Environmental Law Specialist
Farmer
Fetal Physiologist
Flavorist
Food Processing Technologist
Food Production Manager
Food Quality Control Inspector
Flower Grower
Forest Ecologist
Forest Economist
Forest Engineer
Forest Geneticist
Forest Manager

Forest Pathologist
Forest Plantation Manager
Forest Products Technologist
Forest Protection Expert
Forest Soils Analyst
Forester
Forestry Information Specialist
Freeze-Dry Engineer
Fresh Water Biologist
Grant Proposal Writer
Health Administrator
Health Inspector
Health Scientist
Hospital Administrator
Hydrologist
Illustrator
Immunochemist
Immunodiagnostic Assay Developer
Inflammation Technologist
Landscape Architect
Landscape Designer
Legislative Aid
Lepidopterist
Liaison Scientist, Library of Medicine Computer Biologist
Life Science Computer Technologist
Lipid Biochemist
Livestock Inspector
Lumber Inspector
Medical Assistant
Medical Imaging Technician
Medical Officer
Medical Products Developer
Medical Writer
Microbial Physiologist
Microbiologist
Mine Reclamation Scientist
Molecular Endocrinologist
Molecular Neurobiologist
Molecular Parasitologist
Molecular Toxicologist
Molecular Virologist
Morphologist
Natural Products Chemist
Natural Resources Manager
Nature Writer
Nematode Control Biologist
Nematode Specialist
Nematologist
Neuroanatomist
Neurobiologist
Neurophysiologist
Neuroscientist
Nucleic Acids Chemist
Nursing Aid
Nutritionist
Occupational Health Officer
Ornamental Horticulturist
Paleontologist
Paper Chemist
Parasitologist
Pathologist
Peptide Biochemist
Pharmaceutical Writer
Pharmaceutical Sales
Pharmacologist
Physiologist
Planning Consultant
Plant Pathologist
Plant Physiologist
Production Agronomist
Protein Biochemist
Protein Structure & Design Technician
Purification Biochemist
Quantitative Geneticist
Radiation Biologist
Radiological Scientist
Regional Planner
Regulatory Biologist
Renal Physiologist
Renal Toxicologist
Reproductive Toxicologist
Research and Development Director
Research Technician
Research Liaison Scientist
Research Products Designer
Research Proposal Writer
Safety Assessment Sanitarian
Scientific Illustrator
Scientific Photographer
Scientific Reference Librarian
Scientific Writer
Soil Microbiologist
Space Station Life Support Technician
Spectroscopist
Sports Product Designer
Steroid Health Assessor
Taxonomic Biologist
Teacher
Technical Analyst
Technical Science Project Writer
Textbook Editor
Theoretical Ecologist
Timber Harvester
Toxicologist
Toxic Waste Treatment Specialist
Urban Planner
Water Chemist
Water Resources Biologist
Wood Chemist
Wood Fuel Technician
Zoning and Planning Manager
Zoologist
Zoo Animal Breeder
Zoo Animal Behaviorist
Zoo Designer
Zoo Inspector

*Results of one survey of California State University graduates. Some careers may require advanced degrees

Glossary

A

Abiotic Environment Components of ecosystems that include all nonliving factors. (41)

Abscisic Acid (ABA) A plant hormone that inhibits growth and causes stomata to close. ABA may not be commonly involved in leaf drop. (21)

Abscission Separation of leaves, fruit, and flowers from the stem. (21)

Acclimation A physiological adjustment to environmental stress. (28)

Acetyl CoA Acetyl coenzyme A. A complex formed when acetic acid binds to a carrier coenzyme forming a bridge between the end products of glycolysis and the Krebs cycle in respiration. (9)

Acetylcholine Neurotransmitter released by motor neurons at neuromuscular junctions and by some interneurons. (23)

Acid Rain Occurring in polluted air, rain that has a lower pH than rain from areas with unpolluted air. (40)

Acids Substances that release hydrogen ions (H^+) when dissolved in water. (3)

Acid Snow Occurring in polluted air, snow that has a lower pH than snow from areas with unpolluted air. (40)

Acoelomates Animals that lack a body cavity between the digestive cavity and body wall. (39)

Acquired Immune Deficiency Syndrome (AIDS) Disease caused by infection with HIV (Human Immunodeficiency Virus) that destroys the body's ability to mount an immune response due to destruction of its helper T cells. (30, 36)

Actin A contractile protein that makes up the major component of the thin filaments of a muscle cell and the microfilaments of nonmuscle cells. (26)

Action Potential A sudden, dramatic reversal of the voltage (potential difference) across the plasma membrane of a nerve or muscle cell due to the opening of the sodium channels. The basis of a nerve impulse. (23)

Activation Energy Energy required to initiate chemical reaction. (6)

Active Site Region on an enzyme that binds its specific substrates, making them more reactive. (6)

Active Transport Movement of substances into or out of cells against a concentration gradient, i.e., from a region of lower concentration to a region of higher concentration. The process requires an expenditure of energy by the cell. (7)

Adaptation A hereditary trait that improves an organism's chances of survival and/or reproduction. (33)

Adaptive Radiation The divergence of many species from a single ancestral line. (33)

Adenosine Triphosphate (ATP) The molecule present in all living organisms that provides energy for cellular reactions in its phosphate bonds. ATP is the universal energy currency of cells. (6)

Adenylate Cyclase An enzyme activated by hormones that converts ATP to cyclic AMP, a molecule that activates resting enzymes. (25)

Adrenal Cortex Outer layer of the adrenal glands. It secretes steroid hormones in response to ACTH. (25)

Adrenal Medulla An endocrine gland that controls metabolism, cardiovascular function, and stress responses. (25)

Adrenocorticotropic Hormone (ACTH) An anterior pituitary hormone that stimulates the cortex of the adrenal glands to secrete cortisol and other steroid hormones. (25)

Adventitious Root System Secondary roots that develop from stem or leaf tissues. (18)

Aerobe An organism that requires oxygen to release energy from food molecules. (9)

Aerobic Respiration Pathway by which glucose is completely oxidized to CO_2 and H_2O, requiring oxygen and an electron transport system. (9)

Afferent (Sensory) Neurons Neurons that conduct impulses from the sense organs to the central nervous system. (23)

Age-Sex Structure The number of individuals of a certain age and sex within a population. (43)

Aggregate Fruits Fruits that develop from many pistils in a single flower. (20)

AIDS See Acquired Immune Deficiency Syndrome.

Albinism A genetic condition characterized by an absence of epidermal pigmentation that can result from a deficiency of any of a variety of enzymes involved in pigment formation. (12)

Alcoholic Fermentation The process in which electrons removed during glycolysis are transferred from NADH to form alcohol as an end product. Used by yeast during the commercial process of ethyl alcohol production. (9)

Aldosterone A hormone secreted by the adrenal cortex that stimulates reabsorption of sodium from the distal tubules and collecting ducts of the kidneys. (23)

Algae Any unicellular or simple colonial photosynthetic eukaryote. (37,38)

Algin A substance produced by brown algae harvested for human application because of its ability to regulate texture and consistency of products. Found in ice cream, cosmetics, marshmellows, paints, and dozens of other products. (38)

Allantois Extraembryonic membrane that serves as a repository for nitrogenous wastes. In placental mammals, it helps form the vascular connections between mother and fetus. (32)

Allele Alternative form of a gene at a particular site, or locus, on the chromosome. (12)

Allele Frequency The relative occurrence of a certain allele in individuals of a population. (33)

Allelochemicals Chemicals released by some plants and animals that deter or kill a predator or competitor. (42)

Allelopathy A type of interaction in which one organism releases allelochemicals that harm another organism. (42)

Allergy An inappropriate response by the immune system to a harmless foreign substance leading to symptoms such as itchy eyes, runny nose, and congested airways. If the reaction occurs throughout the body (anaphylaxis) it can be life threatening. (30)

Allopatric Speciation Formation of new species when gene flow between parts of a population is stopped by geographic isolation. (33)

Alpha Helix Portion of a polypeptide chain organized into a defined spiral conformation. (4)

Alternation of Generations Sequential change during the life cycle of a plant in which a haploid (1N) multicellular stage (gametophyte) alternates with a diploid (2N) multicellular stage (sporophyte). (38)

Alternative Processing When a primary RNA transcript can be processed to form more than one mRNA depending on conditions. (15)

Altruism The performance of a behavior that benefits another member of the species at some cost to the one who does the deed. (44)

Alveolus A tiny pouch in the lung where gas is exchanged between the blood and the air; the functional unit of the lung where CO_2 and O_2 are exchanged. (29)

Alzheimer's Disease A degenerative disease of the human brain, particularly affecting acetylcholine-releasing neurons and the hippocampus, characterized by the presence of tangled fibrils within the cytoplasm of neurons and amyloid plaques outside the cells. (23)

Amino Acids Molecules containing an amino group ($-NH_2$) and a carboxyl group ($-COOH$) attached to a central carbon atom. Amino acids are the subunits from which proteins are constructed. (4)

Amniocentesis A procedure for obtaining fetal cells by withdrawing a sample of the fluid that surrounds a developing fetus (amniotic fluid) using a hypodermic needle and syringe. (17)

Amnion Extraembryonic membrane that envelops the young embryo and encloses the amniotic fluid that suspends and cushions it. (32)

Amoeba A protozoan that employs pseudopods for motility. (37)

Amphibia A vertebrate class grouped into three orders: Caudata (tailed amphibians); Anura (tail-less amphibians); Apoda (rare wormlike, burrowing amphibians). (39)

Anabolic Steroids Steroid hormones, such as testosterone, which promote biosynthesis (anabolism), especially protein synthesis. (25)

Anabolism Biosynthesis of complex molecules from simpler compounds. Anabolic pathways are endergonic, i.e., require energy. (6)

Anaerobe Organism that does not require oxygen to release energy from food molecules. (9)

Anaerobic Respiration Pathway by which glucose is completely oxidized, using an electron transport system but requiring a terminal electron acceptor other than oxygen. (Compare with fermentation.) (9)

Analogous Structures (Homoplasies) Structures that perform a similar function, such as wings in birds and insects, but did not originate from the same structure in a common ancestor. (33)

Anaphase Stage of mitosis when the kinetochores split and the sister chromatids (now termed chromosomes) move to opposite poles of the spindle. (10)

Anatomy Study of the structural characteristics of an organism. (18)

Angiosperm (Anthophyta) Any plant having its seeds surrounded by fruit tissue formed from the mature ovary of the flowers. (38)

Animal A mobile, heterotrophic, multicellular organism, classified in the Animal kingdom. (39)

Anion A negatively charged ion. (3)

Annelida The phylum which contains segmented worms (earthworms, leeches, and bristleworms). (39)

Annuals Plants that live for one year or less. (18)

Annulus A row of specialized cells encircling each sporangium on the fern frond; facilitates rupture of the sporangium and dispersal of spors. (38)

Antagonistic Muscles Pairs of muscles whose contraction bring about opposite actions as illustrated by the biceps and triceps, which bends or straightens the arm at the elbow, respectively. (26)

Antenna Pigments Components of photosystems that gather light energy of different wavelengths and then channel the absorbed energy to a reaction center. (8)

Anterior In anatomy, at or near the front of an animal; the opposite of posterior. (39)

Anterior Pituitary A true endocrine gland manufacturing and releasing six hormones when stimulated by releasing factors from the hypothalamus. (25)

Anther The swollen end of the stamen (male reproductive organ) of a flowering plant. Pollen grains are produced inside the anther lobes in pollen sacs. (20)

Antibiotic A substance produced by a fungus or bacterium that is capable of preventing the growth of bacteria. (2)

Antibodies Proteins produced by plasma cells. They react specifically with the antigen that stimulated their formation. (30)

Anticodon Triplet of nucleotides in tRNA that recognizes and base pairs with a particular codon in mRNA. (14)

Antidiuretic Hormone (ADH) One of the two hormones released by the posterior pituitary. ADH increases water reabsorption in the kidney, which then produces a more concentrated urine. (25)

Antigen Specific foreign agent that triggers an immune response. (30)

Aorta Largest blood vessel in the body through which blood leaves the heart and enters the systemic circulation. (28)

Apical Dominance The growth pattern in plants in which axillary bud growth is inhibited by the hormone auxin, present in high concentrations in terminal buds. (21)

Apical Meristems Centers of growth located at the tips of shoots, axillary buds, and roots. Their cells divide by mitosis to produce new cells for primary growth in plants. (18)

Aposematic Coloring Warning coloration which makes an organism stand out from its surroundings. (42)

Appendicular Skeleton The bones of the appendages and of the pectoral and pelvic girdles. (26)

Aquatic Living in water. (40)

Archaebacteria Members of the kingdom Monera that differ from typical bacteria in the structure of their membrane lipids, their cell walls, and some characteristics that resemble those of eukaryotes. Their lack of a true nucleus, however, accounts for their assignment to the Moneran kingdom. (36)

Archenteron In gastrulation, the hollow core of the gastrula that becomes an animal's digestive tract. (32)

Arteries Large, thick-walled vessels that carry blood away from the heart. (28)

Arterioles The smallest arteries, which carry blood toward capillary beds. (28)

Arthropoda The most diverse phylum on earth, so called from the presence of jointed limbs. Includes insects, crabs, spiders, centipedes. (39)

Ascospores Sexual fungal spore borne in a sac. Produced by the sac fungi, Ascomycota. (37)

Asexual Reproduction Reproduction without the union of male and female gametes. (31)

Association In ecological communities, a major organization characterized by uniformity and two or more dominant species. (41)

Asymmetric Referring to a body form that cannot be divided to produce mirror images. (39)

Atherosclerosis Condition in which the inner walls of arteries contain a buildup of cholesterol-containing plaque that tends to occlude the channel and act as a site for the formation of a blood clot (thrombus). (7)

Atmosphere The layer of air surrounding the Earth. (40)

Atom The fundamental unit of matter that can enter into chemical reactions; the smallest unit of matter that possesses the qualities of an element. (3)

Atomic Mass Combined number of protons and neutrons in the nucleus of an atom. (3)

Atomic Number The number of protons in the nucleus of an atom. (3)

ATP (see **Adenosine Triphosphate**)

ATPase An enzyme that catalyzes a reaction in which ATP is hydrolyzed. These enzymes are typically involved in reactions where energy stored in ATP is used to drive an energy-requiring reaction, such as active transport or muscle contractility. (7, 26)

ATP Synthase A large protein complex present in the plasma membrane of bacteria, the inner membrane of mitochondria, and the thylakoid membrane of chloroplasts. This complex consists of a baseplate in the membrane, a channel across the membrane through which protons can pass, and a spherical head (F_1 particle) which contains the site where ATP is synthesized from ADP and P_i (8, 9)

Atrioventricular (AV) Node A neurological center of the heart, located at the top of the ventricles. (28)

Atrium A contracting chamber of the heart which forces blood into the ventricle. There are two atria in the hearts of all vertebrates, except fish which have one atrium. (28)

Atrophy The shrinkage in size of structure, such as a bone or muscle, usually as a result of disuse. (26)

Autoantibodies Antibodies produced against the body's own tissue. (30)

Autoimmune Disease Damage to a body tissue due to an attack by autoantibodies. Examples include thyroiditis, multiple sclerosis, and rheumatic fever. (30)

Autonomic Nervous System The nerves that control the involuntary activities of the internal organs. It is composed of the parasympathetic system, which functions during normal activity, and the sympathetic system, which operates in times of emergency or prolonged exertion. (23)

Autosome Any chromosome that is not a sex chromosome. (13)

Autotrophs Organisms that satisfy their own nutritional needs by building organic molecules photosynthetically or chemosynthetically from inorganic substances. (8)

Auxins Plant growth hormones that promote cell elongation by softening cell walls. (21)

Axial Skeleton The bones aligned along the long axis of the body, including the skull, vertebral column, and ribcage. (26)

Axillary Bud A bud that is directly above each leaf on the stem. It can develop into a new stem or a flower. (18)

Axon The long, sometimes branched extension of a neuron which conducts impulses from the cell body to the synaptic knobs. (23)

◀ B ▶

Bacteriophage A virus attacking specific bacteria that multiplies in the bacterial host cell and usually destroys the bacterium as it reproduces. (36)

Balanced Polymorphism The maintenance of two or more alleles for a single trait at fairly high frequencies. (33)

Bark Common term for the periderm. A collective term for all plant tissues outside the secondary xylem. (18)

Base Substance that removes hydrogen ions (H^+) from solutions. (3)

Basidiospores Sexual spores produced by basidiomycete fungi. Often found by the millions on gills in mushrooms. (37)

Basophil A phagocytic leukocyte which also releases substances, such as histamine, that trigger an inflammatory response. (28)

Batesian Mimicry The resemblance of a good-tasting or harmless species to a species with unpleasant traits. (42)

Bathypelagic Zone The ocean zone beneath the mesopelagic zone, characterized by no light; inhabited by heterotrophic bacteria and benthic scavengers. (40)

B Cell A lymphocyte that becomes a plasma cell and produces antibodies when stimulated by an antigen. (30)

Benthic Zone The deepest ocean zone; the ocean floor, inhabited by bottom dwelling organisms. (40)

Bicarbonate Ion $HCO_{-3.}$ (3, 29)

Biennials Plants that live for two years. (18)

Bilateral Symmetry The quality possessed by organisms whose body can be divided into mirror images by only one median plane. (39)

Bile Salts Detergentlike molecules produced by the liver and stored by the gallbladder that function in lipid emulsification in the small intestine. (27)

Binomial A term meaning "two names" or "two words". Applied to the system of nomenclature for categorizing living things with a genus and species name that is unique for each type of organism. (1)

Biochemicals Organic molecules produced by living cells. (4)

Bioconcentration The ability of an organism to accumulate substances within its' body or specific cells. (41)

Biodiversity Biological diversity of species, including species diversity, genetic diversity, and ecological diversity. (43)

Biogeochemical Cycles The exchanging of chemical elements between organisms and the abiotic environment. (41)

Biological Control Pest control through the use of naturally occurring organisms such as predators, parasites, bacteria, and viruses. (41)

Biological Magnification An increase in concentration of slowly degradable chemicals in organisms at successively higher trophic levels; for example, DDT or PCB's. (41)

Bioluminescence The capability of certain organisms to utilize chemical energy to produce light in a reaction catalyzed by the enzyme luciferase. (9)

Biomass The weight of organic material present in an ecosystem at any one time. (41)

Biome Broad geographic region with a characteristic array of organisms. (40)

Biosphere Zone of the earth's soil, water, and air in which living organisms are found. (40)

Biosynthesis Construction of molecular components in the growing cell and the replacement of these compounds as they deteriorate. (6)

Biotechnology A new field of genetic engineering; more generally, any practical application of biological knowledge. (16)

Biotic Environment Living components of the environment. (40)

Biotic Potential The innate capacity of a population to increase tremendously in size were it not for curbs on growth; maximum population growth rate. (43)

Blade Large, flattened area of a leaf; effective in collecting sunlight for photosynthesis. (18)

Blastocoel The hollow fluid-filled space in a blastula. (32)

Blastocyst Early stage of a mammalian embryo, consisting of a mass of cells enclosed in a hollow ball of cells called the trophoblast. (32)

Blastodisk In bird and reptile development, the stage equivalent to a blastula. Because of the large amount of yolk, cleavage produces two flattened layers of cells with a blastocoel between them. (32)

Blastomeres The cells produced during embryonic cleavage. (32)

Blastopore The opening of the archenteron that is the embryonic predecessor of the anus in vertebrates and some other animals. (32)

Blastula An early developmental stage in many animals. It is a ball of cells that encloses a cavity, the blastocoel. (32)

Blood A type of connective tissue consisting of red blood cells, white blood cells, platelets, and plasma. (28)

Blood Pressure Positive pressure within the cardiovascular system that propels blood through the vessels. (28)

Blooms are massive growths of algae that occur when conditions are optimal for algae proliferation. (37)

Body Plan The general layout of a plant's or animal's major body parts. (39)

Bohr Effect Increased release of O_2 from hemoglobin molecules at lower pH. (29)

Bone A tissue composed of collagen fibers, calcium, and phosphate that serves as a means of support, a reserve of calcium and phosphate, and an attachment site for muscles. (26)

Botany Branch of biology that studies the life cycles, structure, growth, and classification of plants. (18)

Bottleneck A situation in which the size of a species' population drops to a very small number of individuals, which has a major impact on the likelihood of the population recovering its earlier genetic diversity. As occurred in the cheetah population. (33)

Bowman's Capsule A double-layered container that is an invagination of the proximal end of the renal tubule that collects molecules and wastes from the blood. (28)

Brain Mass of nerve tissue composing the main part of the central nervous system. (23)

Brainstem The central core of the brain, which coordinates the automatic, involuntary body processes. (23)

Bronchi The two divisions of the trachea through which air enters each of the two lungs. (29)

Bronchioles The smallest tubules of the respiratory tract that lead into the alveoli of the lungs where gas exchange occurs. (29)

Bryophyta Division of non-vascular terrestrial plants that include liverworts, mosses, and hornworts. (38)

Budding Asexual process by which offspring develop as an outgrowth of a parent. (39)

Buffers Chemicals that couple with free hydrogen and hydroxide ions thereby resisting changes in pH. (3)

Bundle Sheath Parenchyma cells that surround a leaf vein which regulate the uptake and release of materials between the vascular tissue and the mesophyll cells. (18)

◀ C ▶

C_3 Synthesis The most common pathway for fixing CO_2 in the synthesis reactions of photosynthesis. It is so named because the first detectable organic molecule into which CO_2 is incorporated is a 3-carbon molecule, phosphoglycerate (PGA). (8)

C_4 Synthesis Pathway for fixing CO_2 during the light-independent reactions of photosynthesis. It is so named because the first detectable organic molecule into which CO_2 is incorporated is a 4-carbon molecule. (8)

Calcitonin A thyroid hormone which regulates blood calcium levels by inhibiting its release from bone. (25)

Calorie Energy (heat) necessary to elevate the temperature of one gram of water by one degree Centigrade (1° C). (6)

Calvin Cycle The cyclical pathway in which CO_2 is incorporated into carbohydrate. See C_3 synthesis. (8)

Calyx The outermost whorl of a flower, formed by the sepals. (20)

CAM Crassulacean acid metabolism. A variation of the photosynthetic reactions in plants, biochemically identical to C_4 synthesis except that all reactions occur in the same cell and are separated by time. Because CAM plants open their stomates at night, they have a competitive advantage in hot, dry climates. (8)

Cambium A ring or cluster of meristematic cells that increase the width of stems and roots when they divide to produce secondary tissues. (18)

Camouflage Adaptations of color, shape and behavior that make an organism more difficult to detect. (42)

Cancer A disease resulting from uncontrolled cell divisions. (10,13)

Capillaries The tiniest blood vessels consisting of a single layer of flattened cells. (28)

Capillary Action Tendency of water to be pulled into a small-diameter tube. (3)

Carbohydrates A group of compounds that includes simple sugars and all larger molecules constructed of sugar subunits, e.g. polysaccharides. (4)

Carbon Cycle The cycling of carbon in different chemical forms, from the environment to organisms and back to the environment. (41)

Carbon Dioxide Fixation In photosynthesis, the combination of CO_2 with carbon-accepting molecules to form organic compounds. (8)

Carcinogen A cancer-causing agent. (13)

Cardiac Muscle One of the three types of muscle tissue; it forms the muscle of the heart. (26)

Cardiovascular System The organ system consisting of the heart and the vessels through which blood flows. (28)

Carnivore An animal that feeds exclusively on other animals. (42)

Carotenoid A red, yellow, or orange plant pigment that absorbs light in 400-500 nm wavelengths. (8)

Carpels Central whorl of a flower containing the female reproductive organs. Each separate carpel, or each unit of fused carpels, is called a pistil. (20)

Carrier Proteins Proteins within the plasma membrane that bind specific substances and facilitate their movement across the membrane. (7)

Carrying Capacity The size of a population that can be supported indefinitely in a given environment. (43)

Cartilage A firm but flexible connective tissue. In the human, most cartilage originally present in the embryo is transformed into bones. (26)

Casparian Strip The band of waxy suberin that surrounds each endodermal cell of a plant's root tissue. (18)

Catabolism Metabolic pathways that degrade complex compounds into simpler molecules, usually with the release of the chemical energy that held the atoms of the larger molecule together. (6)

Catalyst A chemical substance that accelerates a reaction or causes a reaction to occur but remains unchanged by the reaction. Enzymes are biological catalysts. (6)

Cation A positively charged ion. (3)

Cecum A closed-ended sac extending from the intestine in grazing animals lacking a rumen (e.g., horses) that enables them to digest cellulose. (27)

Cell The basic structural unit of all organisms. (5)

Cell Body Region of a neuron that contains most of the cytoplasm, the nucleus, and other organelles. It relays impulses from the dendrites to the axon. (23)

Cell Cycle Complete sequence of stages from one cell division to the next. The stages are denoted G_1, S, G_2, and M phase. (10)

Cell Differentiation The process by which the internal contents of a cell become assembled into a structure that allows the cell to carry out a specific set of activities, such as secretion of enzymes or contraction. (32)

Cell Division The process by which one cell divides into two. (10)

Cell Fusion Technique whereby cells are caused to fuse with one another producing a large cell with a common cytoplasm and plasma membrane. (5, 10)

Cell Plate In plants, the cell wall material deposited midway between the daughter cells during cytokinesis. Plate material is deposited by small Golgi vesicles. (5, 10)

Cell Sap Solution that fills a plant vacuole. In addition to water, it may contain pigments, salts, and even toxic chemicals. (5)

Cell Theory The fundamental theory of biology that states: 1) all organisms are composed of one or more cells, 2) the cell is the basic organizational unit of life, 3) all cells arise from pre-existing cells. (5)

Cellular Respiration (See **Aerobic respiration**)

Cellulose The structural polysaccharide comprising the bulk of the plant cell wall. It is the most abundant polysaccharide in nature. (4, 5)

Cell Wall Rigid outer-casing of cells in plants and other organisms which gives support, slows dehydration, and prevents a cell from bursting when internal pressure builds due to an influx of water. (5)

Central Nervous System In vertebrates, the brain and spinal cord. (23)

Centriole A pinwheel-shaped structure at each pole of a dividing animal cell. (10)

Centromere Indented region of a mitotic chromosome containing the kinetochore. (10)

Cephalization The clustering of neural tissues and sense organs at the anterior (leading) end of the animal. (39)

Cerebellum A bulbous portion of the vertebrate brain involved in motor coordination. Its prominence varies greatly among different vertebrates . (23)

Cerebral Cortex The outer, highly convoluted layer of the cerebrum. In the human, this is the center of higher brain functions, such as speech and reasoning. (23)

Cerebrospinal Fluid Fluid present within the ventricles of the brain, central canal of the spinal cord, and which surrounds and cushions the central nervous system. (23)

Cerebrum The most dominant part of the human forebrain, composed of two cerebral hemispheres, generally associated with higher brain functions. (23)

Cervix The lower tip of the uterus. (31)

Chapparal A type of shrubland in California, characterized by drought-tolerant and fire-adapted plants. (40)

Character Displacement Divergence of a physical trait in closely related species in response to competition. (42)

Chemical Bonds Linkage between atoms as a result of electrons being shared or donated. (3)

Chemical Evolution Spontaneous synthesis of increasingly complex organic compounds from simpler molecules. (35)

Chemical Reaction Interaction between chemical reactants. (6)

Chemiosmosis The process by which a pH gradient drives the formation of ATP. (8, 9)

Chemoreceptors Sensory receptors that respond to the presence of specific chemicals. (24)

Chemosynthesis An energy conversion process in which inorganic substances (H, N, Fe, or S) provide energized electrons and hydrogen for carbohydrate formation (9, 36)

Chiasmata Cross-shaped regions within a tetrad, occurring at points of crossing over or genetic exchange. (11)

Chitin Structural polysaccharide that forms the hard, strong external skeleton of many arthropods and the cell walls of fungi. (4)

Chlamydia Obligate intracellular parasitic bacteria that lack a functional ATP-generating system. (36)

Chlorophyll Pigments Major light-absorbing pigments of photosynthesis. (8)

Chlorophyta Green algae, the largest group of algae; members of this group were very likely the ancestors of the modern plant kingdom. (38)

Chloroplasts An organelle containing chlorophyll found in plant cells in which photosynthesis occurs. (5, 8)

Cholecystokinin (CCK) Hormone secreted by endocrine cells in the wall of the small intestine that stimulates the release of digestive products by the pancreas. (27)

Chondrocytes Living cartilage cells embedded within the protein-polysaccharide matrix they manufacture. (26)

Chordamesoderm In vertebrates, the block of mesoderm that underlies the dorsal ectoderm of the gastrula, induces the formation of the nervous system, and gives rise to the notochord. (32)

Chordate A member of the phylum Chordata possessing a skeletal rod of tissue called a notochord, a dorsal hollow nerve cord, gill slits, and a post-anal tail at some stage of its development. (39)

Chorion The outermost of the four extraembryonic membranes. In placental mammals, it forms the embryonic portion of the placenta. (32)

Chorionic Villus Sampling (CVS) A procedure for obtaining fetal cells by removing a small sample of tissue from the developing placenta of a pregnant woman. (17)

Chromatid Each of the two identical subunits of a replicated chromosome. (10)

Chromatin DNA-protein fibers which, during prophase, condense to form the visible chromosomes. (5, 10)

Chromatography A technique for separating different molecules on the basis of their solubility in a particular solvent. The mixture of substances is spotted on a piece of paper or other material, one end of which is then placed in the solvent. As the solvent moves up the paper by capillary action, each substance in the mixture is carried a particular distance depending on its solubility in the moving solvent. (8)

Chromosomes Dark-staining structures in which the organism's genetic material (DNA) is organized. Each species has a characteristic number of chromosomes. (5, 10)

Chromosome Aberrations Alteration in the structure of a chromosome from the normal state. Includes chromosome deletions, duplications, inversions, and translocations. (13)

Chromosome Puff A site on an insect polytene chromosome where the DNA has unraveled and is being transcribed. (15)

Cilia Short, hairlike structures projecting from the surfaces of some cells. They beat in coordinated ways, are usually found in large numbers, and are densely packed. (5)

Ciliated Mucosa Layer of ciliated epithelial cells lining the respiratory tract. The beating of cilia propels an associated mucous layer and trapped foreign particles. (29)

Circadian Rhythm Behavioral patterns that cycle during approximately 24 hour intervals.

Circulatory System The system that circulates internal fluids throughout an organism to deliver oxygen and nutrients to cells and to remove metabolic wastes. (28)

Class (Taxonomic) A level of the taxonomic hierarchy that groups together members of related orders. (1)

Classical Conditioning A form of learning in which an animal develops a response to a new stimulus by repeatedly associating the new stimulus with a stimulus that normally elicits the response. (44)

Cleavage Successive mitotic divisions in the early embryo. There is no cell growth between divisions. (32)

Cleavage Furow Constriction around the middle of a dividing cell caused by constriction of microfilaments. (10)

Climate The general pattern of average weather conditions over a long period of time in a specific region, including precipitation, temperature, solar radiation, and humidity. (40)

Climax Final or stable community of successional stages, that is more or less in equilibrium with existing environmental conditions for a long period of time. (41)

Climax Community Community that remains essentially the same over long periods of time; final stage of ecological succession. (41)

Clitoris A protrusion at the point where the labia minora merge; rich in sensory neurons and erectile tissue. (31)

Clonal Selection Mechanism The mechanism by which the body can synthesize antibodies specific for the foreign substance (antigen) that stimulated their production. (30)

Clones Offspring identical to the parent, produced by asexual processes. (15)

Closed Circulatory System Circulatory system in which blood travels throughout the body in a continuous network of closed tubes. (Compare with open circulatory system). (28)

Clumped Pattern Distribution of individuals of a population into groups, such as flocks or herds. (43)

Cnidaria A phylum that consists of radial symmetrical animals that have two cell layers. There are three classes: 1) Hydrozoa (hydra), 2) Scyphozoa (jellyfish), 3) Anthozoa (sea anemones, corals). Most are marine forms that live in warm, shallow water. (39)

Cnidocytes Specialized stinging cells found in the members of the phylum Cnidaria. (39)

Coastal Waters Relatively warm, nutrient-rich shallow water extending from the high-tide mark on land to the sloping continental shelf. The greatest concentration of marine life are found in coastal waters. (40)

Coated Pits Indentations at the surfaces of cells that contain a layer of bristly protein (called clathrin) on the inner surface of the plasma membrane. Coated pits are sites where cell receptors become clustered. (7)

Cochlea Organ within the inner ear of mammals involved in sound reception. (24)

Codominance The simultaneous expression of both alleles at a genetic locus in a heterozygous individual. (12)

Codon Linear array of three nucleotides in mRNA. Each triplet specifies a particular amino acid during the process of translation. (14)

Coelomates Animals in which the body cavity is completely lined by mesodermally-derived tissues. (39)

Coenzyme An organic cofactor, typically a vitamin or a substance derived from a vitamin. (6)

Coevolution Evolutionary changes that result from reciprocal interactions between two species, e.g., flowering plants and their insect pollinators. (33)

Cofactor A non-protein component that is linked covalently or noncovalently to an enzyme and is required by the enzyme to catalyze the reaction. Cofactors may be organic molecules (coenzymes) or metals. (6)

Cohesion The tendency of different parts of a substance to hold together because of forces acting between its molecules. (3)

Coitus Sexual union in mammals. (31)

Coleoptile Sheath surrounding the tip of the monocot seedling, protecting the young stem and leaves as they emerge from the soil. (21)

Collagen The most abundant protein in the human body. It is present primarily in the extracellular space of connective tissues such as bone, cartilage, and tendons. (26)

Collenchyma Living plant cells with irregularly thickened primary cell walls. A supportive cell type often found inside the epidermis of stems with primary growth. Angular, lacunar and laminar are different types of collenchyma cells. (18)

Commensalism A form of symbiosis in which one organism benefits from the union while the other member neither gains nor loses. (42)

Community The populations of all species living in a given area. (41)

Compact Bone The solid, hard outer regions of a bone surrounding the honey-combed mass of spongy bone. (26)

Companion Cell Specialized parenchyma cell associated with a sieve-tube member in phloem. (18)

Competition Interaction among organisms that require the same resource. It is of two types: 1) intraspecific (between members of the same species); 2) interspecific (between members of different species). (42)

Competitive Exclusion Principle (Gause's Principle) Competition in which a winner species captures a greater share of resources, increasing its survival and reproductive capacity. The other species is gradually displaced. (42)

Competitive Inhibition Prevention of normal binding of a substrate to its enzyme by the presence of an inhibitory compound that competes with the substrate for the active site on the enzyme. (6)

Complement Blood proteins with which some antibodies combine following attachment to antigen (the surface of microorganisms). The bound complement punches the tiny holes in the plasma membrane of the foreign cell, causing it to burst. (28)

Complementarity The relationship between the two strands of a DNA molecule determined by the base pairing of nucleotides on the two strands of the helix. A nucleotide with guanine on one strand always pairs with a nucleotide having cytosine on the other strand; similarly with adenine and thymine. (14)

Complete Digestive Systems Systems that have a digestive tract with openings at both ends—a mouth for entry and an anus for exit. (27)

Complete Flower A flower containing all four whorls of modified leaves—sepels, petals, stamen, and carpels. (20)

Compound Chemical substances composed of atoms of more than one element. (3)

Compound Leaf A leaf that is divided into leaflets, with two or more leaflets attached to the petiole. (18)

Concentration Gradient Regions in a system of differing concentration representing potential energy, such as exist in a cell and its environment, that cause molecules to move from areas of higher concentration to lower concentration. (7)

Conditioned Reflex A reflex ("automatic") response to a stimulus that would not normally have elicited the response. Conditioned reflexes develop by repeated association of a new stimulus with an old stimulus that normally elicits the response. (44)

Conformation The three-dimensional shape of a molecule as determined by the spatial arrangement of its atoms. (4)

Conformational Change Change in molecular shape (as occurs, for example, in an enzyme as it catalyzes a reaction, or a myosin molecule during contraction). (6)

Conjugation A method of reproduction in single-celled organisms in which two cells link and exchange nuclear material. (11)

Connective Tissues Tissues that protect, support, and hold together the internal organs and other structures of animals. Includes bone, cartilage, tendons, and other tissues, all of which have large amounts of extracellular material. (22)

Consumers Heterotrophs in a biotic environment that feed on other organisms or organic waste. (41)

Continental Drift The continuous shifting of the earth's land masses explained by the theory of plate tectonics. (35)

Continuous Variation An inheritance pattern in which there is graded change between the two extremes in a phenotype (compare with discontinuous variation). (12)

Contraception The prevention of pregnancy. (31)

Contractile Proteins Actin and myosin, the protein filaments that comprise the bulk of the muscle mass. During contraction of skeletal muscle, these filaments form a temporary association and slide past each other, generating the contractile force. (26)

Control (Experimental) A duplicate of the experiment identical in every way except for the one variable being tested. Use of a control is necessary to demonstrate cause and effect. (2)

Convergent Evolution The evolution of similar structures in distantly related organisms in response to similar environments. (33)

Cork Cambium In stems and roots of perennials, a secondary meristem that produces the outer protective layer of the bark. (18)

Coronary Arteries Large arteries that branch immediately from the aorta, providing oxygen-rich blood to the cardiac muscle. (28)

Corpus Callosum A thick cable composed of hundreds of millions of neurons that connect the right and left cerebral hemispheres of the mammalian brain. (23)

Corpus Luteum In the mammalian ovary, the structure that develops from the follicle after release of the egg. It secretes hormones that prepare the uterine endometrium to receive the developing embryo. (31)

Cortex In the stem or root of plants, the region between the epidermis and the vascular tissues. Composed of ground tissue. In animals, the outermost portion of some organs. (18)

Cotyledon The seed leaf of a dicot embryo containing stored nutrients required for the germinated seed to grow and develop, or a food digesting seed leaf in a monocot embryo. (20)

Countercurrent Flow Mechanism for increasing the exchange of substances or heat from one stream of fluid to another by having the two fluids flow in opposite directions. (29)

Covalent Bonds Linkage between two atoms which share the same electrons in their outermost shells. (3)

Cranial Nerves Paired nerves which emerge from the central stalk of the vertebrate brain and innervate the body. Humans have 12 pairs of cranial nerves. (23)

Cranium The bony casing which surrounds and protects the vertebrate brain. (23)

Cristae The convolutions of the inner membrane of the mitochondrion. Embedded within them are the components of the electron transport system and proton channels for chemiosmosis. (9)

Crossing Over During synapsis, the process by which homologues exchange segments with each other. (11)

Cryptic Coloration A form of camouflage wherein an organism's color or patterning helps it resemble its background. (42)

Cutaneous Respiration The uptake of oxygen across virtually the entire outer body surface. (29)

Cuticle 1) Waxy layer covering the outer cell walls of plant epidermal cells. It retards water vapor loss and helps prevent dehydration. (18) 2) Outer protective, nonliving covering of some animals, such as the exoskeleton of anthropods. (26, 39)

Cyanobacteria A type of prokaryote capable of photosynthesis using water as a source of electrons. Cyanobacteria were responsible for initially creating an O_2-containing atmosphere on earth. (35, 36)

Cyclic AMP (Cyclic adenosine monophosphate) A ring-shaped molecular version of an ATP minus two phosphates. A regulatory molecule formed by the enzyme adenylate cyclase which converts ATP to cAMP. A second messenger. (25)

Cyclic Pathways Metabolic pathways in which the intermediates of the reaction are regenerated while assisting the conversion of the substrate to product. (9)

Cyclic Photophosphorylation A pathway that produces ATP, but not NADPH, in the light reactions of photosynthesis. Energized electrons are shuttled from a reaction center, along a molecular pathway, back to the original reaction center, generating ATP en route. (8)

Cysts Protective, dormant structure formed by some protozoa. (37)

Cytochrome Oxidase A complex of proteins that serves as the final electron carrier in the mitochondrial electron transport system, transferring its electrons to O_2 to form water. (9)

Cytokinesis Final event in eukaryotic cell division in which the cell's cytoplasm and the new nuclei are partitioned into separate daughter cells. (10)

Cytokinins Growth-producing plant hormones which stimulate rapid cell division. (21)

Cytoplasm General term that includes all parts of the cell, except the plasma membrane and the nucleus. (5)

Cytoskeleton Interconnecting network of microfilaments, microtubules, and intermediate filaments that serves as a cell scaffold and provides the machinery for intracellular movements and cell motility. (5)

Cytotoxic (Killer) T Cells A class of T cells capable of recognizing and destroying foreign or infected cells. (30)

◄ D ►

Day Neutral Plants Plants that flower at any time of the year, independent of the relative lengths of daylight and darkness. (21)

Deciduous Trees or shrubs that shed their leaves in a particular season, usually autumn, before entering a period of dormancy. (40)

Deciduous Forest Forests characterized by trees that drop their leaves during unfavorable conditions, and leaf out during warm, wet seasons. Less dense than tropical rain forests. (40)

Decomposers (Saprophytes) Organisms that obtain nutrients by breaking down organic compounds in wastes and dead organisms. Includes fungi, bacteria, and some insects. (41)

Deletion Loss of a portion of a chromosome, following breakage of DNA. (13)

Denaturation Change in the normal folding of a protein as a result of heat, acidity, or alkalinity. Such changes result in a loss of enzyme functioning. (4)

Dendrites Cytoplasmic extensions of the cell body of a neuron. They carry impulses from the area of stimulation to the cell body. (23)

Denitrification The conversion by denitrifying bacteria of nitrites and nitrates into nitrogen gas. (41)

Denitrifying Bacteria Bacteria which take soil nitrogen, usable to plants, and convert it to unusable nitrogen gas. (41)

Density-Dependent Factors Factors that control population growth which are influenced by population size. (43)

Density-Independent Factors Factors that control population growth which are not affected by population size. (43)

Deoxyribonucleic Acid (DNA) Double-stranded polynucleotide comprised of deoxyribose (a sugar), phosphate, and four bases (adenine, guanine, cytosine, and thymine). Encoded in the sequence of nucleotides are the instructions for making proteins. DNA is the genetic material in all organisms except certain viruses. (14)

Depolarization A decrease in the potential difference (voltage) across the plasma membrane of a cell typically due to an increase in the movement of sodium ions into the cell. Acts to excite a target cell. (23)

Dermal Bone Bones of vertebrates that form within the dermal layer of the skin, such as the scales of fishes and certain bones of the skull. (26)

Dermal Tissue System In plants, the epidermis in primary growth, or the periderm in secondary growth. (18)

Dermis In animals, layer of cells below the epidermis in which connective tissue predominates. Embedded within it are vessels, various glands, smooth muscle, nerves, and follicles. (26)

Desert Biome characterized by intense solar radiation, very little rainfall, and high winds. (40)

Detrivore Organism that feeds on detritus, dead organisms or their parts, and living organisms' waste. (41)

Deuterostome One path of development exhibited by coelomate animals (e.g., echinoderms and chordates). (39)

Diabetes Mellitus A disease caused by a deficiency of insulin or its receptor, preventing glucose from being absorbed by the cells. (25)

Diaphragm A sheet of muscle that separates the thoracic cavity from the abdominal wall. (29)

Diastolic Pressure The second number of a blood pressure reading; the lowest pressure in the arteries just prior to the next heart contraction. (28)

Diatoms are golden-brown algae that are distinguished most dramatically by their intricate silica shells. (37)

Dicotyledonae (Dicots) One of the two classes of flowering plants, characterized by having seeds with two cotyledons, flower parts in 4s or 5s, net-veined leaves, one main root, and vascular bundles in a circular array within the stem. (Compare with Monocotylenodonae). (18)

Diffusion Tendency of molecules to move from a region of higher concentration to a region of lower concentration, until they are uniformly dispersed. (7)

Digestion The process by which food particles are disassembled into molecules small enough to be absorbed into the organism's cells and tissues. (27)

Digestive System System of specialized organs that ingests food, converts nutrients to a form that can be distributed throughout the animal's body, and eliminates undigested residues. (27)

Digestive Tract Tubelike channel through which food matter passes from its point of ingestion at the mouth to the elimination of indigestible residues from the anus. (27)

Dihybrid Cross A mating between two individuals that differ in two genetically-determined traits. (12)

Dimorphism Presence of two forms of a trait within a population, resulting from diversifying selection. (33)

Dinoflagellates Single-celled photosynthesizers that have two flagella. They are members of the pyrophyta, phosphorescent algae that sometimes cause red tide, often synthesizing a neurotoxin that accumulates in plankton eaters, causing paralytic shellfish poisoning in people who eat the shellfish. (37)

Dioecious Plants that produce either male or female reproductive structures but never both. (38)

Diploid Having two sets of homologous chromosomes. Often written 2N. (10, 13)

Directional Selection The steady shift of phenotypes toward one extreme. (33)

Discontinuous Variation An inheritance pattern in which the phenomenon of all possible phenotypes fall into distinct categories. (Compare with continuous variation). (12)

Displays The signals that form the language by which animals communicate. These signals are species specific and stereotyped and may be visual, auditory, chemical, or tactile. (44)

Disruptive Coloration Coloration that disguises the shape of an organism by breaking up its outline. (42)

Disruptive Selection The steady shift toward more than one extreme phenotype due to the elimination of intermediate phenotypes as has occurred among African swallowtail butterflies whose members resemble more than one species of distasteful butterfly. (33)

Divergent Evolution The emergence of new species as branches from a single ancestral lineage. (33)

Diversifying Selection The increasing frequency of extreme phenotypes because individuals with average phenotypes die off. (33)

Diving Reflex Physiological response that alters the flow of blood in the body of diving mammals that allows the animal to maintain high levels of activity without having to breathe. (29)

Division (or Phylum) A level of the taxonomic hierarchy that groups together members or related classes. (1)

DNA (see **Deoxyribonucleic Acid**)

DNA Cloning The amplification of a particular DNA by use of a growing population of bacteria. The DNA is initially taken up by a bacterial cell—usually as a plasmid—and then replicated along with the bacteria's own DNA. (16)

DNA Fingerprint The pattern of DNA fragments produced after treating a sample of DNA with a particular restriction enzyme and separating the fragments by gel electrophoresis. Since different members of a population have DNA with a different nucleotide sequence, the pattern of DNA fragments produced by this method can be used to identify a particular individual. (16)

DNA Ligase The enzyme that covalently joins DNA fragments into a continuous DNA strand. The enzyme is used in a cell during replication to seal newly-synthesized fragments and by biotechnologists to form recombinant DNA molecules from separate fragments. (14, 16)

DNA Polymerase Enzyme responsible for replication of DNA. It assembles free nucleotides, aligning them with the complementary ones in the unpaired region of a single strand of DNA template. (14)

Dominant The form of an allele that masks the presence of other alleles for the same trait. (12)

Dormancy A resting period, such as seed dormancy in plants or hibernation in animals, in which organisms maintain reduced metabolic rates. (21)

Dorsal In anatomy, the back of an animal. (39)

Double Blind Test A clinical trial of a drug in which neither the human subjects or the researchers know who is receiving the drug or placebo. (2)

Down Syndrome Genetic disorder in humans characterized by distinct facial appearance and mental retardation, resulting from an extra copy of chromosome number 21 (trisomy 21) in each cell. (11, 17)

Duodenum First part of the human small intestine in which most digestion of food occurs. (27)

Duplication The repetition of a segment of a chromosome. (13)

◄ E ►

Ecdysis Molting process by which an arthropod periodically discards its exoskeleton and replaces it with a larger version. The process is controlled by the hormone ecydysone. (39)

Ecdysone An insect steroid hormone that triggers molting and metamorphosis. (15)

Echinodermata A phylum composed of animals having an internal skeleton made of many small calcium carbonate plates which have jutting spines. Includes sea stars, sea urchins, etc. (39)

Echolocation The use of reflected sound waves to help guide an animal through its environment and/or locate objects. (24)

Ecological Equivalent Organisms that occupy similar ecological niches in different regions or ecosystems of the world. (41)

Ecological Niche The habitat, functional role(s), requirements for environmental resources and tolerance ranges for each abiotic condition in relation to an organism. (41)

Ecological Pyramid Illustration showing the energy content, numbers of organisms, or biomass at each trophic level. (41)

Ecology The branch of biology that studies interactions among organisms as well as the interactions of organisms and their physical environment. (40)

Ecosystem Unit comprised of organisms interacting among themselves and with their physical environment. (41)

Ecotypes Populations of a single species with different, genetically fixed tolerance ranges. (41)

Ectoderm In animals, the outer germ cell layer of the gastrula. It gives rise to the nervous system and integument. (32)

Ectotherms Animals that lack an internal mechanism for regulating body temperature. "Cold-blooded" animals. (28)

Edema Swelling of a tissue as the result of an accumulation of fluid that has moved out of the blood vessels. (28)

Effectors Muscle fibers and glands that are activated by neural stimulation. (23)

Efferent (Motor) Nerves The nerves that carry messages from the central nervous system to the effectors, the muscles, and glands. They are divided into two systems: somatic and autonomic. (23)

Egg Female gamete, also called an ovum. A fertilized egg is the product of the union of female and male gametes (egg and sperm cells). (32)

Electrocardiogram (EKG) Recording of the electrical activity of the heart, which is used to diagnose various types of heart problems. (28)

Electron Acceptor Substances that are capable of accepting electrons transferred from an electron donor. For example, molecular oxygen (O_2) is the terminal electron acceptor during respiration. Electron acceptors also receive electrons from chlorophyll during photosynthesis. Electron acceptors may act as part of an electron transport system by transferring the electrons they receive to another substance. (8, 9)

Electron Carrier Substances (such as NAD^+ and FAD) that transport electrons from one step of a metabolic pathway to the next or from metabolic reactions to biosynthetic reactions. (8, 9)

Electrons Negatively charged particles that orbit the atomic nucleus. (3)

Electron Transport System Highly organized assembly of cytochromes and other proteins which transfer electrons. During transport, which occurs within the inner membranes of mitochondria and chloroplasts, the energy extracted from the electrons is used to make ATP. (8, 9)

Electrophoresis A technique for separating different molecules on the basis of their size and/or electric charge. There are various ways the technique is used. In gel electrophoresis, proteins or DNA fragments are driven through a porous gel by their charge, but become separated according to size; the larger the molecule, the slower it can work its way through the pores in the gel, and the less distance it travels along the gel. (16)

Element Substance composed of only one type of atom. (3)

Embryo An organism in the early stages of development, beginning with the first division of the zygote. (32)

Embryo Sac The fully developed female gametophyte within the ovule of the flower. (20)

Emigration Individuals permanently leaving an area or population. (43)

Endergonic Reactions Chemical reactions that require energy input from another source in order to occur. (6)

Endocrine Glands Ductless glands, which secrete hormones directly into surrounding tissue fluids and blood vessels for distribution to the rest of the body by the circulatory system. (25)

Endocytosis A type of active transport that imports particles or small cells into a cell. There are two types of endocytic processes: phagocytosis, where large particles are ingested by the cell, and pinocytosis, where small droplets are taken in. (7)

Endoderm In animals, the inner germ cell layer of the gastrula. It gives rise to the digestive tract and associated organs and to the lungs. (32)

Endodermis The innermost cylindrical layer of cortex surrounding the vascular tissues of the root. The closely pressed cells of the endodermis have a waxy band, forming a waterproof layer, the Casparian strip. (18)

Endogenous Plant responses that are controlled internally, such as biological clocks controlling flower opening. (21)

Endometrium The inner epithelial layer of the uterus that changes markedly with the uterine (menstrual) cycle in preparation for implantation of an embryo. (31)

Endoplasmic Reticulum (ER) An elaborate system of folded, stacked and tubular membranes contained in the cytoplasm of eukaryotic cells. (5)

Endorphins (Endogenous Morphinelike Substances) A class of peptides released from nerve cells of the limbic system of the brain that can block perceptions of pain and produce a feeling of euphoria. (23)

Endoskeleton The internal support structure found in all vertebrates and a few invertebrates (sponges and sea stars). (26)

Endosperm Nutritive tissue in plant embryos and seeds. (20)

Endosperm Mother Cell A binucleate cell in the embryo sac of the female gametophyte, occurring in the ovule of the ovary in angiosperms. Each nucleus is haploid; after fertilization, nutritive endosperm develops. (20)

Endosymbiosis Theory A theory to explain the development of complex eukaryotic cells by proposing that some organelles once were free-living prokaryotic cells that then moved into another larger such cell, forming a beneficial union with it. (5)

Endotherms Animals that utilize metabolically produced heat to maintain a constant, elevated body temperature. "Warm-blooded" animals. (28)

End Product The last product in a metabolic pathway. Typically a substance, such as an amino acid or a nucleotide, that will be used as a monomer in the formation of macromolecules. (6)

Energy The ability to do work. (6)

Entropy Energy that is not available for doing work; measure of disorganization or randomness. (6)

Environmental Resistance The factors that eventually limit the size of a population. (43)

Enzyme Biological catalyst; a protein molecule that accelerates the rate of a chemical reaction. (6)

Eosiniphil A type of phagocytic white blood cell. (28)

Epicotyl The portion of the embryo of a dicot plant above the cotyledons. The epicotyl gives rise to the shoot. (20)

Epidermis In vertebrates, the outer layer of the skin, containing superficial layers of dead cells produced by the underlying living epithelial cells. In plants, the outer layer of cells covering leaves, primary stem, and primary root. (26, 18)

Epididymis Mass of convoluted tubules attached to each testis in mammals. After leaving the testis, sperm enter the tubules where they finish maturing and acquire motility. (31)

Epiglottis A flap of tissue that covers the glottis during swallowing to prevent food and liquids from entering the lower respiratory tract. (29)

Epinephrine (Adrenalin) Substance that serves both as an excitatory neurotransmitter released by certain neurons of the CNS and as a hormone released by the adrenal medulla that increases the body's ability to combat a stressful situation. (25)

Epipelagic Zone The lighted upper ocean zone, where photosynthesis occurs; large populations of phytoplankton occur in this zone. (40)

Epiphyseal Plates The action centers for ossification (bone formation). (26)

Epistasis A type of gene interaction in which a particular gene blocks the expression of another gene at another locus. (12)

Epithelial Tissue Continuous sheets of tightly packed cells that cover the body and line its tracts and chambers. Epithelium is a fundamental tissue type in animals. (22)

Erythrocytes Red blood cells. (28)

Erythropoietin A hormone secreted by the kidney which stimulates the formation of erythrocytes by the bone marrow. (28)

Essential Amino Acids Eight amino acids that must be acquired from dietary protein. If even one is missing from the human diet, the synthesis of proteins is prevented. (27)

Essential Fatty Acids Linolenic and linoleic acids, which are required for phospholipid construction and must be acquired from a dietary source. (27)

Essential Nutrients The 16 minerals essential for plant growth, divided into two groups: macronutrients, which are required in large quantities, and micronutrients, which are needed in small amounts. (19)

Estrogen A female sex hormone secreted by the ovaries when stimulated by pituitary gonadotrophins. (31)

Estuaries Areas found where rivers and streams empty into oceans, mixing fresh water with salt water. (40)

Ethology The study of animal behavior. (44)

Ethylene Gas A plant hormone that stimulates fruit ripening. (21)

Etiolation The condition of rapid shoot elongation, small underdeveloped leaves, bent shoot-hook, and lack of chlorophyll, all due to lack of light. (21)

Eubacteria Typical procaryotic bacteria with peptidoglycan in their cell walls. The majority of monerans are eubacteria. (36)

Eukaryotic Referring to organisms whose cellular anatomy includes a true nucleus with a nuclear envelope, as well as other membrane-bound organelles. (5)

Eusocial Species Social species that have sterile workers, cooperative care of the young, and an overlap of generations so that the colony labor is a family affair. (44)

Eutrophication The natural aging process of lakes and ponds, whereby they become marshes and, eventually, terrestrial environments.

Evolution A process whereby the characteristics of a species change over time, eventually leading to the formation of new species that go about life in new ways. (33)

Evolutionarily Stable Strategy (ESS) A behavioral strategy or course of action that depends on what other members of the population are doing. By definition, an ESS cannot be replaced by any other strategy when most of the members of the population have adopted it. (44)

Excitatory Neurons Neurons that stimulate their target cells into activity. (23)

Excretion Removal of metabolic wastes from an organism. (28)

Excretory System The organ system that eliminates metabolic wastes from the body. (28)

Exergonic Reactions Chemical reactions that occur spontaneously with the release of energy. (6)

Exocrine Glands Glands which secrete their products through ducts directly to their sites of action, e.g., tear glands. (26)

Exocytosis A form of active transport used by cells to move molecules, particles, or other cells contained in vesicles across the plasma membrane to the cell's environment. (5)

Exogenous Plant responses that are controlled externally, or by environmental conditions. (21)

Exons Structural gene segments that are transcribed and whose genetic information is subsequently translated into protein. (15)

Exoskeletons Hard external coverings found in some animals (e.g., lobsters, insects) for protection, support, or both. Such organisms grow by the process of molting. (26)

Exploitative Competition A competition in which one species manages to get more of a resource, thereby reducing supplies for a competitor. (42)

Exponential Growth An increase by a fixed percentage in a given time period; such as population growth per year. (43)

Extensor Muscle A muscle which, when contracted, causes a part of the body to straighten at a joint. (26)

External Fertilization Fertilization of an egg outside the body of the female parent. (31)

Extinction The loss of a species. (33)

Extracellular Digestion Digestion occurring outside the cell; occurs in bacteria, fungi, and multicellular animals. (27)

Extracellular Matrix Layer of extracellular material residing just outside a cell. (5)

◀ F ▶

F_1 First filial generation. The first generation of offspring in a genetic cross. (12)

F_2 Second filial generation. The offspring of an F_1 cross. (12)

Facilitated Diffusion The transport of molecules into cells with the aid of "carrier" proteins embedded in the plasma membrane. This carrier-assisted transport does not require the expenditure of energy by the cell. (7)

FAD Flavin adenine dinucleotide. A coenzyme that functions as an electron carrier in metabolic reactions. When it is reduced to $FADH_2$, this molecule becomes a cellular energy source. (9)

Family A level of the taxonomic hierarchy that groups together members of related genera. (1)

Fast-Twitch Fibers Skeletal muscle fibers that depend on anaerobic metabolism to produce ATP rapidly, but only for short periods of time before the onset of fatigue. Fast-twitch fibers generate greater forces for shorter periods than slow-twitch fibers. (9)

Fat A triglyceride consisting of three fatty acids joined to a glycerol. (4)

Fatty Acid A long unbranched hydrocarbon chain with a carboxyl group at one end. Fatty acids lacking a double bond are said to be saturated. (4)

Fauna The animals in a particular region.

Feedback Inhibition (Negative Feedback) A mechanism for regulating enzyme activity by temporarily inactivating a key enzyme in a biosynthetic pathway when the concentration of the end product is elevated. (6)

Fermentation The direct donation of the electrons of NADH to an organic compound without their passing through an electron transport system. (9)

Fertility Rate In humans, the average number of children born to each woman between 15 and 44 years of age. (43)

Fertilization The process in which two haploid nuclei fuse to form a zygote. (32)

Fetus The term used for the human embryo during the last seven months in the uterus. During the fetal stage, organ refinement accompanies overall growth. (32)

Fibrinogen A rod-shaped plasma protein that, converted to fibrin, generates a tangled net of fibers that binds a wound and stops blood loss until new cells replace the damaged tissue. (28)

Fibroblasts Cells found in connective tissues that secrete the extracellular materials of the connective tissue matrix. These cells are easily isolated from connective tissues and are widely used in cell culture. (22)

Fibrous Root System Many approximately equal-sized roots; monocots are characterized by a fibrous root system. Also called diffuse root system. (18)

Filament The stalk of a stamen of angiosperms, with the anther at its tip. Also, the threadlike chain of cells in some algae and fungi. (20)

Filamentous Fungus Multicellular members of the fungus kingdom comprised mostly of living threads (hyphae) that grow by division of cells at their tips (see molds). (37)

Filter Feeders Aquatic animals that feed by straining small food particles from the surrounding water. (27, 39)

Fitness The relative degree to which an individual in a population is likely to survive to reproductive age and to reproduce. (33)

Fixed Action Patterns Motor responses that may be triggered by some environmental stimulus, but once started can continue to completion without external stimuli. (44)

Flagella Cellular extensions that are longer than cilia but fewer in number. Their undulations propel cells like sperm and many protozoans, through their aqueous environment. (5)

Flexor Muscle A muscle which, when contracted, causes a part of the body to bend at a joint. (26)

Flora The plants in a particular region. (21)

Florigen Proposed A chemical hormone that is produced in the leaves and stimulates flowering. (21)

Fluid Mosaic Model The model proposes that the phospholipid bilayer has a viscosity similar to that of light household oil and that globular proteins float like icebergs within this bilayer. The now favored explanation for the architecture of the plasma membrane. (5)

Follicle (Ovarian) A chamber of cells housing the developing oocytes. (31)

Food Chain Transfers of food energy from organism to organism, in a linear fashion. (41)

Food Web The map of all interconnections between food chains for an ecosystem. (41)

Forest Biomes Broad geographic regions, each with characteristic tree vegetation: 1) tropical rain forests (lush forests in a broad band around the equator), 2) deciduous forests (trees and shrubs drop their leaves during unfavorable seasons), 3) coniferous forest (evergreen conifers). (40)

Fossil Record An entire collection of remains from which paleontologists attempt to reconstruct the phylogeny, anatomy, and ecology of the preserved organisms. (34)

Fossils The preserved remains of organisms from a former geologic age. (34)

Fossorial Living underground.

Founder Effect The potentially dramatic difference in allele frequency of a small founding population as compared to the original population. (33)

Founder Population The individuals, usually few, that colonize a new habitat. (33)

Frameshift Mutation The insertion or deletion of nucleotides in a gene that throws off the reading frame. (14)

Free Radical Atom or molecule containing an unpaired electron, which makes it highly reactive. (3)

Freeze-Fracture Technique in which cells are frozen into a block which is then struck with a knife blade that fractures the block in two. Fracture planes tend to expose the center of membranes for EM examination. (5)

Fronds The large leaf-like structures of ferns. Unlike true leaves, fronds have an apical meristem and clusters of sporangia called sori. (38)

Fruit A mature plant ovary (flower) containing seeds with plant embryos. Fruits protect seeds and aid in their dispersal. (20)

Fruiting Body A spore-producing structure that extends upward in an elevated position from the main mass of a mold or slime mold. (37)

FSH Follicle stimulating hormone. A hormone secreted by the anterior pituitary that prepares a female for ovulation by stimulating the primary follicle to ripen or stimulates spermatogenesis in males. (31)

Functional Groups Accessory chemical entities (e.g., —OH, $—NH_2$, $—CH_3$), which help determine the identity and chemical properties of a compound. (4)

Fundamental Niche The potential ecological niche of a species, including all factors affecting that species. The fundamental niche is usually never fully utilized. (41)

Fungus Yeast, mold, or large filamentous mass forming macroscopic fruiting bodies, such as mushrooms. All fungi are eukaryotic nonphotosynthetic heterotrophics with cell walls. (37)

◄ G ►

G_1 Stage The first of three consecutive stages of interphase. During G_1, cell growth and normal functions occur. The duration of this stage is most variable. (10)

G_2 Stage The final stage of interphase in which the final preparations for mitosis occur. (10)

Gallbladder A small saclike structure that stores bile salts produced by the liver. (27)

Gamete A haploid reproductive cell—either a sperm or an egg. (10)

Gas Exchange Surface Surface through which gases must pass in order to enter or leave the body of an animal. It may be the plasma membrane of a protistan or the complex tissues of the gills or the lungs in multicellular animals. (29)

Gastrovascular Cavity In cnidarians and flatworms, the branched cavity with only one opening. It functions in both digestion and transport of nutrients. (39)

Gastrula The embryonic stage formed by the inward migration of cells in the blastula. (32)

Gastrulation The process by which the blastula is converted into a gastrula having three germ layers (ectoderm, mesoderm, and endoderm). (32)

Gated Ion Channels Most passageways through a plasma membrane that allow ions to pass contain "gates" that can occur in either an open or a closed conformation. (7, 23)

Gel Electrophoresis (See **Electrophoresis**)

Gene Pool All the genes in all the individuals of a population. (33)

Gene Regulatory Proteins Proteins that bind to specific sites in the DNA and control the transcription of nearby genes. (15)

Genes Discrete units of inheritance which determine hereditary traits. (12, 14)

Gene Therapy Treatment of a disease by alteration of the person's genotype, or the genotype of particular affected cells. (17)

Genetic Carrier A heterozygous individual who shows no evidence of a genetic disorder but, because they possess a recessive allele for a disorder, can pass the mutant gene on to their offspring. (17)

Genetic Code The correspondence between the various mRNA triplets (codons, e.g., UGC) and the amino acid that the triplet specifies (e.g., cysteine). The genetic code includes 64 possible three-letter words that constitute the genetic language for protein synthesis. (14)

Genetic Drift Random changes in allele frequency that occur by chance alone. Occurs primarily in small populations. (33)

Genetic Engineering The modification of a cell or organism's genetic composition according to human design. (16)

Genetic Equilibrium A state in which allele frequencies in a population remain constant from generation to generation. (33)

Genetic Mapping Determining the locations of specific genes or genetic markers along particular chromosomes. This is typically accomplished using crossover frequencies; the more often alleles of two genes are separated during crossing over, the greater the distance separating the genes. (13)

Genetic Recombination The reshuffling of genes on a chromosome caused by breakage of DNA and its reunion with the DNA of a homologoue. (11)

Genome The information stored in all the DNA of a single set of chromosomes. (17)

Genotype An individual's genetic makeup. (12)

Genus Taxonomic group containing related species. (1)

Geologic Time Scale The division of the earth's 4.5 billion-year history into eras, periods, and epochs based on memorable geologic and biological events. (35)

Germ Cells Cells that are in the process of or have the potential to undergo meiosis and form gametes. (11, 31)

Germination The sprouting of a seed, beginning with the radicle of the embryo breaking through the seed coat. (21)

Germ Layers Collective name for the endoderm, ectoderm, and mesoderm, from which all the structures of the mature animal develop. (32)

Gibberellins More than 50 compounds that promote growth by stimulating both cell elongation and cell division. (21)

Gills Respiratory organs of aquatic animals. (29)

Globin The type of polypeptide chains that make up a hemoglobin molecule.

Glomerular Filtration The process by which fluid is filtered out of the capillaries of the glomerulus into the proximal end of the nephron. Proteins and blood cells remain behind in the bloodstream. (28)

Glomerulus A capillary bundle embedded in the double-membraned Bowman's capsule, through which blood for the kidney first passes. (28)

Glottis Opening leading to the larynx and lower respiratory tract. (29)

Glucagon A hormone secreted by the Islets of Langerhans that promotes glycogen breakdown to glucose. (25)

Glucocorticoids Steroid hormones which regulate sugar and protein metabolism. They are secreted by the adrenal cortex. (25)

Glycogen A highly branched polysaccharide consisting of glucose monomers that serves as a storage of chemical energy in animals. (4)

Glycolysis Cleavage, releasing energy, of the six-carbon glucose molecule into two molecules of pyruvic acid, each containing three carbons. (9)

Glycoproteins Proteins with covalently-attached chains of sugars. (5)

Glycosidic Bond The covalent bond between individual molecules in carbohydrates. (4)

Golgi Complex A system of flattened membranous sacs, which package substances for secretion from the cell. (5)

Gonadotropin-Releasing Hormone (GnRH) Hypothalmic hormone that controls the secretion of the gonadotropins FSH and LH. (31)

Gonadotropins Two anterior pituitary hormones which act on the gonads. Both FSH (follicle-stimulating hormone) and LH (luteinizing hormone) promote gamete development and stimulate the gonads to produce sex hormones. (25)

Gonads Gamete-producing structures in animals: ovaries in females, testes in males. (31)

Grasslands Areas of densely packed grasses and herbaceous plants. (40)

Gravitropisms (Geotropisms) Changes in plant growth caused by gravity. Growth away from gravitational force is called negative gravitropism; growth toward it is positive. (21)

Gray Matter Gray-colored neural tissue in the cerebral cortex of the brain and in the butterfly-shaped interior of the spinal cord. Composed of nonmyelinated cell bodies and dendrites of neurons. (23)

Greenhouse Effect The trapping of heat in the Earth's troposphere, caused by increased levels of carbon dioxide near the Earth's surface; the carbon dioxide is believed to act like glass in a greenhouse, allowing light to reach the Earth, but not allowing heat to escape. (41)

Ground Tissue System All plant tissues except those in the dermal and vascular tissues. (18)

Growth An increase in size, resulting from cell division and/or an increase in the volume of individual cells. (10)

Growth Hormone (GH) Hormone produced by the anterior pituitary; stimulates protein synthesis and bone elongation. (25)

Growth Ring In plants with secondary growth, a ring formed by tracheids and/or vessels with small lumens (late wood) during periods of unfavorable conditions; apparent in cross section. (18)

Guard Cells Specialized epidermal plant cells that flank each stomated pore of a leaf. They regulate the rate of gas diffusion and transpiration. (18)

Guild Group of species with similar ecological niches. (41)

Guttation The forcing of water and mineral completely out to the tips of leaves as a result of positive root pressure. (19)

Gymnosperms The earliest seed plants, bearing naked seeds. Includes the pines, hemlocks, and firs. (38)

◀ H ▶

Habitat The place or region where an organism lives. (41)

Habituation The phenomenon in which an animal ceases to respond to a repetitive stimulus. (23, 44)

Hair Cells Sensory receptors of the inner ear that respond to sound vibration and bodily movement. (24)

Half-Life The time required for half the mass of a radioactive element to decay into its stable, non-radioactive form. (3)

Haplodiploidy A genetic pattern of sex determination in which fertilized eggs develop into females and non-fertilized eggs develop into males (as occurs among bees and wasps). (44)

Haploid Having one set of chromosomes per cell. Often written as 1N. (10)

Hardy-Weinberg Law The maintenance of constant allele frequencies in a population from one generation to the next when certain conditions are met. These conditions are the absence of mutation and migration, random mating, a large population, and an equal chance of survival for all individuals. (33)

Haversian Canals A system of microscopic canals in compact bone that transport nutrients to and remove wastes from osteocytes. (26)

Heart An organ that pumps blood (or hemolymph in arthropods) through the vessels of the circulatory system. (28)

Helper T Cells A class of T cells that regulate immune responses by recognizing and activating B cells and other T cells. (30)

Hemocoel In arthropods, the unlined spaces into which fluid (hemolymph) flows when it leaves the blood vessels and bathes the internal organs. (28)

Hemoglobin The iron-containing blood protein that temporarily binds O_2 and releases it into the tissues. (4, 29)

Hemophilia A genetic disorder determined by a gene on the X chromosome (an X-linked trait) that results from the failure of the blood to form clots. (13)

Herbaceous Plants having only primary growth and thus composed entirely of primary tissue. (18)

Herbivore An organism, usually an animal, that eats primary producers (plants). (42)

Herbivory The term for the relationship of a secondary consumer, usually an animal, eating primary producers (plants). (42)

Heredity The passage of genetic traits to offspring which consequently are similar or identical to the parent(s). (12)

Hermaphrodites Animals that possess gonads of both the male and the female. (31)

Heterosporous Higher vascular plants producing two types of spores, a megaspore which grows into a female gametophyte and a microspore which grows into a male gametophyte. (38)

Heterozygous A term applied to organisms that possess two different alleles for a trait. Often, one allele (A) is dominant, masking the presence of the other (a), the recessive. (12)

High Intertidal Zone In the intertidal zone, the region from mean high tide to around just below sea level. Organisms are submerged about 10% of the time. (40)

Histones Small basic proteins that are complexed with DNA to form nucleosomes, the basic structural components of the chromatin fiber. (14)

Homeobox That part of the DNA sequence of homeotic genes that is similar (homologous) among diverse animal species. (32)

Homeostasis Maintenance of fairly constant internal conditions (e.g., blood glucose level, pH, body temperature, etc.) (22)

Homeotic Genes Genes whose products act during embryonic development to affect the spatial arrangement of the body parts. (32)

Hominids Humans and the various groups of extinct, erect-walking primates that were either our direct ancestors or their relatives. Includes the various species of *Homo* and *Australopithecus*. (34)

Homo the genus that contains modern and extinct species of humans. (34)

Homologous Structures Anatomical structures that may have different functions but develop from the same embryonic tissues, suggesting a common evolutionary origin. (34)

Homologues Members of a chromosome pair, which have a similar shape and the same sequence of genes along their length. (10)

Homoplasy (see **Analogous Structures**)

Homosporous Plants that manufacture only one type of spore, which develops into a gametophyte containing both male and female reproductive structures. (38)

Homozygous A term applied to an organism that has two identical alles for a particular trait. (12)

Hormones Chemical messengers secreted by ductless glands into the blood that direct tissues to change their activities and correct imbalances in body chemistry. (25)

Host The organism that a parasite lives on and uses for food. (42)

Human Chorionic Gonadotropin (HCG) A hormone that prevents the corpus luteum from degenerating, thereby maintaining an adequate level of progesterone during pregnancy. It is produced by cells of the early embryo. (25)

Human Immunodeficiency Virus (HIV) The infectious agent that causes AIDS, a disease in which the immune system is seriously disabled. (30, 36)

Hybrid An individual whose parents possess different genetic traits in a breeding experiment or are members of different species. (12)

Hybridization Occurs when two distinct species mate and produce hybrid offspring. (33)

Hybridoma A cell formed by the fusion of a malignant cell (a myeloma) and an antibody-producing lymphocyte. These cells proliferate indefinitely and produce monoclonal antibodies. (30)

Hydrogen Bonds Relatively weak chemical bonds formed when two molecules share an atom of hydrogen. (3)

Hydrologic Cycle The cycling of water, in various forms, through the environment, from Earth to atmosphere and back to Earth again. (41)

Hydrolysis Splitting of a covalent bond by donating the H^+ or OH^- of a water molecule to the two components. (4)

Hydrophilic Molecules Polar molecules that are attracted to water molecules and readily dissolve in water. (3)

Hydrophobic Interaction When nonpolar molecules are "forced" together in the presence of a polar solvent, such as water. (3)

Hydrophobic Molecules Nonpolar substances, insoluble in water, which form aggregates to minimize exposure to their polar surroundings. (3)

Hydroponics The science of growing plants in liquid nutrient solutions, without a solid medium such as soil. (19)

Hydrosphere That portion of the Earth composed of water. (40)

Hydrostatic Skeletons Body support systems found usually in underwater animals (e.g., marine worms). Body shape is protected against gravity and other physical forces by internal hydrostatic pressure produced by contracting muscles encircling their closed, fluid-filled chambers. (26)

Hydrothermal Vents Fissures in the ocean floor where sea water becomes superheated. Chemosynthetic bacteria that live in these vents serve as the autotrophs that support a diverse community of ocean-dwelling organisms. (8)

Hyperpolarization An increase in the potential difference (voltage) across the plasma membrane of a cell typically due to an increase in the movement of potassium ions out of the cell. Acts to inhibit a target cell. (23)

Hypertension High blood pressure (above about 130/90). (28)

Hypertonic Solutions Solutions with higher solute concentrations than found inside the cell. These cause a cell to lose water and shrink. (7)

Hypervolume In ecology, a multidimensional area which includes all factors in an organism's ecological niche, or its' potential niche. (41)

Hypocotyl Portion of the plant embryo below the cotyledons. The hypocotyl gives rise to the root and, very often, to the lower part of the stem. (20)

Hypothalamus The area of the brain below the thalamus that regulates body temperature, blood pressure, etc. (25)

Hypothesis A tentative explanation for an observation or a phenomenon, phrased so that it can be tested by experimentation. (2)

Hypotonic Solutions Solutions with lower solute concentrations than found inside the cell. These cause a cell to accumulate water and swell. (7)

◀ I ▶

Immigration Individuals permanently moving into a new area or population. (43)

Immune System A system in vertebrates for the surveillance and destruction of disease-causing microorganisms and cancer cells. Composed of lymphocytes, particularly B cells and T cells, and triggered by the introduction of antigens into the body which makes the body, upon their destruction, resistant to a recurrence of the same disease. (30)

Immunoglobulins (IGs) Antibody molecules. (30)

Imperfect Flowers Flowers that contain either stamens or carpels, making them male or female flowers, respectively. (20)

Imprinting A type of learning in which an animal develops an association with an object after exposure to the object during a critical period early in its life. (44)

Inbreeding When individuals mate with close relatives, such as brothers and sisters. May occur when population sizes drastically shrink and results in a decrease in genetic diversity. (33)

Incomplete Digestive Tract A digestive tract with only one opening through which food is taken in and residues are expelled. (27)

Incomplete (Partial) Dominance A phenomenon in which heterozygous individuals are phenotypically distinguishable from either homozygous type. (12)

Incomplete Flower Flowers lacking one or more whorls of sepals, petals, stamen, or pistils. (20)

Independent Assortment The shuffling of members of homologous chromosome pairs in meiosis I. As a result, there are new chromosome combinations in the daughter cells, which later produce offspring with random mixtures of traits from both parents. (11, 12)

Indoleatic Acid (IAA) An auxin responsible for many plant growth responses including apical dominance, a growth pattern in which shoot tips prevent axillary buds from sprouting. (21)

Induction The process in which one embryonic tissue induces another tissue to differentiate along a pathway that it would not otherwise have taken. (32) Stimulation of transcription of a gene in an operon. Occurs when the repressor protein is unable to bind to the operator. (15)

Inflammation A body strategy initiated by the release of chemicals following injury or infection which brings additional blood with its protective cells to the injured area. (30)

Inhibitory Neurons Neurons that oppose a response in the target cells. (23)

Inhibitory Neurotransmitters Substances released from inhibitory neurons where they synapse with the target cell. (23)

Innate Behavior Actions that are under fairly precise genetic control, typically species-specific, highly stereotyped, and that occur in a complete form the first time the stimulus is encountered. (44)

Insight Learning The sudden solution to a problem without obvious trial-and-error procedures. (44)

Insulin One of the two hormones secreted by endocrine centers called Islets of Langerhans; promotes glucose absorption, utilization, and storage. Insulin is secreted by them when the concentration of glucose in the blood begins to exceed the normal level. (25)

Integumentary System The body's protective external covering, consisting of skin and subcutaneous tissue. (26)

Integuments Protective covering of the ovule. (20)

Intercellular Junctions Specialized regions of cell-cell contact between animal cells. (5)

Intercostal Muscles Muscles that lie between the ribs in humans whose contraction expands the thoracic cavity during breathing. (29)

Interference Competition One species' direct interference by another species for the same limited resource; such as aggressive animal behavior. (42)

Internal Fertilization Fertilization of an egg within the body of the female. (31)

Interneurons Neurons situated entirely within the central nervous system. (23)

Internodes The portion of a stem between two nodes. (18)

Interphase Usually the longest stage of the cell cycle during which the cell grows, carries out normal metabolic functions, and replicates its DNA in preparation for cell division. (10)

Interstitial Cells Cells in the testes that produce testosterone, the major male sex hormone. (31)

Interstitial Fluid The fluid between and surrounding the cells of an animal; the extracellular fluid. (28)

Intertidal Zone The region of beach exposed to air between low and high tides. (40)

Intracellular Digestion Digestion occurring inside cells within food vacuoles. The mode of digestion found in protists and some filter-feeding animals (such as sponges and clams). (27)

Intraspecific Competition Individual organisms of one species competing for the same limited resources in the same habitat, or with overlapping niches. (42)

Intrinsic Rate of Increase (r_m) the maximum growth rate of a population under conditions of maximum birth rate and minimum death rate. (43)

Introns Intervening sequences of DNA in the middle of structural genes, separating exons. (15)

Invertebrates Animals that lack a vertebral column, or backbone. (39)

Ion An electrically charged atom created by the gain or loss of electrons. (3)

Ionic Bond The noncovalent linkage formed by the attraction of oppositely charged groups. (3)

Islets of Langerhans Clusters of endocrine cells in the pancreas that produce insulin and glucagon. (25)

Isolating Mechanisms Barriers that prevent gene flow between populations or among segments of a single population. (33)

Isotopes Atoms of the same element having a different number of neutrons in their nucleus. (3)

Isotonic Solutions Solutions in which the solute concentration outside the cell is the same as that inside the cell. (7)

◄ J ►

Joints Structures where two pieces of a skeleton are joined. Joints may be flexible, such as the knee joint of the human leg or the joints between segments of the exoskeleton of the leg of an insect, or inflexible, such as the joints (sutures) between the bones of the skull. (26)

J-Shaped Curve A curve resulting from exponential growth of a population. (43)

◄ K ►

Karyotype A visual display of an individual's chromosomes. (10)

Kidneys Paired excretory organs which, in humans, are fist-sized and attached to the lower spine. In vertebrates, the kidneys remove nitrogenous wastes from the blood and regulate ion and water levels in the body. (28)

Killer T Cells A type of lymphocyte that functions in the destruction of virus-infected cells and cancer cells. (30)

Kinases Enzymes that catalyze reactions in which phosphate groups are transferred from ATP to another molecule. (6)

Kinetic Energy Energy in motion. (6)

Kinetochore Part of a mitotic (or meiotic) chromosome that is situated within the centromere and to which the spindle fibers attach. (10)

Kingdom A level of the taxonomic hierarchy that groups together members of related phyla or divisions. Modern taxonomy divides all organisms into five Kingdoms: Monera, Protista, Fungi, Plantae, and Animalia. (1)

Klinefelter Syndrome A male whose cells have an extra X chromosome (XXY). The syndrome is characterized by underdeveloped male genitalia and feminine secondary sex characteristics. (17)

Krebs Cycle A circular pathway in aerobic respiration that completely oxidizes the two pyruvic acids from glycolysis. (9)

K-Selected Species Species that produce one or a few well-cared for individuals at a time. (43)

◄ L ►

Lacteal Blind lymphatic vessel in the intestinal villi that receives the absorbed products of lipid digestion. (27)

Lactic Acid Fermentation The process in which electrons removed during glycolysis are transferred from NADH to pyruvic acid to form lactic acid. Used by various prokaryotic cells under oxygen-deficient conditions and by muscle cells during strenuous activity. (9)

Lake Large body of standing fresh water, formed in natural depressions in the Earth. Lakes are larger than ponds. (40)

Lamella In bone, concentric cylinders of calcified collagen deposited by the osteocytes. The laminated layers produce a greatly strengthened structure. (26)

Large Intestine Portion of the intestine in which water and salts are reabsorbed. It is so named because of its large diameter. The large instestine, except for the rectum, is called the colon. (27)

Larva A self-feeding, sexually, and developmentally immature form of an animal. (32)

Larynx The short passageway connecting the pharynx with the lower airways. (29)

Latent (hidden) Infection Infection by a microorganism that causes no symptoms but the microbe is well-established in the body. (36)

Lateral Roots Roots that arise from the pericycle of older roots; also called branch roots or secondary roots. (18)

Law of Independent Assortment Alleles on nonhomologous chromosomes segregate independently of one another. (12)

Law of Segregation During gamete formation, pairs of alleles separate so that each sperm or egg cell has only one gene for a trait. (12)

Law of the Minimum The ecological principle that a species' distribution will be limited by whichever abiotic factor is most deficient in the environment. (41)

Laws of Thermodynamics Physical laws that describe the relationship of heat and mechanical energy. The first law states that energy cannot be created or destroyed, but one form can change into another. The second law states that the total energy the universe decreases as energy conversions occur and some energy is lost as heat. (6)

Leak Channels Passageways through a plasma membrane that do not contain gates and, therefore, are always open for the limited diffusion of a specific substance (ion) through the membrane. (7, 23)

Learning A process in which an animal benefits from experience so that its behavior is better suited to environmental conditions. (44)

Lenticels Loosely packed cells in the periderm of the stem that create air channels for transferring CO_2, H_2O, and O_2. (18)

Leukocytes White blood cells. (28)

LH Luteinizing hormone. A hormone secreted by the anterior pituitary that stimulates testosterone production in males and triggers ovulation and the transformation of the follicle into the corpus luteum in females. (31)

Lichen Symbiotic associations between certain fungi and algae. (37)

Life Cycle The sequence of events during the lifetime of an organism from zygote to reproduction. (39)

Ligaments Strong straps of connective tissue that hold together the bones in articulating joints or support an organ in place. (26)

Light-Dependent Reactions First stage of photosynthesis in which light energy is converted to chemical energy in the form of energy-rich ATP and NADPH. (8)

Light-Independent Reactions Second stage of photosynthesis in which the energy stored in ATP and NADPH formed in the light reactions is used to drive the reactions in which carbon dioxide is converted to carbohydrate. (8)

Limb Bud A portion of an embryo that will develop into either a forelimb or hindlimb. (32)

Limbic System A series of an interconnected group of brain structures, including the thalamus and hypothalamus, controlling memory and emotions. (23)

Limiting Factors The critical factors which impose restraints of the distribution, health, or activities of an organism. (41)

Limnetic Zone Open water of lakes, through which sunlight penetrates and photosynthesis occurs. (40)

Linkage The tendency of genes of the same chromosome to stay together rather than to assort independently. (13)

Linkage Groups Groups of genes located on the same chromosome. The genes of each linkage group assort independently of the genes of other linkage groups. In all eukaryotic organisms, the number of linkage groups is equal to the haploid number of chromosomes. (13)

Lipids A diverse group of biomolecules that are insoluble in water. (4)

Lithosphere The solid outer zone of the Earth; composed of the crust and outermost portion of the mantle. (40)

Littoral Zone Shallow, nutrient-rich waters of a lake, where sunlight reaches the bottom; also the lakeshore. Rooted vegetation occurs in this zone. (40)

Locomotion The movement of an organism from one place to another. (26)

Locus The chromosomal location of a gene. (13)

Logistic Growth Population growth producing a sigmoid, or S-shaped, growth curve. (43)

Long-Day Plants Plants that flower when the length of daylight exceeds some critical period. (21)

Longitudinal Fission The division pattern in flagellated protozoans, where division is along the length of the cell.

Loop of Henle An elongated section of the renal tubule that dips down into the kidney's medulla and then ascends back out to the cortex. It separates the proximal and distal convoluted tubules and is responsible for forming the salt gradient on which water reabsorption in the kidney depends. (28)

Low Density Lipoprotein (LDL) Particles that transport cholesterol in the blood. Each particle consists of about 1,500 cholesterol molecules surrounded by a film of phospholipids and protein. LDLs are taken into cells following their binding to cell surface LDL receptors. (7)

Low Intertidal Zone In the intertidal zone, the region which is uncovered by "minus" tides only. Organisms are submerged about 90% of the time. (40)

Lumen A space within an hollow organ or tube. (28)

Luminescence (see **Bioluminescence**)

Lungs The organs of terrestrial animals where gas exchange occurs. (29)

Lymph The colorless fluid in lymphatic vessels. (28)

Lymphatic System Network of fluid-carrying vessels and associated organs that participate in immunity and in the return of tissue fluid to the main circulation. (28)

Lymphocytes A group of non-phagocytic white blood cells which combat microbial invasion, fight cancer, and neutralize toxic chemicals. The two classes of lymphocytes, B cells and T cells, are the heart of the immune system. (28, 30)

Lymphoid Organs Organs associated with production of blood cells and the lymphatic system, including the thymus, spleen, appendix, bone marrow, and lymph nodes. (30)

Lysis (1) To split or dissolve. (2) Cell bursting.

Lysomes A type of storage vesicle produced by the Golgi complex, containing hydrolytic (digestive) enzymes capable of digesting many kinds of macromolecules in the cell. The membrane around them keeps them sequestered. (5)

◀ M ▶

M Phase That portion of the cell cycle during which mitosis (nuclear division) and cytokinesis (cytoplasmic division) takes place. (10)

Macroevolution Evolutionary changes that lead to the appearance of new species. (33)

Macrofungus Filamentous fungus so named for the large size of its fleshy sexual structures; a mushroom, for example. (37)

Macromolecules Large polymers, such as proteins, nucleic acids, and polysaccharides. (4)

Macronutrients Nutrients required by plants in large amounts: carbon, oxygen, hydrogen, nitrogen, potassium, calcium, phosphorus, magnesium, and sulfur. (19)

Macrophages Phagocytic cells that develop from monocytes and present antigen to lymphocytes. (30)

Macroscopic Referring to biological observations made with the naked eye or a hand lens.

Mammals A class of vertebrates that possesses skin covered with hair and that nourishes their young with milk from mammary glands. (39)

Mammary Glands Glands contained in the breasts of mammalian mothers that produce breast milk. (39)

Marsupials Mammals with a cloaca whose young are born immature and complete their development in an external pouch in the mother's skin. (39)

Mass Extinction The simultaneous extinction of a multitude of species as the result of a drastic change in the environment. (33, 35)

Maternal Chromosomes The set of chromosomes in an individual that were inherited from the mother. (11)

Mechanoreceptors Sensory receptors that respond to mechanical pressure and detect motion, touch, pressure, and sound. (24)

Medulla The center-most portion of some organs. (23)

Medusa The motile, umbrella-shaped body form of some members of the phylum Cnidaria, with mouth and tentacles on the lower, concave service. (Compare with polyp.) (39)

Megaspores Spores that divide by mitosis to produce female gametophytes that produce the egg gamete. (20)

Meiosis The division process that produces cells with one-half the number of chromosomes in each somatic cell. Each resulting daughter cell is haploid (1N) (11)

Meiosis I A process of reductional division in which homologous chromosomes pair and then segregate. Homologues are partitioned into separate daughter cells. (11)

Meiosis II Second meiotic division. A division process resembling mitosis, except that the haploid number of chromosomes is present. After the chromosomes line up at the meta-phase plate, the two sister chromatids separate. (11)

Melanin A brown pigment that gives skin and hair its color (12)

Melanoma A deadly form of skin cancer that develops from pigment cells in the skin and is promoted by exposure to the sun. (14)

Memory Cells Lymphocytes responsible for active immunity. They recall a previous exposure to an antigen and, on subsequent exposure to the same antigen, proliferate rapidly into plasma cells and produce large quantities of antibodies in a short time. This protection typically lasts for many years. (30)

Mendelian Inheritance Transmission of genetic traits in a manner consistent with the principles discovered by Gregor Mendel. Includes traits controlled by simple dominant or recessive alleles; more complex patterns of transmission are referred to as Nonmendelian inheritance. (12)

Meninges The thick connective tissue sheath which surrounds and protects the vertebrate brain and spinal cord. (23)

Menstrual Cycle The repetitive monthly changes in the uterus that prepare the endometrium for receiving and supporting an embryo. (31)

Meristematic Region New cells arise from this undifferentiated plant tissue; found at root or shoot apical meristems, or lateral meristems. (18)

Meristems In plants, clusters of cells that retain their ability to divide, thereby producing new cells. One of the four basic tissues in plants. (18)

Mesoderm In animals, the middle germ cell layer of the gastrula. It gives rise to muscle, bone, connective tissue, gonads, and kidney. (32)

Mesopelagic Zone The dimly lit ocean zone beneath the epipelagic zone; large fishes, whales and squid occupy this zone; no phytoplankton occur in this zone. (40)

Mesophyll Layers of cells in a leaf between the upper and lower epidermis; produced by the ground meristem. (18)

Messenger RNA (mRNA) The RNA that carries genetic information from the DNA in the nucleus to the ribosomes in the cytoplasm, where the sequence of bases in the mRNA is translated into a sequence of amino acids. (14)

Metabolic Intermediates Compounds produced as a substrate are converted to end product in a series of enzymatic reactions. (6)

Metabolic Pathways Set of enzymatic reactions involved in either building or dismantling complex molecules. (6)

Metabolic Rate A measure of the level of activity of an organism usually determined by measuring the amount of oxygen consumed by an individual per gram body weight per hour. (22)

Metabolic Water Water produced as a product of metabolic reactions. (28)

Metabolism The sum of all the chemical reactions in an organism; includes all anabolic and catabolic reactions. (6)

Metamorphosis Transformation from one form into another form during development. (32)

Metaphase The stage of mitosis when the chromosomes line-up along the metaphase plate, a plate that usually lies midway between the spindle poles. (10)

Metaphase Plate Imaginary plane within a dividing cell in which the duplicated chromosomes become aligned during metaphase. (10)

Microbes Microscopic organisms. (36)

Microbiology The branch of biology that studies microorganisms. (36)

Microevolution Changes in allele frequency of a species' gene pool which has not generated new species. Exemplified by changes in the pigmentation of the peppered moth and by the acquisition of pesticide resistance in insects. (33)

Microfibrils Bundles formed from the intertwining of cellulose molecules, i.e., long chains of glucose molecules in the cell walls of plants. (5)

Microfilaments Thin actin-containing protein fibers that are responsible for maintenance of cell shape, muscle contraction and cyclosis. (5)

Micrometer One millionth (1/1,000,000) of a meter.

Micronutrients Nutrients required by plants in small amounts: iron, chlorine, copper, manganese, zinc, molybdenum, and boron. (19)

Micropyle A small opening in the integuments of the ovule through which the pollen tube grows to deliver sperm. (21)

Microspores Spores within anthers of flowers. They divide by mitosis to form pollen grains, the male gametophytes that produce the plant's sperm. (20)

Microtubules Thin, hollow tubes in cells; built from repeating protein units of tubulin. Microtubules are components of cilia, flagella, and the cytoskeleton. (5)

Microvilli The small projections on the cells that comprise each villus of the intestinal wall, further increasing the absorption surface area of the small intestine. (27)

Middle Intertidal Zone In the intertidal zone, the region which is covered and uncovered twice a day, the zero of tide tables. Organisms are submerged about 50% of the time. (40)

Migration Movements of a population into or out of an area. (44)

Mimicry A defense mechanism where one species resembles another in color, shape, behavior, or sound. (42)

Mineralocorticoids Steroid hormones which regulate the level of sodium and potassium in the blood. (25)

Mitochondria Organelles that contain the biochemical machinery for the Krebs cycle and the electron transport system of aerobic respiration. They are composed of two membranes, the inner one forming folds, or cristae. (9)

Mitosis The process of nuclear division producing daughter cells with exactly the same number of chromosomes as in the mother cell. (10)

Mitosis Promoting Factor (MPF) A protein that appears to be a universal trigger of cell division in eukaryotic cells. (10)

Mitotic Chromosomes Chromosomes whose DNA-protein threads have become coiled into microscopically visible chromosomes, each containing duplicated chromatids ready to be separated during mitosis. (10)

Molds Filamentous fungi that exist as colonies of threadlike cells but produce no macroscopic fruiting bodies. (37)

Molecule Chemical substance formed when two or more atoms bond together; the smallest unit of matter that possesses the qualities of a compound. (3)

Mollusca A phylum, second only to Arthropoda in diversity. Composed of three main classes: 1) Gastropoda (spiral-shelled), 2) Bivalvia (hinged shells), 3) Cephalopoda (with tentacles or arms and no, or very reduced shells). (39)

Molting (Ecdysis) Shedding process by which certain arthropods lose their exoskeletons as their bodies grow larger. (39)

Monera The taxonomic kingdom comprised of single-celled prokaryotes such as bacteria, cyanobacteria, and archebacteria. (36)

Monoclonal Antibodies Antibodies produced by a clone of hybridoma cells, all of which descended from one cell. (30)

Monocotyledae (Monocots) One of the two divisions of flowering plants, characterized by seeds with a single cotyledon, flower parts in 3s, parallel veins in leaves, many roots of approximately equal size, scattered vascular bundles in its stem anatomy, pith in its root anatomy, and no secondary growth capacity. (18)

Monocytes A type of leukocyte that gives rise to macrophages. (28)

Monoecious Both male and female reproductive structures are produced on the same sporophyte individual. (20, 38)

Monohybrid Cross A mating between two individuals that differ only in one genetically-determined trait. (12)

Monomers Small molecular subunits which are the building blocks of macromolecules. The macromolecules in living systems are constructed of some 40 different monomers. (4)

Monotremes A group of mammals that lay eggs from which the young are hatched. (39)

Morphogenesis The formation of form and internal architecture within the embryo brought about by such processes as programmed cell death, cell adhesion, and cell movement. (32)

Morphology The branch of biology that studies form and structure of organisms.

Mortality Deathrate in a population or area. (43)

Motile Capable of independent movement.

Motor Neurons Nerve cells which carry outgoing impulses to their effectors, either glands or muscles. (23)

Mucosa The cell layer that lines the digestive tract and secretes a lubricating layer of mucus. (27)

Mullerian Mimicry Resemblance of different species, each of which is equally obnoxious to predators. (42)

Multicellular Consisting of many cells. (35)

Multichannel Food Chain Where the same primary producer supplies the energy for more than one food chain. (41)

Multiple Allele System Three or more possible alleles for a given trait, such as ABO blood groups in humans. (12)

Multiple Fission Division of the cell's nucleus without a corresponding division of cytoplasm.

Multiple Fruits Fruits that develop from pistils of separate flowers. (20)

Muscle Fiber A muiltinucleated skeletal muscle cell that results from the fusion of several pre-muscle cells during embryonic development. (26)

Muscle Tissue Bundles and sheets of contractile cells that shorten when stimulated, providing force for controlled movement. (26)

Mutagens Chemical or physical agents that induce genetic change. (14)

Mutation Random heritable changes in DNA that introduce new alleles into the gene pool. (14)

Mutualism The symbiotic interaction in which both participants benefit. (42)

Mycology The branch of biology that studies fungi. (37)

Mycorrhizae An association between soil fungi and the roots of vascular plants, increasing the plant's ability to extract water and minerals from the soil. (19)

Myelin Sheath In vertebrates, a jacket which covers the axons of high-velocity neurons, thereby increasing the speed of a neurological impulse. (23)

Myofibrils In striated muscle, the banded fibrils that lie parallel to each other, constituting the bulk of the muscle fiber's interior and powering contraction. (26)

Myosin A contractile protein that makes up the major component of the thick filaments of a muscle cell and is also present in nonmuscle cells. (26)

◄ N ►

NADPH Nicotinamide adenine dinucleotide phosphate. NADPH is formed by reduction of $NADP^+$, and serves as a store of electrons for use in metabolism (see Reducing Power). (9)

NAD^+ Nicotinamide adenine dinucleotide. A coenzyme that functions as an electron carrier in metabolic reactions. When reduced to NADH, the molecule becomes a cellular energy source. (9)

Natality Birthrate in a population or area. (43)

Natural Killer (NK) Cells Nonspecific, lymphocytelike cells which destroy foreign cells and cancer cells. (30)

Natural Selection Differential survival and reproduction of organisms with a resultant increase in the frequency of those best adapted to the environment. (33)

Neanderthals A subspecies of Homo sapiens different from that of modern humans that were characterized by heavy bony skeletons and thick bony ridges over the eyes. They disappeared about 35,000 years ago. (34)

Nectary Secretory gland in flowering plants containing sugary fluid that attracts pollinators as a food source. Usually located at the base of the flower. (20)

Negative Feedback Any regulatory mechanism in which the increased level of a substance inhibits further production of that substance, thereby preventing harmful accumulation. A type of homeostatic mechanism. (22, 25)

Negative Gravitropism In plants, growth against gravitational forces, or shoot growth upward. (21)

Nematocyst Within the stinging cell (cnidocyte) of cnidarians, a capsule that contains a coiled thread which, when triggered, harpoons prey and injects powerful toxins. (39)

Nematoda The widespread and abundant animal phylum containing the roundworms. (39)

Nephridium A tube surrounded by capillaries found in an organism's excretory organs that removes nitrogenous wastes and regulates the water and chemical balance of body fluids. (28)

Nephron The functional unit of the vertebrate kidney, consisting of the glomerulus, Bowman's capsule, proximal and distal convoluted tubules, and loop of Henle. (28)

Nerve Parallel bundles of neurons and their supporting cells. (23)

Nerve Impulse A propagated action potential. (23)

Nervous Tissue Excitable cells that receive stimuli and, in response, transmit an impulse to another part of the animal. (23)

Neural Plate In vertebrates, the flattened plate of dorsal ectoderm of the late gastrula that gives rise to the nervous system. (32)

Neuroglial Cells Those cells of a vertebrate nervous system that are not neurons. Includes a variety of cell types including Schwann cells. (23)

Neuron A nerve cell. (23)

Neurosecretory Cells Nervelike cells that secrete hormones rather than neurotransmitter substances when a nerve impulse reaches the distal end of the cell. In vertebrates, these cells arise from the hypothalamus. (25)

Neurotoxins Substances, such as curare and tetanus toxin, that interfere with the transmission of neural impulses. (23)

Neurotransmitters Chemicals released by neurons into the synaptic cleft, stimulating or inhibiting the post-synaptic target cell. (23)

Neurulation Formation by embryonic induction of the neural tube in a developing vertebrate embryo. (32)

Neutrons Electrically neutral (uncharged) particles contained within the nucleus of the atom. (3)

Neutrophil Phagocytic leukocyte, most numerous in the human body. (28)

Niche An organism's habitat, role, resource requirements, and tolerance ranges for each abiotic condition. (42)

Niche Breadth Relative size and dimension of ecological niches; for example, broad or narrow niches. (41)

Niche Overlap Organisms that have the same habitat, role, environmental requirements, or needs. (41)

Nitrogen Fixation The conversion of atmospheric nitrogen gas N_2 into ammonia (NH_3) by certain bacteria and cyanobacteria. (19)

Nitrogenous Wastes Nitrogen-containing metabolic waste products, such as ammonia or urea, that are produced by the breakdown of proteins and nucleic acids. (28)

Nodes The attachment points of leaves to a stem. (18)

Nodes of Ranvier Uninsulated (nonmyelinated) gaps along the axon of a neuron. (23)

Noncovalent Bonds Linkages between two atoms that depend on an attraction between positive and negative charges between molecules or ions. Includes ionic and hydrogen bonds. (3)

Non-Cyclic Photophosphorylation The pathway in the light reactions of photosynthesis in which electrons pass from water, through two photosystems, and then ultimately to $NADP^+$. During the process, both ATP and NADPH are produced. It is so named because the electrons do not return to their reaction center. (8)

Nondisjunction Failure of chromosomes to separate properly at meiosis I or II. The result is that one daughter will receive an extra chromosome and the other gets one less. (11, 13)

Nonpolar Molecules Molecules which have an equal charge distribution throughout their structure and thus lack regions with a localized positive or negative charge. (3)

Notochord A flexible rod that is below the dorsal surface of the chordate embryo, beneath the nerve cord. In most chordates, it is replaced by the vertebral column. (32)

Nuclear Envelope A double membrane pierced by pores that separates the contents of the nucleus from the rest of the eukaryotic cell. (5)

Nucleic Acids DNA and RNA; linear polymers of nucleotides, responsible for the storage and expression of genetic information. (4, 14)

Nucleoid A region in the prokaryotic cell that contains the genetic material (DNA). It is unbounded by a nuclear membrane. (36)

Nucleoplasm The semifluid substance of the nucleus in which the particulate structures are suspended. (5)

Nucleosomes Nuclear protein complex consisting of a length of DNA wrapped around a central cluster of 8 histones. (14)

Nucleotides Monomers of which DNA and RNA are built. Each consists of a 5-carbon sugar, phosphate, and a nitrogenous base. (4)

Nucleous (pl. nucleoli) One or more darker regions of a nucleus where each ribosomal subunit is assembled from RNA and protein. (5)

Nucleus The large membrane-enclosed organelle that contains the DNA of eukaryotic cells. (5)

Nucleus, Atomic The center of an atom containing protons and neutrons. (3)

◀ O ▶

Obligate Symbiosis A symbiotic relationship between two organisms that is necessary for the survival or both organisms. (42)

Olfaction The sense of smell. (24)

Oligotrophic Little nourished, as a young lake that has few nutrients and supports little life. (40)

Omnivore An animal that obtains its nutritional needs by consuming plants and other animals. (42)

Oncogene A gene that causes cancer, perhaps activated by mutation or a change in its chromosomal location. (10)

Oocyte A female germ cell during any of the stages of meiosis. (31)

Oogenesis The process of egg production. (31)

Oogonia Female germ cells that have not yet begun meiosis. (31)

Open Circulatory System Circulatory system in which blood travels from vessels to tissue spaces, through which it percolates prior to returning to the main vessel (compare with closed circulatory system). (28)

Operator A regulatory gene in the operon of bacteria. It is the short DNA segment to which the repressor binds, thus preventing RNA polymerase from attaching to the promoter. (15)

Operon A regulatory unit in prokaryotic cells that controls the expression of structural genes. The operon consists of structural genes that produce enzymes for a particular metabolic pathway, a regulator region composed of a promoter and an operator, and R (regulator) gene that produces a repressor. (15)

Order A level of the taxonomic hierarchy that groups together members of related families. (1)

Organ Body part composed of several tissues that performs specialized functions. (22)

Organelle A specialized part of a cell having some particular function. (5)

Organic Compounds Chemical compounds that contain carbon. (4)

Organism A living entity able to maintain its organization, obtain and use energy, reproduce, grow, respond to stimuli, and display homeostatis. (1)

Organogenesis Organ formation in which two or more specialized tissue types develop in a precise temporal and spatial relationship to each other. (32)

Organ System Group of functionally related organs. (22)

Osmoregulation The maintenance of the proper salt and water balance in the body's fluids. (28)

Osmosis The diffusion of water through a differentially permeable membrane into a hypertonic compartment. (7)

Ossification Synthesis of a new bone. (26)

Osteoclast A type of bone cell which breaks down the bone, thereby releasing calcium into the bloodstream for use by the body. Osteoclasts are activated by hormones released by the parathyroid glands. (26)

Osteocytes Living bone cells embedded within the calcified matrix they manufacture. (26)

Osteoporosis A condition present predominantly in postmenopausal women where the bones are weakened due to an increased rate of bone resorption compared to bone formation. (26)

Ovarian Cycle The cycle of egg production within the mammalian ovary. (31)

Ovarian Follicle In a mammalian ovary, a chamber of cells in which the oocyte develops. (31)

Ovary In animals, the egg-producing gonad of the female. In flowering plants, the enlarged base of the pistil, in which seeds develop. (20)

Oviduct (Fallopian Tube) The tube in the female reproductive organ that connects the ovaries and uterus and where fertilization takes place. (31)

Ovulation The release of an egg (ovum) from the ovarian follicle. (31)

Ovule In seed plants, the structure containing the female gametophyte, nucellus, and integuments. After fertilization, the ovule develops into a seed. (20, 38)

Ovum An unfertilized egg cell; a female gamete. (31)

Oxidation The removal of electrons from a compound during a chemical reaction. For a carbon atom, the fewer hydrogens bonded to a carbon, the greater the oxidation state of the atom. (6)

Oxidative Phosphorylation The formation of ATP from ADP and inorganic phosphate that occurs in the electron-transport chain of cellular respiration. (8, 9)

Oxyhemoglobin A complex of oxygen and hemoglobin, formed when blood passes through the lungs and is dissociated in body tissues, where oxygen is released. (29)

Oxytocin A female hormone released by the posterior pituitary which triggers uterine contractions during childbirth and the release of milk during nursing. (25)

◀ P ▶

P680 Reaction Center (P = Pigment) Special chlorophyll molecule in Photosystem II that traps the energy absorbed by the other pigment molecules. It absorbs light energy maximally at 680 nm. (8)

Palisade Parenchyma In dicot leaves, densely packed, columnar shaped cells functioning in photosynthesis. Found just beneath the upper epidermis. (18)

Pancreas In vertebrates, a large gland that produces digestive enzymes and hormones. (27)

Parallel Evolution When two species that have descended from the same ancestor independently acquire the same evolutionary adaptations. (33)

Parapatric Speciation The splitting of a population into two species' populations under conditions where the members of each population reside in adjacent areas. (33)

Parasite An organism that lives on or inside another called a host, on which it feeds. (39, 42)

Parasitism A relationship between two organisms where one organism benefits, and the other is harmed. (42)

Parasitoid Parasitic organisms, such as some insect larvae, which kill their host. (42)

Parasympathetic Nervous System Part of the autonomic nervous system active during relaxed activity. (23)

Parathyroid Glands Four glands attached to the thyroid gland which secrete parathyroid hormone (PTH). When blood calcium levels are low, PTH is secreted, causing calcium to be released from bone. (25)

Parenchyma The most prevalent cell type in herbaceous plants. These thin-walled, polygonal-shaped cells function in photosynthesis and storage. (18)

Parthenogenesis Process by which offspring are produced without egg fertilization. (31)

Passive Immunity Immunity achieved by receiving antibodies from another source, as occurs with a newborn infant during nursing. (30)

Paternal Chromosomes The set of chromosomes in an individual that were inherited from the father. (11)

Pathogen A disease-causing microorganism. (36)

Pectoral Girdle In humans, the two scapulae (shoulder blades) and two clavicles (collarbones) which support and articulate with the bones of the upper arm. (26)

Pedicel A shortened stem carrying a flower. (20)

Pedigree A diagram showing the inheritance of a particular trait among the members of the family. (13)

Pelagic Zone The open oceans, divided into three layers: 1) photo- or epipelagic (sunlit), 2) mesopelagic (dim light), 3) aphotic or bathypelagic (always dark). (40)

Pelvic Girdle The complex of bones that connect a vertebrate's legs with its backbone. (26)

Penis An intrusive structure in the male animal which releases male gametes into the female's sex receptacle. (31)

Peptide Bond The covalent bond between the amino group of one amino acid and the carboxyl group of another. (4)

Peptidoglycan A chemical component of the prokaryotic cell wall. (36)

Percent Annual Increase A measure of population increase; the number of individuals (people) added to the population per 100 individuals. (43)

Perennials Plants that live longer than two years. (18)

Perfect Flower Flowers that contain both stamens and pistils. (20)

Perforation Plate In plants, that portion of the wall of vessel members that is perforated, and contains an area with neither primary nor secondary cell wall; a "hole" in the cell wall. (18)

Pericycle One or more layers of cells found in roots, with phloem or xylem to its' inside, and the endodermis to its' outside. Functions in producing lateral roots and formation of the vascular cambium in roots with secondary growth. (18)

Periderm Secondary tissue that replaces the epidermis of stems and roots. Consists of cork, cork cambium, and an internal layer of parenchyma cells. (18)

Peripheral Nervous System Neurons, excluding those of the brain and spinal cord, that permeate the rest of the body. (23)

Peristalsis Sequential waves of muscle contractions that propel a substance through a tube. (27)

Peritoneum The connective tissue that lines the coelomic cavities. (39)

Permeability The ability to be penetrable, such as a membrane allowing molecules to pass freely across it. (7)

Petal The second whorl of a flower, often brightly colored to attract pollinators; collectively called the corolla. (20)

Petiole The stalk leading to the blade of a leaf. (18)

pH A scale that measures the concentration of hydrogen ions in a solution. The pH scale extends from 0 to 14. Acidic solutions have a pH of less than 7; alkaline solutions have a pH above 7; neutral solutions have a pH equal to 7. (3)

Phagocytosis Engulfing of food particles and foreign cells by amoebae and white blood cells. A type of endocytosis. (5)

Pharyngeal Pouches In the vertebrate embryo, outgrowths from the walls of the pharynx that break through the body surface to form gill slits. (32)

Pharynx The throat; a portion of both the digestive and respiratory system just behind the oral cavity. (29)

Phenotype An individual's observable characteristics that are the expression of its genotype. (12)

Pheromones Chemicals that, when released by an animal, elicit a particular behavior in other animals of the same species. (44)

Phloem The vascular tissue that transports sugars and other organic molecules from sites of photosynthesis and storage to the rest of the plant. (18)

Phloem Loading The transfer of assimilates to phloem conducting cells, from photosynthesizing source cells. (19)

Phloem Unloading The transfer of assimilates to storage (sink) cells, from phloem conducting cells. (19)

Phospholipids Lipids that contain a phosphate and a variable organic group that form polar, hydrophilic regions on an otherwise nonpolar, hydrophobic molecule. They are the major structural components of membranes. (4)

Phosphorylation A chemical reaction in which a phosphate group is added to a molecule or atom. (6)

Photoexcitation Absorption of light energy by pigments, causing their electrons to be raised to a higher energy level. (8)

Photolysis The splitting of water during photosynthesis. The electrons from water pass to Photosystem II, the protons enter the lumen of the thylakoid and contribute to the proton gradient across the thylakoid membrane, and the oxygen is released into the atmosphere. (8)

Photon A particle of light energy. (8)

Photoperiod Specific lengths of day and night which control certain plant growth responses to light, such as flowering or germination. (21)

Photoperiodism Changes in the behavior and physiology of an organism in response to the relative lengths of daylight and darkness, i.e., the photoperiod. (21)

Photoreceptors Sensory receptors that respond to light. (24)

Photorespiration The phenomenon in which oxygen binds to the active site of a CO_2-fixing enzyme, thereby competing with CO_2 fixation, and lowering the rate of photosynthesis. (8)

Photosynthesis The conversion by plants of light energy into chemical energy stored in carbohydrate. (8)

Photosystems Highly organized clusters of photosynthetic pigments and electron/hydrogen carriers embedded in the thylakoid membranes of chloroplasts. There are two photosystems, which together carry out the light reactions of photosynthesis. (8)

Photosystem I Photosystem with a P700 reaction center; participates in cyclic photophosphorylation as well as in noncyclic photophosphorylation. (8)

Photosystem II Photosystem activated by a P680 reaction center; participates only in noncyclic photophosphorylation and is associated with photolysis of water. (8)

Phototropism The growth responses of a plant to light. (21)

Phyletic Evolution The gradual evolution of one species into another. (33)

Phylogeny Evolutionary history of a species. (35)

Phylum The major taxonomic divisions in the Animal kingdom. Members of a phylum share common, basic features. The Animal kingdom is divided into approximately 35 phyla. (39)

Physiology The branch of biology that studies how living things function. (22)

Phytochrome A light-absorbing pigment in plants which controls many plant responses, including photoperiodism. (21)

Phytoplankton Microscopic photosynthesizers that live near the surface of seas and bodies of fresh water. (37)

Pineal Gland An endocrine gland embedded within the brain that secretes the hormone melatonin. Hormone secretion is dependent on levels of environmental light. In amphibians and reptiles, melatonin controls skin coloration. In humans, pineal secretions control sexual maturation and daily rhythms. (25)

Pinocytosis Uptake of small droplets and dissolved solutes by cells. A type of endocytosis. (5)

Pistil The female reproductive part and central portion of a flower, consisting of the ovary, style and stigma. May contain one carpel, or one or more fused carpels. (20)

Pith A plant tissue composed of parenchyma cells, found in the central portion of primary growth stems of dicots, and monocot roots. (18)

Pith Ray Region between vascular bundles in vascular plants. (18)

Pituitary Gland (see **Posterior and Anterior Pituitary).**

Placenta In mammals (exclusive of marsupials and monotremes), the structure through which nutrients and wastes are exchanged between the mother and embryo/fetus. Develops from both embryonic and uterine tissues. (32)

Plant Multicellular, autotrophic organism able to manufacture food through photosynthesis. (38)

Plasma In vertebrates, the liquid portion of the blood, containing water, proteins (including fibrinogen), salts, and nutrients. (28)

Plasma Cells Differentiated antibody-secreting cells derived from B lymphocytes. (30)

Plasma Membrane The selectively permeable, molecular boundary that separates the cytoplasm of a cell from the external environment. (5)

Plasmid A small circle of DNA in bacteria in addition to its own chromosome. (16)

Plasmodesmata Openings between plant cell walls, through which adjacent cells are connected via cytoplasmic threads. (19)

Plasmodium Genus of protozoa that causes malaria. (37)

Plasmodium A huge multinucleated "cell" stage of a plasmodial slime mold that feeds on dead organic matter. (37)

Plasmolysis The shrinking of a plant cell away from its cell wall when the cell is placed in a hypertonic solution. (7)

Platelets Small, cell-like fragments derived from special white blood cells. They function in clotting. (28)

Plate Tectonics The theory that the earth's crust consists of a number of rigid plates that rest on an underlying layer of semimolten rock. The movement of the earth's plates results from the upward movement of molten rock into the solidified crust along ridges within the ocean floor. (35)

Platyhelminthes The phylum containing simple, bilaterally symmetrical animals, the flatworms. (39)

Pleiotropy Where a single mutant gene produces two or more phenotypic effects. (12)

Pleura The double-layered sac which surrounds the lungs of a mammal. (29, 39)

Pneumocytis Pneumonia (PCP) A disease of the respiratory tract caused by a protozoan that strikes persons with immunodeficiency diseases, such as AIDS. (30)

Point Mutations Changes that occur at one point within a gene, often involving one nucleotide in the DNA. (14)

Polar Body A haploid product of meiosis of a female germ cell that has very little cytoplasm and disintegrates without further function. (31)

Polar Molecule A molecule with an unequal charge distribution that creates distinct positive and negative regions or poles. (3)

Pollen The male gametophyte of seed plants, comprised of a generative nucleus and a tube nucleus surrounded by a tough wall. (20)

Pollen Grain The male gametophyte of conifers and angiosperms, containing male gametes. In angiosperms, pollen grains are contained in the pollen sacs of the anther of a flower. (20)

Pollination The transfer of pollen grains from the anther of one flower to the stigma of another. The transfer is mediated by wind, water, insects, and other animals. (20)

Polygenic Inheritance An inheritance pattern in which a phenotype is determined by two or more genes at different loci. In humans, examples include height and pigmentation. (12)

Polymer A macromolecule formed of monomers joined by covalent bonds.. Includes proteins, polysaccharides, and nucleic acids. (4)

Polymerase Chain Reaction (PCR) Technique to amplify a specific DNA molecule using a temperature-sensitive DNA polymerase obtained from a heat-resistant bacterium. Large numbers of copies of the initial DNA molecule can be obtained in a short period of time, even when the starting material is present in vanishingly small amounts, as for example from a blood stain left at the scene of a crime. (16)

Polymorphic Property of some protozoa to produce more than one stage of organism as they complete their life cycles. (37)

Polymorphic Genes Genes for which several different alleles are known, such as those that code for human blood type. (17)

Polyp Stationary body form of some members of the phylum Cnidaria, with mouth and tentacles facing upward. (Compare with medusa.) (39)

Polypeptide An unbranched chain of amino acids covalently linked together and assembled on a ribosome during translation. (4)

Polyploidy An organism or cell containing three or more complete sets of chromosomes. Polyploidy is rare in animals but common in plants. (33)

Polysaccharide A carbohydrate molecule consisting of monosaccharide units. (4)

Polysome A complex of ribosomes found in chains, linked by mRNA. Polysomes contain the ribosomes that are actively assembling proteins. (14)

Polytene Chromosomes Giant banded chromosomes found in certain insects that form by the repeated duplication of DNA. Because of the multiple copies of each gene in a cell, polytene chromosomes can generate large amounts of a gene product in a short time. Transcription occurs at sites of chromosome puffs. (13)

Pond Body of standing fresh water, formed in natural depressions in the Earth. Ponds are smaller than lakes. (40)

Population Individuals of the same species inhabiting the same area. (43)

Population Density The number of individual species living in a given area. (43)

Positive Gravitropism In plants, growth with gravitational forces, or root growth downward. (21)

Posterior Pituitary A gland which manufactures no hormones but receives and later releases hormones produced by the cell bodies of neurons in the hyopthalamus. (25)

Potential Energy Stored energy, such as occurs in chemical bonds. (6)

Preadaptation A characteristic (adaptation) that evolved to meet the needs of an organism in one type of habitat, but fortuitously allows the organism to exploit a new habitat. For example, lobed fins and lungs evolved in ancient fishes to help them live in shallow, stagnant ponds, but also facilitated the evolution of terrestrial amphibians. (33, 39)

Precells Simple forerunners of cells that, presumably, were able to concentrate organic molecules, allowing for more frequent molecular reactions. (35)

Predation Ingestion of prey by a predator for energy and nutrients. (42)

Predator An organism that captures and feeds on another organism (prey). (42)

Pressure Flow In the process of phloem loading and unloading, pressure differences resulting from solute increases in phloem conducting cells and neighboring xylem cells cause the flow of water to phloem. A concentration gradient is created between xylem and phloem cells. (19)

Prey An organism that is captured and eaten by another organism (predator). (42)

Primary Consumer Organism that feeds exclusively on producers (plants). Herbivores are primary consumers. (41)

Primary Follicle In the mammalian ovary, a structure composed of an oocyte and its surrounding layer of follicle cells. (31)

Primary Growth Growth from apical meristems, resulting in an increase in the lengths of shoots and roots in plants. (18)

Primary Immune Response Process of antibody production following the first exposure to an antigen. There is a lag time from exposure until the appearance in the blood of protective levels of antibodies. (30)

Primary Oocyte Female germ cell that is either in the process of or has completed the first meiotic division. In humans, germ cells may remain in this stage in the ovary for decades. (31)

Primary Producers All autotrophs in a biotic environment that use sunlight or chemical energy to manufacture food from inorganic substances. (41)

Primary Sexual Characteristics Gonads, reproductive tracts, and external genitals. (31)

Primary Spermatocyte Male germ cell that is either in the process of or has completed the first meiotic division. (31)

Primary Succession The development of a community in an area previously unoccupied by any community; for example, a "bare" area such as rock, volcanic material, or dunes. (41)

Primary Tissues Tissues produced by primary meristems of a plant, which arise from the shoot and root apical meristems. In general, primary tissues are a result of an increase in plant length. (18)

Primary Transcript An RNA molecule that has been transcribed but not yet subjected to any type of processing. The primary transcript corresponds to the entire stretch of DNA that was transcribed. (15)

Primates Order of mammals that includes humans, apes, monkeys, and lemurs. (39)

Primitive An evolutionary early condition. Primitive features are those that were also present in an early ancestor, such as five digits on the feet of terrestrial vertebrates. (34)

Prions An infectious particle that contains protein but no nucleic acid. It causes slow diseases of animals, including neurological disease of humans. (36)

Processing-Level Control Control of gene expression by regulating the pathway by which a primary RNA transcript is processed into an mRNA. (15)

Products In a chemical reaction, the compounds into which the reactants are transformed. (6)

Profundal Zone Deep, open water of lakes, where it is too dark for photosynthesis to occur. (40)

Progesterone A hormone produced by the corpus luteum within the ovary. It prepares and maintains the uterus for pregnancy, participates in milk production, and prevents the ovary from releasing additional eggs late in the cycle or during pregnancy. (25)

Prokaryotic Referring to single-celled organisms that have no membrane separating the DNA from the cytoplasm and lack membrane-enclosed organelles. Prokaryotes are confined to the kingdom Monera; they are all bacteria. (36)

Prokaryotic Fission The most common type of cell division in bacteria (prokaryotes). Duplicated DNA strands are attached to the plasma membrane and become separated into two cells following membrane growth and cell wall formation. (10, 36)

Prolactin A hormone produced by the anterior pituitary, stimulating milk production by mammary glands. (25)

Promoter A short segment of DNA to which RNA polymerase attaches at the start of transcription. (15)

Prophase Longest phase of mitosis, involving the formation of a spindle, coiling of chromatin fibers into condensed chromosomes, and movement of the chromosomes to the center of the cell. (10)

Prostaglandins Hormones secreted by endocrine cells scattered throughout the body responsible for such diverse functions as contraction of uterine muscles, triggering the inflammatory response, and blood clotting. (25)

Prostate Gland A muscular gland which produces and releases fluids that make up a substantial portion of the semen. (31)

Proteins Long chains of amino acids, linked together by peptide bonds. They are folded into specific shapes essential to their functions. (4)

Prothallus The small, heart-shaped gametophyte of a fern. (38)

Protists A member of the kingdom Protista; simple eukaryotic organisms that share broad taxonomic similarities. (36, 37)

Protocooperation Non-compulsory interactions that benefit two organisms, e.g., lichens. (42)

Proton Gradient A difference in hydrogen ion (proton) concentration on opposite sides of a membrane. Proton gradients are formed during photosynthesis and respiration and serve as a store of energy used to drive ATP formation. (8, 9)

Protons Positively charged particles within the nucleus of an atom. (3)

Protostomes One path of development exhibited by coelomate animals (e.g., mollusks, annelids, and arthropods). (39)

Protozoa Member of protist kingdom that is unicellular and eukaryotic; vary greatly in size, motility, nutrition and life cycle. (37)

Provirus DNA copy of a virus' nucleic acid that becomes integrated into the host cell's chromosome. (36)

Pseudocoelamates Animals in which the body cavity is not lined by cells derived from mesoderm. (39)

Pseudopodia (psuedo = false, pod = foot). Pseudopodia are fingerlike extensions of cytoplasm that flow forward from the "body" of an amoeba; the rest of the cell then follows. (37)

Puberty Development of reproductive capacity, often accompanied by the appearance of secondary sexual characteristics. (31)

Pulmonary Circulation The loop of the circulatory system that channels blood to the lungs for oxygenation. (28)

Punctuated Equilibrium Theory A theory to explain the phenomenon of the relatively sudden appearance of new species, followed by long periods of little or no change. (33)

Punnett Square Method A visual method for predicting the possible genotypes and their expected ratios from a cross. (12)

Pupa In insects, the stage in metamorphosis between the larva and the adult. Within the pupal case, there is dramatic transformation in body form as some larval tissues die and others differentiate into those of the adult. (32)

Purine A nitrogenous base found in DNA and RNA having a double ring structure. Adenine and guanine are purines. (14)

Pyloric Sphincter Muscular valve between the human stomach and small intestine. (27)

Pyrimidine A nitrogenous base found in DNA and RNA having a single ring structure. Cytosine, thymine, and uracil are pyrimidines. (14)

Pyramid of Biomass Diagrammatic representation of the total dry weight of organisms at each trophic level in a food chain or food web. (41)

Pyramid of Energy Diagrammatic representation of the flow of energy through trophic levels in a food chain or food web. (41)

Pyramid of Numbers Similar to a pyramid of energy, but with numbers of producers and consumers given at each trophic level in a food chain or food web. (41)

◄ Q ►

Quiescent Center The region in the apical meristem of a root containing relatively inactive cells. (18)

◄ R ►

R-Group The variable portion of a molecule. (4)

r-Selected Species Species that possess adaptive strategies to produce numerous offspring at once. (43)

Radial Symmetry The quality possessed by animals whose bodies can be divided into mirror images by more than one median plane. (39)

Radicle In the plant embryo, the tip of the hypocotyl that eventually develops into the root system. (20)

Radioactivity A property of atoms whose nucleus contains an unstable combination of particles. Breakdown of the nucleus causes the emission of particles and a resulting change in structure of the atom. Biologists use this property to track labeled molecules and to determine the age of fossils. (3)

Radiodating The use of known rates of radioactive decay to date a fossil or other ancient object. (3, 34)

Radioisotope An isotope of an element that is radioactive. (3)

Radiolarian A prozoan member of the protistan group Sarcodina that secretes silicon shells through which it captures food.

Rainshadow The arid, leeward (downwind) side of a mountain range. (40)

Random Distribution Distribution of individuals of a population in a random manner; environmental conditions must be similar and individuals do not affect each other's location in the population. (43)

Reactants Molecules or atoms that are changed to products during a chemical reaction. (6)

Reaction A chemical change in which starting molecules (reactants) are transformed into new molecules (products). (6)

Reaction Center A special chlorophyll molecule in a photosystem (P_{700} in Photosystem I, P_{680} in Photosystem II). (8)

Realized Niche Part of the fundamental niche of an organism that is actually utilized. (41)

Receptacle The base of a flower where the flower parts are attached; usually a widened area of the pedicel. (20)

Receptor-Mediated Endocytosis The uptake of materials within a cytoplasmic vesicle (endocytosis) following their binding to a cell surface receptor. (7)

Receptor Site A site on a cell's plasma membrane to which a chemical such as a hormone binds. Each surface site permits the attachment of only one kind of hormone. (5)

Recessive An allele whose expression is masked by the dominant allele for the same trait. (12)

Recombinant DNA A DNA molecule that contains DNA sequences derived from different biological sources that have been joined together in the laboratory. (16)

Recombination The rejoining of DNA pieces with those of a different strand or with the same strand at a point different from where the break occurred. (11, 13)

Red Marrow The soft tissue in the interior of bones that produces red blood cells. (26)

Red Tide Growth of one of several species of reddish brown dinoflagellate algae so extensive that it tints the coastal waters and inland lakes a distinctive red color. Often associated with paralytic shellfish poisoning (see dinoflagellates). (37)

Reducing Power A measure of the cell's ability to transfer electrons to substrates to create molecules of higher energy content. Usually determined by the available store of NADPH, the molecule from which electrons are transferred in anabolic (synthetic) pathways. (6)

Reduction The addition of electrons to a compound during a chemical reaction. For a carbon atom, the more hydrogens that are bonded to the carbon, the more reduced the atom. (6)

Reduction Division The first meiotic division during which a cell's chromosome number is reduced in half. (11)

Reflex An involuntary response to a stimulus. (23)

Reflex Arc The simplest example of central nervous system control, involving a sensory neuron, motor neuron, and usually an interneuron. (23)

Regeneration Ability of certain animals to replace injured or lost limbs parts by growth and differentiation of undifferentiated stem cells. (15)

Region of Elongation In root tips, the region just above the region of cell division, where cells elongate and the root length increases. (18)

Region of Maturation In root tips, the region above the region of elongation; cells differentiate and root hairs occur in this region. (18)

Regulatory Genes Genes whose sole function is to control the expression of structural genes. (15)

Releaser A sign stimulus that is given by an individual to another member of the same species, eliciting a specific innate behavior. (44)

Releasing Factors Hormones secreted by the tips of hypothalmic neurosecretory cells that stimulate the anterior pituitary to release its hormones. GnRH, for example, stimulates the release of gonadotropins. (25)

Renal Referring to the kidney. (28)

Replication Duplication of DNA, usually prior to cell division. (14)

Replication Fork The site where the two strands of a DNA helix are unwinding during replication. (14)

Repression Inhibition of transcription of a gene which, in an operon, occurs when repressor protein binds to the operator. (15)

Repressor Protein encoded by a bacterial regulatory gene that binds to an operator site of an operon and inhibits transcription. (15)

Reproduction The process by which an organism produces offspring. (31)

Reproductive Isolation Phenomenon in which members of a single population become split into two populations that no longer interbreed. (33)

Reproductive System System of specialized organs that are utilized for the production of gametes and, in some cases, the fertilization and/or development of an egg. (31)

Reptiles Members of class Reptilia, scaly, air-breathing, egg-laying vertebrates such as lizards, snakes, turtles, and crocodiles. (39)

Resolving Power The ability of an optical instrument (eye, microscopes) to discern whether two very close objects are separate from each other. (APP.)

Resource Partitioning Temporal or spatial sharing of a resource by different species. (42)

Respiration Process used by organisms to exchange gases with the environment; the source of oxygen required for metabolism. The process organisms use to oxidize glucose to CO_2 and H_2O using an electron transport system to extract energy from electrons and store it in the high-energy bonds of ATP. (29)

Respiratory System The specialized set of organs that function in the uptake of oxygen from the environment. (29)

Resting Potential The electrical potential (voltage) across the plasma membrane of a neuron when the cell is not carrying an impulse. Results from a difference in charge across the membrane. (23)

Restriction Enzyme A DNA-cutting enzyme found in bacteria. (16)

Restriction Fragment Length Polymorphism (RFLP) Certain sites in the DNA tend to have a highly variable sequence from one individual to another. Because of these differences, restriction enzymes cut the DNA from different individuals into fragments of different length. Variations in the length of particular fragments (RFLPs) can be used as genetic signposts for the identification of nearby genes of interest. (17)

Restriction Fragments The DNA fragments generated when purified DNA is treated with a particular restriction enzyme. (16)

Reticular Formation A series of interconnected sites in the core of the brain (brainstem) that selectively arouse conscious activity. (23)

Retroviruses RNA viruses that reverse the typical flow of genetic information; within the infected cell, the viral DNA serves as a template for synthesis of a DNA copy. Examples include HIV, which causes AIDS, and certain cancer viruses. (36)

Reverse Genetics Determining the amino acid sequence and function of a polypeptide from the nucleotide sequence of the gene that codes for that polypeptide. (17)

Reverse Transcriptase An enzyme present in retroviruses that transcriibes a strand of DNA, using viral RNA as the template. (36)

Rhizoids Slender cells that resemble roots but do not absorb water or minerals. (36)

Rhodophyta Red algae; seaweeds that can absorb deeper penetrating light rays than most aquatic photosynthesizers. (36)

Rhyniophytes Ancient plants having vascular tissue which thrived in marshy areas during the Silurian period.

Ribonucleic Acid (RNA) Single-stranded chain of nucleotides each comprised of ribose (a sugar), phosphate, and one of four bases (adenine, guanine, cytosine, and uracil). The sequence of nucleotides in RNA is dictated by DNA, from which it is transcribed. There are three classes of RNA: mRNA, tRNA, and rRNA, all required for protein synthesis. (4, 14)

Ribosomal RNA (rRNA) RNA molecules that form part of the ribosome. Included among the rRNAs is one that is thought to catalyze peptide bond formation. (14)

Ribosomes Organelles involved in protein synthesis in the cytoplasm of the cell. (14)

Ribozymes RNAs capable of catalyzing a chemical reaction, such as peptide bond formation or RNA cutting and splicing. (15)

Rickettsias A group of obligate intracellular parasites, smaller than the typical prokaryote. They cause serious diseases such as typhus. (36)

River Flowing body of surface fresh water; rivers are formed from the convergence of streams. (40)

RNA Polymerase The enzyme that directs transcription and assembling RNA nucleotides in the growing chain. (14)

RNA Processing The process by which the intervening (noncoding) portions of a primary RNA transcript are removed and the remaining (coding) portions are spliced together to form an mRNA. (15)

Root Cap A protective cellular helmet at the tip of a root that surrounds delicate meristematic cells and shields them from abrasion and directs the growth downward. (18)

Root Hairs Elongated surface cells near the tip of each root for the absorption of water and minerals. (18)

Root Nodules Knobby structures on the roots of certain plants. They house nitrogen-fixing bacteria which supply nitrogen in a form that can be used by the plant. (19)

Root Pressure A positive pressure as a result of continuous water supply to plant roots that assists (along with transpirational pull) the pushing of water and nutrients up through the xylem. (19)

Root System The below-ground portion of a plant, consisting of main roots, lateral roots, root hairs, and associated structures and systems such as root nodules or mycorrhizae. (18)

Rough ER (RER) Endoplasmatic reticulum with many ribosomes attached. As a result, they appear rough in electron micrographs. (5)

Ruminant Grazing mammals that possess an additional stomach chamber called rumen which is heavily fortified with cellulose-digesting microorganisms. (27)

◄ S ►

S Phase The second stage of interphase in which the materials needed for cell division are synthesized and an exact copy of cell's DNA is made by DNA replication. (10)

Sac Body The body plan of simple animals, like cnidarians, where there is a single opening leading to and from a digestion chamber.

Saltatory Conduction The "hopping" movement of an impulse along a myelinated neuron from one Node of Ranvier to the next one. (23)

Sap Fluid found in xylem or sieve of phloem. (20)

Saprophyte Organisms, mainly fungi and bacteria, that get their nutrition by breaking down organic wastes and dead organisms, also called decomposers. (42)

Saprobe Organism that obtains its nutrients by decomposing dead organisms. (37)

Sarcolemma The plasma membrane of a muscle fiber. (26)

Sarcomere The contractile unit of a myofibril in skeletal muscle. (26)

Sarcoplasmic Reticulum (SR) In skeletal muscle, modified version of the endoplasmic reticulum that stores calcium ions. (26)

Savanna A grassland biome with alternating dry and rainy seasons. The grasses and scattered trees support large numbers of grazing animals. (40)

Scaling Effect A property that changes disproportionately as the size of organisms increase. (22)

Scanning Electron Microscope (SEM) A microscope which operates by showering electrons back and forth across the surface of a specimen prepared with a thin metal coating. The resultant image shows three-dimensional features of the specimen's surface. (APP.)

Schwann Cells Cells which wrap themselves around the axons of neurons forming an insulating myelin sheath composed of many layers of plasma membrane. (23)

Sclereids Irregularly-shaped sclerenchyma cells, all having thick cell walls; a component of seed coats and nuts. (18)

Sclerenchyma Component of the ground tissue system of plants. They are thick walled cells of various shapes and sizes, providing support or protection. They continue to function after the cell dies. (18)

Sclerenchyma Fibers Non-living elongated plant cells with tapering ends and thick secondary walls. A supportive cell type found in various plant tissues. (18)

Sebaceous Glands Exocrine glands of the skin that produce a mixture of lipids (sebum) that oil the hair and skin. (26)

Secondary Cell Wall An additional cell wall that improves the strength and resiliency of specialized plant cells, particularly those cells found in stems that support leaves, flowers, and fruit. (5)

Secondary Consumer Organism that feeds exclusively on primary consumers; mostly animals, but some plants. (41)

Secondary Growth Growth from cambia in perennials; results in an increase in the diameter of stems and roots. (18)

Secondary Meristems (vascular cambium, cork cambrium) Rings or clusters of meristematic cells that increase the width of stems and roots when the divide. (18)

Secondary Sex Characteristics Those characteristics other than the gonads and reproductive tract that develop in response to sex hormones. For example, breasts and pubic hair in women and a deep voice and pubic hair in men. (31)

Secondary Succession The development of a community in an area previously occupied by a community, but which was disturbed in some manner; for example, fire, development, or clear-cutting forests. (41)

Secondary Tissues Tissues produced to accommodate new cell production in plants with woody growth. Secondary tissues are produced from cambia, which produce vascular and cork tissues, leading to an increase in plant girth. (18)

Second Messenger Many hormones, such as glucagon and thyroid hormone, evoke a response by binding to the outer surface of a target cell and causing the release of another substance (which is the second messenger). The best-studied second messenger is cyclic AMP which is formed by an enzyme on the inner surface of the plasma membrane following the binding of a hormone to the outer surface of the membrane. The cyclic AMP diffuses into the cell and activates a protein kinase. (25)

Secretion The process of exporting materials produced by the cell. (5)

Seed A mature ovule consisting of the embryo, endosperm, and seed coat. (20)

Seed Dormancy Metabolic inactivity of seeds until favorable conditions promote seed germination. (20)

Secretin Hormone secreted by endocrine cells in the wall of the intestine that stimulates the release of digestive products from the pancreas. (25)

Segmentation A condition in which the body is constructed, at least in part, from a series of repeating parts. Segmentation occurs in annelids, arthropods, and vertebrates (as revealed during embryonic development). (39)

Selectively Permeable A term applied to the plasma membrane because membrane proteins control which molecules are transported. Enables a cell to import and accumulate the molecules essential for normal metabolism. (7)

Semen The fluid discharged during a male orgasm. (31)

Semiconsevative Replication The manner in which DNA replicates; half of the original DNA strand is conserved in each new double helix. (14)

Seminal Vesicles The organs which produce most of the ejaculatory fluid. (31)

Seminiferous Tubules Within the testes, highly coiled and compacted tubules, lined with a self-perpetuating layer of spermatogonia, which develop into sperm. (31)

Senescence Aging and eventual death of an organism, organ or tissue. (3, 18)

Sense Strand The one strand of a DNA double helix that contains the information that encodes the amino sequence of a polypeptide. This is the strand that is selectively transcribed by RNA polymerase forming an mRNA that can be properly translated. (14)

Sensory Neurons Neurons which relay impulses to the central nervous system. (23)

Sensory Receptors Structures that detect changes in the external and internal environment and transmit the information to the nervous system. (24)

Sepal The outermost whorl of a flower, enclosing the other flower parts as a flower bud; collectively called the calyx. (20)

Sessile Sedentary, incapable of independent movement. (39)

Sex Chromosomes The one chromosomal pair that is not identical in the karyotypes of males and females of the same animal species. (10, 13)

Sex Hormones Steroid hormones which influence the production of gametes and the development of male or female sex characteristics. (25)

Sexual Dimorphism Differences in the appearance of males and females in the same species. (33)

Sexual Reproduction The process by which haploid gametes are formed and fuse during fertilization to form a zygote. (31)

Sexual Selection The natural selection of adaptations that improve the chances for mating and reproducing. (33)

Shivering Involuntary muscular contraction for generating metabolic heat that raises body temperature. (28)

Shoot In angiosperms, the system consisting of stems, leaves, flowers and fruits. (18)

Shoot System The above-ground portion of an angiosperm plant consisting of stems with nodes, including branches, leaves, flowers and fruits. (18)

Short-Day Plants Plants that flower in late summer or fall when the length of daylight becomes shorter than some critical period. (21)

Shrubland A biome characterized by densely growing woody shrubs in mediterranean type climate; growth is so dense that understory plants are not usually present. (40)

Sickle Cell Anemia A genetic (recessive autosomal) disorder in which the beta globin genes of adult hemoglobin molecules contain an amino acid substitution which alters the ability of hemoglobin to transport oxygen. During times of oxygen stress, the red blood cells of these individuals may become sickle shaped, which interferes with the flow of the cells through small blood vessels. (4, 17)

Sieve Plate Found in phloem tissue in plants, the wall between sieve-tube members, containing perforated areas for passage of materials. (18)

Sieve-Tube Member A living, food-conducting cell found in phloem tissue of plants; associated with a companion cell. (18)

Sigmoid Growth Curve An S-shaped curve illustrating the lag phase, exponential growth, and eventual approach of a population to its carrying capacity. (43)

Sign Stimulus An object or action in the environment that triggers an innate behavior. (44)

Simple Fruits Fruits that develop from the ovary of one pistil. (20)

Simple Leaf A leaf that is undivided; only one blade attached to the petiole. (18)

Sinoatrial (SA) Node A collection of cells that generates an action potential regulating heart beat; the heart's pacemaker. (28)

Skeletal Muscles Separate bundles of parallel, striated muscle fibers anchored to the bone, which they can move in a coordinated fashion. They are under voluntary control. (26)

Skeleton A rigid form of support found in most animals either surrounding the body with a protective encasement or providing a living girder system within the animal. (26)

Skull The bones of the head, including the cranium. (26)

Slow-Twitch Fibers Skeletal muscle fibers that depend on aerobic metabolism for ATP production. These fibers are capable of undergoing contraction for extended periods of time without fatigue, but generate lesser forces than fast-twitch fibers. (9)

Small Intestine Portion of the intestine in which most of the digestion and absorption of nutrients takes place. It is so named because of its narrow diameter. There are three sections: duodenum, jejunum, and ilium. (27)

Smell Sense of the chemical composition of the environment. (24)

Smooth ER (SER) Membranes of the endoplasmic reticulum that have no ribosomes on their surface. SER is generally more tubular than the RER. Often acts to store calcium or synthesize steroids. (5)

Smooth Muscle The muscles of the internal organs (digestive tract, glands, etc.). Composed of spindle-shaped cells that interlace to form sheets of visceral muscle. (26)

Social Behavior Behavior among animals that live in groups composed of individuals that are dependent on one another and with whom they have evolved mechanisms of communication. (44)

Social Learning Learning of a behavior from other members of the species. (44)

Social Parasitism Parasites that use behavioral mechanisms of the host organism to the parasite's advantage, thereby harming the host. (42)

Solute A substance dissolved in a solvent. (3)

Solution The resulting mixture of a solvent and a solute. (3)

Solvent A substance in which another material dissolves by separating into individual molecules or ions. (3)

Somatic Cells Cells that do not have the potential to form reproductive cells (gametes). Includes all cells of the body except germ cells. (11)

Somatic Nervous System The nerves that carry messages to the muscles that move the skeleton either voluntarily or by reflex. (23)

Somatic Sensory Receptors Receptors that respond to chemicals, pressure, and temperature that are present in the skin, muscles, tendons, and joints. Provides a sense of the physiological state of the body. (24)

Somites In the vertebrate embryo, blocks of mesoderm on either side of the notochord that give rise to muscles, bones, and dermis. (32)

Speciation The formation of new species. Occurs when one population splits into separate populations that diverge genetically to the point where they become separate species. (33)

Species Taxonomic subdivisions of a genus. Each species has recognizable features that distinguish it from every other species. Members of one species generally will not interbreed with members of other species. (33)

Specific Epithet In taxonomy, the second term in an organism's scientific name identifying its species within a particular genus. (1)

Spermatid Male germ cell that has completed meiosis but has not yet differentiated into a sperm. (31)

Spermatogenesis The production of sperm. (31)

Spermatogonia Male germ cells that have not yet begun meiosis. (31)

Spermatozoa (Sperm) Male gametes. (31)

Sphinctors Circularly arranged muscles that close off the various tubes in the body.

Spinal Cord A centralized mass of neurons for processing neurological messages and linking the brain with that part of peripheral nervous system not reached by the cranial nerves. (23)

Spinal Nerves Paired nerves which emerge from the spinal cord and innervate the body. Humans have 31 pairs of spinal nerves. (23)

Spindle Apparatus In dividing eukaryotic cells, the complex rigging, made of microtubules, that aligns and separates duplicated chromosomes. (10)

Splash Zone In the intertidal zone, the uppermost region receiving splashes and sprays of water to the mean of high tides. (40)

Spleen One of the organs of the lymphatic system that produces lymphocytes and filters blood; also produces red blood cells in the human fetus. (28)

Splicing The step during RNA processing in which the coding segments of the primary transcript are covalently linked together to form the mRNA. (15)

Spongy Parenchyma In monocot and dicot leaves, loosely arranged cells functioning in photosynthesis. Found above the lower epidermis and beneath the palisade parenchyma in dicots, and between the upper and lower epidermis in monocots. (18)

Spontaneous Generation Disproven concept that living organisms can arise directly from inanimate materials. (2)

Sporangiospores Black, asexual spores of the zygomycete fungi. (37)

Sporangium A hollow structure in which spores are formed. (37)

Spores In plants, haploid cells that develop into the gametophyte generation. In fungi, an asexual or sexual reproductive cell that gives rise to a new mycelium. Spores are often lightweight for their dispersal and adapted for survival in adverse conditions. (37)

Sporophyte The diploid spore producing generation in plants. (38)

Stabilizing Selection Natural selection favoring an intermediate phenotype over the extremes. (33)

Starch Polysaccharides used by plants to store energy. (4)

Stamen The flower's male reproductive organ, consisting of the pollen-producing anther supported by a slender stalk, the filament. (20)

Stem In plants, the organ that supports the leaves, flowers, and fruits. (18)

Stem Cells Cells which are undifferentiated and capable of giving rise to a variety of different types of differentiated cells. For example, hematopoietic stem cells are capable of giving rise to both red and white blood cells. (17)

Steroids Compounds classified as lipids which have the basic four-ringed molecular skeleton as represented by cholesterol. Two examples of steroid hormones are the sex hormones; testosterone in males and estrogen in females. (4, 25)

Stigma The sticky area at the top of each pistil to which pollen adheres. (20)

Stimulus Any change in the internal or external environment to which an organism can respond. (24)

Stomach A muscular sac that is part of the digestive system where food received from the esophagus is stored and mixed, some breakdown of food occurs, and the chemical degradation of nutrients begins. (27)

Stomates (Pl. Stomata) Microscopic pores in the epidermis of the leaves and stems which allow gases to be exchanged between the plant and the external environment. (18)

Stratified Epithelia Multicellular layered epithelium. (22)

Stream Flowing body of surface fresh water; streams merge together into larger streams and rivers. (40)

Stretch Receptors Sensory receptors embedded in muscle tissue enabling muscles to respond reflexively when stretched. (23, 24)

Striated Referring to the striped appearance of skeletal and cardiac muscle fibers. (26)

Strobilus In lycopids, terminal, cone-like clusters of specialized leaves that produce sporangia.

Stroma The fluid interior of chloroplasts. (8)

Stromatolites Rocks formed from masses of dense bacteria and mineral deposits. Some of these rocky masses contain cells that date back over three billion years revealing the nature of early prokaryotic life forms. (35)

Structural Genes DNA segments in bacteria that direct the formation of enzymes or structural proteins. (15)

Style The portion of a pistil which joins the stigma to the ovary. (20)

Substrate-Level Phosphorylation The formation of ATP by direct transfer of a phosphate group from a substrate, such as a sugar phosphate, to ADP. ATP is formed without the involvement of an electron transport system. (9)

Substrates The reactants which bind to enzymes and are subsequently converted to products. (6)

Succession The orderly progression of communities leading to a climax community. It is one of two types: primary, which occurs in areas where no community existed before; and secondary, which occurs in disturbed habitats where some soil and perhaps some organisms remain after the disturbance. (41)

Succulents Plants having fleshy, water-storing stems or leaves. (40)

Suppressor T Cells A class of T cells that regulate immune responses by inhibiting the activation of other lymphocytes. (30)

Surface Area-to-Volume Ratio The ratio of the surface area of an organism to its volume, which determines the rate of exchange of materials between the organism and its environment. (22)

Surface Tension The resistance of a liquid's surface to being disrupted. In aqueous solutions, it is caused by the attraction between water molecules. (3)

Survivorship Curve Graph of life expectancy, plotted as the number of survivors versus age. (43)

Sweat Glands Exocrine glands of the skin that produce a dilute salt solution, whose evaporation cools in the body. (26)

Symbiosis A close, long-term relationship between two individuals of different species. (42)

Symmetry Referring to a body form that can be divided into mirror image halves by at least one plane through its body. (39)

Sympathetic Nervous System Part of the autonomic nervous system that tends to stimulate bodily activities, particularly those involved with coping with stressful situations. (23)

Sympatric Speciation Speciation that occurs in populations with overlapping distributions. It is common in plants when polyploidy arises within a population. (33)

Synapse Juncture of a neuron and its target cell (another neuron, muscle fiber, gland cell). (23)

Synapsis The pairing of homologous chromosomes during prophase of meiosis I. (11)

Synaptic Cleft Small space between the synaptic knobs of a neuron and its target cell. (23)

Synaptic Knobs The swellings that branch from the end of the axon. They deliver the neurological impulse to the target cell. (23)

Synaptonemal Complex Ladderlike structure that holds homologous chromosomes together as a tetrad during crossing over in prophase I of meiosis. (11)

Synovial Cavities Fluid-filled sacs around joints, the function of which is to lubricate and separate articulating bone surfaces. (26)

Systemic Circulation Part of the circulatory system that delivers oxygenated blood to the tissues and routes deoxygenated blood back to the heart. (28)

Systolic Pressure The first number of a blood pressure reading; the highest pressure attained in the arteries as blood is propelled out of the heart. (28)

◄ T ►

Taiga A biome found south of tundra biomes; characterized by coniferous forests, abundant precipitation, and soils that thaw only in the summer. (40)

Tap Root System Root system of plants having one main root and many smaller lateral roots. Typical of conifers and dicots. (18)

Taste Sense of the chemical composition of food. (24)

Taxonomy The science of classifying and grouping organisms based on their morphology and evolution. (1)

T Cell Lymphocytes that carry out cell-mediated immunity. They respond to antigen stimulation by becoming helper cells, killer cells, and memory cells. (30)

Telophase The final stage of mitosis which begins when the chromosomes reach their spindle poles and ends when cytokinesis is completed and two daughter cells are produced. (10)

Tendon A dense connective tissue cord that connects a skeletal muscle to a bone. (26)

Teratogenic Embryo deforming. Chemicals such as thalidomide or alcohol are teratogenic because they disturb embryonic development and lead to the formation of an abnormal embryo and fetus. (32)

Terminal Electron Acceptor In aerobic respiration, the molecule of O_2 which removes the electron pair from the final cytochrome of the respiratory chain. (9)

Terrestrial Living on land. (40)

Territory (Territoriality) An area that an animal defends against intruders, generally in the protection of some resource. (42, 44)

Tertiary Consumer Animals that feed on secondary consumers (plant or animal) or animals only. (41)

Test Cross An experimental procedure in which an individual exhibiting a dominant trait is crossed to a homozygous recessive to determine whether the first individual is homozygous or heterozygous. (12)

Testis In animals, the sperm-producing gonad of the male. (23)

Testosterone The male sex hormone secreted by the testes when stimulated by pituitary gonadotropins. (31)

Tetrad A unit of four chromatids formed by a synapsed pair of homologous chromosomes, each of which has two chromatids. (11)

Thallus In liverworts, the flat, ground-hugging plant body that lacks roots, stems, leaves, and vascular tissues. (38)

Theory of Tolerance Distribution, abundance and existence of species in an ecosystem are determined by the species' range of tolerance of chemical and physical factors. (41)

Thermoreceptors Sensory receptors that respond to changes in temperature. (24)

Thermoregulation The process of maintaining a constant internal body temperature in spite of fluctuations in external temperatures. (28)

Thigmotropism Changes in plant growth stimulated by contact with another object, e.g., vines climbing on cement walls. (21)

Thoracic Cavity The anterior portion of the body cavity in which the lungs are suspended. (39)

Thylakoids Flattened membrane sacs within the chloroplast. Embedded in these membranes are the light-capturing pigments and other components that carry out the light-dependent reactions of photosynthesis. (8)

Thymus Endocrine gland in the chest where T cells mature. (30)

Thyroid Gland A butterfly-shaped gland that lies just in front of the human windpipe, producing two metabolism-regulating hormones, thyroxin and triodothyronine. (25)

Thyroid Hormone A mixture of two iodinated amino acid hormones (thyroxin and triiodothyronine) secreted by the thyroid gland. (25)

Thyroid Stimulating Hormone (TSH) An anterior pituitary hormone which stimulates secretion by the thyroid gland. (25)

Tissue An organized group of cells with a similar structure and a common function. (22)

Tissue System Continuous tissues organized to perform a specific function in plants. The three plant tissue systems are: dermal, vascular, and ground (fundamental). (18)

Tolerance Range The range between the maximum and minimum limits for an environmental factor that is necessary for an organism's survival. (41)

Totipotent The genetic potential for one type of cell from a multicellular organism to give rise to any of the organism's cell types, even to generate a whole new organism. (15)

Trachea The windpipe; a portion of the respiratory tract between the larynx and bronchii. (29)

Tracheal Respiratory System A network of tubes (tracheae) and tubules (tracheoles) that carry air from the outside environment directly to the cells of the body without involving the circulatory system. (29, 39)

Tracheid A type of conducting cell found in xylem functioning when a cell is dead to transport water and dissolved minerals through its hollow interior. (18)

Tracheophytes Vascular plants that contain fluid-conducting vessels. (38)

Transcription The process by which a strand of RNA assembles along one of the DNA strands. (14)

Transcriptional-Level Control Control of gene expression by regulating whether or not a specific gene is transcribed and how often. (15)

Transduction A type of genetic recombination resulting from transfer of genes from one organism to another by a virus.

Transfer RNA (tRNA) A type of RNA that decodes mRNA's codon message and translates it into amino acids. (14)

Transgenic Organism An organism that possesses genes derived from a different species. For example, a sheep that carries a human gene and secretes the human protein in its milk is a transgenic animal. (16)

Translation The cell process that converts a sequence of nucleotides in mRNA into a sequence of amino acids in a polypeptide. (14)

Translational-Level Control Control of gene expression by regulating whether or not a specific mRNA is translated into a polypeptide. (15)

Translocation The joining of segments of two nonhomologous chromosomes (13)

Transmission Electron Microscope (TEM) A microscope that works by shooting electrons through very thinly sliced specimens. The result is an enormously magnified image, two-dimensional, of remarkable detail. (App.)

Transpiration Water vapor loss from plant surfaces. (19)

Transpiration Pull The principle means of water and mineral transport in plants, initiated by transpiration. (19)

Transposition The phenomenon in which certain DNA segments (mobile genetic elements, or jumping genes) tend to move from one part of the genome to another part. (15)

Transverse Fission The division pattern in ciliated protozoans where the plane of division is perpendicular to the cell's length.

Trimester Each of the three stages comprising the 266-day period between conception and birth in humans. (32)

Triploid Having three sets of chromosomes, abbreviated 3N. (11)

Trisomy Three copies of a particular chromosome per cell. (17)

Trophic Level Each step along a feeding pathway. (41)

Trophozoite The actively growing stage of polymorphic protozoa. (37)

Tropical Rain Forest Lush forests that occur near the equator; characterized by high annual rainfall and high average temperature. (40)

Tropical Thornwood A type of shrubland occurring in tropical regions with a short rainy season. Plants lose their small leaves during dry seasons, leaving sharp thorns. (40)

Tropic Hormones Hormones that act on endocrine glands to stimulate the production and release of other hormones. (25)

Tropisms Changes in the direction of plant growth in response to environmental stimuli, e.g., light, gravity, touch. (21)

True-Breeder Organisms that, when bred with themselves, always produce offspring identical to the parent for a given trait. (12)

Tubular Reabsorption The process by which substances are selectively returned from the fluid in the nephron to the bloodstream. (28)

Tubular Secretion The process by which substances are actively and selectively transported from the blood into the fluid of the nephron. (28)

Tumor-Infiltrating Lymphocytes (TILs) Cytotoxic T cells found within a tumor mass that have the capability to specifically destroy the tumor cells. (30)

Tumor-Suppressor Genes Genes whose products act to block the formation of cancers. Cancers form only when both copies of these genes (one on each homologue) are mutated. (13)

Tundra The marshy, unforested biome in the arctic and at high elevations. Frigid temperatures for most of the year prevent the subsoil from thawing, which produces marshes and ponds. Dominant vegetation includes low growing plants, lichens, and mosses. (40)

Turgor Pressure The internal pressure in a plant cell caused by the diffusion of water into the cell. Because of the rigid cell wall, pressure can increase to where it eventually stops the influx of more water. (7)

Turner Syndrome A person whose cells have only one X chromosome and no second sex chromosome (XO). These individuals develop as immature females. (17)

◄ U ►

Ultimate (Top) Consumer The final carnivore trophic level organism, or organisms that escaped predation; these consumers die and are eventually consumed by decomposers. (41)

Ultracentrifuge An instrument capable of spinning tubes at very high speeds, delivering centrifugal forces over 100,000 times the force of gravity. (9)

Unicellular The description of an organism where the cell is the organism. (35)

Uniform Pattern Distribution of individuals of a population in a uniform arrangement, such as individual plants of one species uniformly spaced across a region. (43)

Urethra In mammals, a tube that extends from the urinary bladder to the outside. (28)

Urinary Tract The structures that form and export urine: kidneys, ureters, urinary bladder, and urethra. (28)

Urine The excretory fluid consisting of urea, other nitrogenous substances, and salts dissolved in water. It is formed by the kidneys. (28)

Uterine (Menstrual) Cycle The repetitive monthly changes in the uterus that prepare the endometrium for receiving and sustaining an embryo. (31)

Uterus An organ in the female reproductive system in which an embryo implants and is maintained during development. (31)

◄ V ►

Vaccines Modified forms of disease-causing microbes which cannot cause disease but retain the same antigens of it. They permit the immune system to build memory cells without diseases developing during the primary immune response. (30)

Vacoconstriction Reduction in the diameter of blood vessels, particularly arterioles. (28)

Vacuole A large organelle found in mature plant cells, occupying most of the cell's volume, sometimes more than 90% of it. (5)

Vagina The female mammal's copulatory organ and birth canal. (31)

Variable (Experimental) A factor in an experiment that is subject to change, i.e., can occur in more than one state. (2)

Vascular Bundles Groups of vascular tissues (xylem and phloem) in the shoot of a plant. (19)

Vascular Cambrium In perennials, a secondary meristem that produces new vascular tissues. (18)

Vascular Cylinder Groups of vascular tissues in the central region of the root. (18)

Vascular Plants Plants having a specialized conducting system of vessels and tubes for transporting water, minerals, food, etc., from one region to another. (18)

Vascular Tissue System All the vascular tissues in a plant, including xylem, phloem, and the vascular cambium or procambium. (18)

Vasodilation Increase in the diameter of blood vessels, particularly arterioles. (28)

Veins In plants, vascular bundles in leaves. In animals, blood vessels that return blood to the heart. (28)

Venation The pattern of vein arrangement in leaf blades. (18)

Ventricle Lower chamber of the heart which pumps blood through arteries. There is one ventricle in the heart of lower vertebrates and two ventricles in the four-chambered heart of birds and mammals. (28)

Venules Small veins that collect blood from the capillaries. They empty into larger veins for return to the heart. (28)

Vertebrae The bones that form the backbone. In the human there are 33 bones arranged in a gracefully curved line along the bone, cushioned from one another by disks of cartilage. (26)

Vertebral Column The backbone, which encases and protects the spinal cord. (26)

Vertebrates Animals with a backbone. (39)

Vesicles Small membrane-enclosed sacs which form from the ER and Golgi complex. Some store chemicals in the cells; others move to the surface and fuse with the plasma membrane to secrete their contents to the outside. (5)

Vessel A tube or connecting duct containing or circulating body fluids. (18)

Vessel Member A type of conducting cell in xylem functioning when the cell is dead to transport water and dissolved minerals through its hollow interior; also called a vessel element. (18)

Vestibular Apparatus A portion of the inner ear of vertebrates that gathers information about the position and movement of the head for use in maintaining balance and equilibrium. (24)

Vestigial Structure Remains of ancestral structures or organs which were, at one time, useful. (34)

Villi Finger-like projections of the intestinal wall that increase the absorption surface of the intestine. (27)

Viroids are associated with certain diseases of plants. Each viroid consists solely of a small single-stranded circle of RNA unprotected by a protein coat. (36)

Virus Minute structures composed of only heredity information (DNA or RNA), surrounded by a protein or protein/lipid coat. After infection, the viral mucleic acid subverts the metabolism of the host cell, which then manufactures new virus particles. (36)

Visible Light The portion of the electromagnetic spectrum producing radiation from 380 nm to 750 nm detectable by the human eye.

Vitamins Any of a group of organic compounds essential in small quantities for normal metabolism. (27)

Vocal Cords Muscular folds located in the larynx that are responsible for sound production in mammals. (29)

Vulva The collective name for the external features of the human female's genitals. (31)

◄ W ►

Water Vascular System A system for locomotion, respiration, etc., unique to echinoderms. (39)

Wavelength The distance separating successive crests of a wave. (8)

Waxes A waterproof material composed of a number of fatty acids linked to a long chain alcohol. (4)

White Matter Regions of the brain and spinal cord containing myelinated axons, which confer the white color. (23)

Wild Type The phenotype of the typical member of a species in the wild. The standard to which mutant phenotypes are compared. (13)

Wilting Drooping of stems or leaves of a plant caused by water loss. (7)

Wood Secondary xylem. (18)

◄ X ►

X Chromosome The sex chromosome present in two doses in cells of a female, and in one dose in the cells of a male. (13)

X-Linked Traits Traits controlled by genes located on the X chromosome. These traits are much more common in males than females. (13)

Xylem The vascular tissue that transports water and minerals from the roots to the rest of the plant. Composed of tracheids and vessel members. (18)

◄ Y ►

Y Chromosome The sex chromosome found in the cells of a male. When Y-carrying sperm unite with an egg, all of which carry a single X chromosome, a male is produced. (13)

Y-Linked Inheritance Genes carried only on Y chromosomes. There are relatively few Y-linked traits; maleness being the most important such trait in mammals. (13)

Yeast Unicellular fungus that forms colonies similar to those of bacteria. (37)

Yolk A deposit of lipids, proteins, and other nutrients that nourishes a developing embryo. (32)

Yolk Sac A sac formed by an extraembryonic membrane. In humans, it manufactures blood cells for the early embryo and later helps to form the umbilical cord. (32)

◄ Z ►

Zero Population Growth In a population, the result when the combined positive growth factors (births and immigration). (43)

Zooplankton Protozoa, small crustaceans and other tiny animals that drift with ocean currents and feed on phytoplankton. (37, 40)

Zygospore The diploid spores of the zygomycete fungi, which include Rhizopus, a common bread mold. After a period of dormancy, the zygospore undergoes meiosis and germinates. (36)

Zygote A fertilized egg. The diploid cell that results from the union of a sperm and egg. (32)

Photo Credits

Part 1 Opener Norbert Wu. **Chapter 1** Fig. 1.1: Jeff Gnass. Fig. 1.2a: Frans Lanting/Minden Pictures, Inc. Fig. 1.2b: David Muench. Fig. 1.3a: Courtesy Dr. Alan Cheetham, National Museum of National History, Smithsonian Institution. Fig. 1.3b: Manfred Kage/Peter Arnold. Fig. 1.3c: Stephen Dalton/NHPA. Fig. 1.3d: Charles Summers, Jr. Fig. 1.3e: Larry West/Photo Researchers. Fig. 1.3f: Bianca Lavies. Fig. 1.3g: Steve Allen/Peter Arnold. Fig. 1.3h: George Grall. Fig. 1.3i: Michio Hoshino/Minden Pictures, Inc. Fig. 1.6a: CNRI/Science Photo Library/Photo Researchers. Fig. 1.6b: M. Abbey/Visuals Unlimited. Fig. 1.6c: Steve Kaufman/Peter Arnold. Fig. 1.6d: Willard Clay. Fig. 1.6e: Jim Bradenburg/Minden Pictures, Inc. Fig. 1.7: (pages 16-17) Charles A. Mauzy; (page 16, top) Anthony Mercieca/Natural Selection; (page 16, center) Wolfgang Bayer/Bruce Coleman, Inc.; (page 16, bottom) Dr. Jeremy Burgess/Science Photo Library/Photo Researchers; (page 17, top) Dr. Eckart Pott/Bruce Coleman, Ltd.; (page 17, bottom) Art Wolfe. Fig. 1..8: Paul Chesley/Photographers Aspen. Fig. 1.9a: Rainbird/Robert Harding Picture Library. Fig. 1.11: Dick Luria/FPG International. **Chapter 2** Fig. 2.2a: Couresy Institut Pasteur. Fig. 2.3a: Topham/The Image Works. Fig. 2.3b: Leonard Lessin/Peter Arnold. Fig. 2.3c: Bettmann Archive. Fig. 2.5a: Laurence Gould/Earth Scenes/Animals Animals. Fig. 2.5b: Ted Horowitz/The Stock Market. **Part 2 Opener:** Nancy Kedersha. **Chapter 3** Fig. 3.1: Franklin Viola. Fig. 3.2: Courtesy Pachyderm Scientific Industries. Bioline: Jacan-Yves Kerban/The Image Bank. Fig. 3.7: Courtesy Stephen Harrison, Harvard Biochemistry Department. Fig. 3.11: Gary Milburn/Tom Stack & Associates. Fig. 3.12: Courtesy R. S. Wilcox, Biology Department, SUNY Binghampton. **Chapter 4** Fig. 4.1: Alastair Black/Tony Stone World Wide. Fig. 4.5a: Don Fawcett/Visuals Unlimited. Fig. 4.5b: Jeremy Burgess/Photo Researchers. Fig. 4.5c: Cabisco/Visuals Unlimited. Fig. 4.6: Tony Stone World Wide. Fig. 4.7: Zig Leszcynski/Animals Animals. Fig. 4.8: Robert & Linda Mitchell. Human Perspective: (a) Photofest (b) Bill Davila/Retna. Fig. 4.12a: Frans Lanting/AllStock, Inc. Fig. 4.12b: Mantis Wildlife Films/Oxford Scientific Films/Animals Animals. Fig. 4.17: Stanley Flegler/Visuals Unlimited. Fig. 4.18a: Courtesy Nelson Max, Lawrence Livermore Laboratory. Fig. 4.18b: Tsuned Hayashida/The Stock Market. **Chapter 5** Fig. 5.1a: The Granger Collection. Fig. 5.1b: Bettmann Archive. Fig. 5.2a: Dr. Jeremey Burgess/Photo Researchers. Fig. 5.2b: CNRI/Photo Researchers. Fig. 5.4: Omikron/Photo Researchers. Fig. 5.6a: Courtesy Richard Chao, California State University at Northridge. Fig. 5.6c: Courtesy Daniel Branton, University of Berkeley. Fig. 5.7: Courtesy G. F. Bohr. Fig. 5.8: Courtesy Michael Mercer, Zoology Department, Arizona State University. Fig. 5.9: Courtesy D. W. Fawcett, Harvard Medical School. Fig. 5.10a: Courtesy U.S. Department of Agriculture, Fig. 5.11: Courtesy Dr. Birgit H. Satir, Albert Einstein College of Medicine. Fig. 5.13: Courtesy Lennart Nilsson, from *A Child Is Born*. Fig. 5.14a: K. R. Porter/Photo Researchers. Fig. 5.14c: Courtesy Lennart Nilsson, BonnierAlba. Fig. 5.14d: Courtesy Lennart Nilsson, From *A Child Is Born*. Fig. 5.15a: Courtesy J. Elliot Weier. Fig. 5.16a: Courtesy J.V. Small. Fig. 5.17a: Peter Parks/Animals Animals. Fig. 5.17b: Courtesy Dr. Manfred Hauser, RUHR-Universitat Rochum. Fig. 5.18: Courtesy Jean Paul Revel, Division of Biology, California Institute of Technology. Fig. 5.19a: Courtesy E. Vivier, from *Paramecium* by W.J. Wagtendonk, Elsevier North Holland Biomedical Press, 1974. Fig. 5.19b: David Phillips/Photo Researchers. Fig. 5.20: Courtesy C.J. Brokaw and T.F. Simonick, *Journal of Cell Biology*, 75:650 (1977). Reproduced with permission. Fig. 5.21: Courtesy L.G. Tilney and K. Fujiwara. Fig. 5.22a: Courtesy W. Gordon Whaley, University of Texas, Austin. Fig. 5.22c: Courtesy R.D. Preston, University of Leeds, London. **Chapter 6** Fig. 6.1a: Alex Kerstitch. Fig. 6.1b: Peter Parks/Earth Scenes. Fig. 6.2: Marty Stouffer/Animals Animals. Fig. 6.3: Kay Chernush/The Image Bank. Fig. 6.4b: Courtesy Computer Graphics Laboratory, University of California, San Francisco. Fig. 6.8: Courtesy Stan Koszelek, Ph.D., University of California, Riverside. Fig. 6.10: Swarthout/The Stock Market. **Chapter 7** Fig. 7.5: Ed Reschke. Fig. 7.10a: Lennart Nilsson, ©Boehringer Ingelheim, International Gmbh; from *The Incredible Machine*. Human Perspective: (Fig. 1a) Martin Rotker/Phototake; (Fig. 1b) Cabisco/Visuals Unlimited. Fig. 7.11a: Courtesy Dr. Ravi Pathak, Southwestern Medical Center, University of Texas. **Chapter 8** Fig. 8.1a: Carr Clifton. Fig. 8.1b: Shizuo Lijima/Tony Stone World Wide. Fig. 8.5: Joe Englander/Viesti Associates, Inc. Fig. 8.6: Courtesy T. Elliot Weier. Fig. 8.10: Courtesy Lawrence Berkeley Laboratory, University of California. Fig. 8.13: Pete Winkel/Atlanta/Stock South. Bioline: Courtesy Woods Hole Oceanographic Institution. **Chapter 9** Fig. 9.1: Stephen Frink/AllStock, Inc. Bioline: (Fig. 1) Frank Oberle/Bruce Coleman, Inc.; (Fig. 2) Movie Stills Archive. Fig. 9.6: Topham/The Image Works. Fig. 9.9: H. Fernandez-Moran. Human Perspective: (top and bottom insets) Courtesy MacDougal; (top) Jerry Cooke; (bottom) Richard Kane/Sportchrome East/West. Fig. 9.11: Peter Parks/NHPA. Fig. 9.12a: Grafton M. Smith/The Image Bank. **Chapter 10** Fig. 10.1: Sipa Press. Fig. 10.2a: Institut Pasteur/CNRI/Phototake. Fig. 10.3a: Dr. R. Vernon/Phototake. Fig. 10.3b: CNRI/Science Photo Library/Photo Researchers. Fig. 10.6: Courtesy Dr. Andrew Bajer, University of Oregon, Fig. 10.7: CNRI/Science Photo Library/Photo Researchers. Fig. 10.8: Dr. G. Shatten/Science Photo Library/Photo Researchers. Fig. 10.9: Courtesy Professor R.G. Kessel. Fig. 10.10: David Phillips/Photo Researchers. **Chapter 11** Fig. 11.3: Courtesy Science Software Systems. Fig. 11.5a: Courtesy Dr. A.J. Solari, from *Chromosoma*, vol. 81, p. 330 (1980), Human Perspective: Donna Zweig/Retna. Fig. 11.7: Cabisco/Visuals Unlimited. **Part 3 Opener:** David Scharf. **Chapter 12** Fig. 12.1: Courtesy Dr. Ing Jaroslav Krizenecky. Fig. 12.6a: Doc Pele/Retna. Fig. 12.6b: FPG International. Fig. 12.9: Sydney Freelance/Gamma Liaison. Fig. 12.11 Hans Reinhard/Bruce Coleman, Inc. **Chapter 13** Fig. 13.2a: Robert Noonan. Fig. 13.6: Biological Photo Service. Bioline: Norbert Wu. Fig. 13.9: Historical Pictures Service. Fig. 13.11a: Courtesy M.L. Barr. Fig. 13.11b: Jean Pragen/Tony Stone World Wide. Fig. 13.12: Courtesy Lawrence Livermore National Laboratory. **Chapter 14** Fig. 14.1b: Lee D. Simon/Science Photo Library/Photo Researchers. Fig. 14.4: David Leah/Science Photo Library/Photo Researchers. Fig. 14.5: Dr. Gopal Murti/Science Photo Library/Photo Researchers. Fig. 14.6b: Fawcett/Olins/Photo Researchers. Fig. 14.7: Courtesy U.K. Laemmli. Fig. 14.10a: From M. Schnos and.R.B. Inman, *Journal of Molecular Biology*, 51:61-73 (1970), ©Academic Press. Fig. 14.10b: Courtesy Professor Joel Huberman, Roswell Park Memorial Institute. Human Perspective: (Fig. 2) Courtesy Skin Cancer Foundation; (Fig. 3) Mark Lewis/Gamma Liaison. Fig. 14.17: Courtesy Dr. O.L. Miller, Oak Ridge National Laboratory. **Chapter 15** Fig. 15.1: Courtesy Richard Goss, Brown University. Fig. 15.2a: (top left) Oxford Scientific Films/Animals Animals; (top right) F. Stuart Westmorland/Tom Stack & Associates. Fig. 15.2b: Courtesy Dr. Cecilio Barrera, Department of Biology, New Mexico State University. Fig. 15.3b: Courtesy Michael Pique, Research Institute of Scripps Clinic. Fig. 15.7b: Courtesy Wen Su and Harrison Echols, University of California, Berkeley. Fig. 15.8a: Courtesy Stephen Case, University of Mississippi Medical Center. Fig. 15.9: Roy Morsch/The Stock Market. **Chapter 16** Fig. 16.1a: David M. Dennis/Tom Stack & Associates. Fig. 16.1b: Courtesy Lakshmi Bhatnagor, Ph.D., Michigan Biotechnology Institute. Fig. 16.2: Art Wolfe/All Stock, Inc. Fig. 16.3: Ken Graham. Fig. 16.4a: Courtesy R.L. Brinster, Laboratory for Reproductive Physiology, University of Pennsylvania. Fig. 16.4b: John Marmaras/Woodfin Camp & Associates. Fig. 16.5: Courtesy Robert Hammer, School of Veterinary Medicine, University of

Pennsylvania. Human Perspective: Courtesy Dr. James Asher, Michigan State University. Fig. 16.6b: Professor Stanley N. Cohen/Photo Researchers. Fig. 16.10: Ted Speigel/Black Star. Fig. 16.11b: Philippe Plailly/Science Photo Library/Photo Researchers. Bioline: Photograph by Anita Corbin and John O'Grady, courtesy the British Council. **Chapter 17** ig. 17.2d: Courtesy Howard Hughs Medical Institution. Fig. 17.3: Tom Raymond/Medichrome/The Stock Shop. Fig. 17.7a: Culver Pictures. Fig. 17.7b: Courtesy Roy Gumpel. Fig. 17.9a: Will & Deni McIntyre/Photo Researchers. Human Perspective: Gamma Liaison. **Part 4 Opener:** J. H. Carmichael, Jr./The Image Bank. **Chapter 18** Fig. 18.1a: Ralph Perry/Black Star. Fig. 18.1b: Jeffrey Hutcherson/DRK Photo. Fig. 18.4: Courtesy Professor Ray Evert, University of Wisconsin, Madison. Fig. 18.5: Walter H. Hodge/Peter Arnold; (inset) Courtesy Carolina Biological Supply Company. Bioline: (Fig. 1) Doug Wilson/West Light; (Fig. 2) Courtesy Gil Brum. Fig. 18.7: Luiz C. Marigo/Peter Arnold. Fig. 18.8: Dr. Jeremy Burgess/Science Photo Library/Photo Researchers. Fig. 18.9: Saul Mayer/The Stock Market; (inset) David Scharf/Peter Arnold. Fig. 18.10a: Carr Clifton. Fig. 18.10b: Ed Reschke. Fig. 18.11: Courtesy Thomas A. Kuster, Forest Products Laboratory/USDA. Fig. 18.12: A.J. Belling/Photo Researchers. Fig. 18.13: Courtesy DSIR Library Centre. Fig. 18.15: Biophoto Associates/Photo Researchers; (inset) Courtesy C.Y. Shih and R.G. Kessel, reproduced from *Living Images*, Science Books International, 1982. Fig. 18.16: Courtesy Industrial Research Ltd, New Zealand. Fig. 18.19: Robert & Linda Mitchell; (inset) From *Botany Principles* by Roy H. Saigo and Barbara Woodworth Saigo, ©1983 Prentice-Hall, Inc. Reproduced with permission. Fig. 18.22a: Jerome Wexler/Photo Researchers. Fig. 18.22b: Fritz Polking/Peter Arnold. Fig. 18.22c: Dwight R. Kuhn. Fig. 18.22d: Richard Kolar/Earth Scenes. Fig. 18.24: Biophoto Associates/Science Source/Photo Researchers. **Chapter 19 Fig. 19.1a: Courtesy** of C.P. Reid, School of Natural Resources, University of Arizona. Fig. 19.3a: Peter Beck/The Stock Market. Fig. 19.3b: Dr. J. Burgess/Photo Researchers. Fig. 19.4: D. Cavagnaro/DRK Photo. Fig. 19.7a: Biophoto Associates/Photo Researchers. Fig. 19.7b: Alfred Pasieka/Peter Arnold. Fig. 19.7c: Courtesy DSIR Library Centre. Fig. 19.8: Martin Zimmerman. **Chapter 20** Fig. 20.1: G.I. Bernard/Earth Scenes/Animals Animals. Fig. 20.2a: Dwight R. Kuhn/DRK Photo. Bioline (Cases 1 and 3) Biological Photo Service; (Case 2) P.H. and S.L. Ward/Natural Science Photos; (Case 4): G.I. Bernard/Animals Animals. Fig. 20.3: E.R. Degginger. Fig. 20,4a: William E. Ferguson. Fig. 20.4b: Phil Degginger. Fig. 20.4c: E.R. Degginger. Fig. 20.5a: Runk/Shoenberger/Grant Heilman Photography. Fig. 20.5b: William E. Ferguson. Fig. 20.6a: Robert Harding Picture Library. Fig. 20,6b Richard Parker/Photo Researchers. Fig. 20.6c: Jack Wilburn/Earth Scenes/Animals Animals. Fig. 20.9a: Visuals Unlimted. Fig. 20.9b: Manfred Kage/Peter Arnold. Human Perspective: Harold Sund/The Image Bank. Fig. 20.15: Courtesy Media Resources, California State Polytechnic University. Fig. 20.16a: John Fowler/Valan Photos. Fig. 20.16b: Richard Kolar/Earth Scenes/Animals Animals. Fig. 20.17: Breck Kent. **Chapter 21** Fig. 21.1a: Gary Milburn/Tom Stack & Associates. Fig. 21.1b: Schafer & Hill/Peter Arnold. Human Perspective: (Fig. 1) Enrico Ferorelli; (Fig. 2) Michael Nichols/Magnum Photos, Inc. Fig. 21.5: Runk/Schoenberger/Grant Heilman Photography. Fig. 21.6: Courtesy Dr. Harlan K. Pratt, University of California, Davis. Fig. 21.7: Scott Camazine/Photo Researchers. Fig. 21.9a: E.R. Degginger. Fig. 21.9b: Fritze Prenze/Earth Scenes/Animals Animals. Fig. 21.12a: Pam Hickman/Valan Photos. Fig. 21.12b: R.F. Head/Earth Scenes/Animals Animals. **Part 5 Opener:** Joe Devenney/The Image Bank. **Chapter 22** Fig. 22.2b: Dr. Mary Notter/Phototake. Fig. 22.3b: Professor P. Motta, Department of Anatomy, University of "La Sapienza", Rome/Photo Researchers. Fig. 22.6a: Fred Bavendam/Peter Arnold. Fig. 22.6b: Michael Fogden/DRK Photo. Fig. 22.6c: Peter Lamberti/Tony Stone World Wide. Fig. 22.6d: Gerry Ellis Nature Photography. Fig. 22.7a: Mike Severns/Tom Stack & Associates. Fig. 22.7b: Norbert Wu. **Chapter 23** ig 23.1:Fawcett/Coggeshall/Photo Researchers. Fig. 23.5: (top left) Vu/©T. Reese and D.W. Fawcett/Visuals Unlimited; (bottom left) Courtesy Lennart Nilsson, *Behold Man*, Little Brown & Co., Boston. Fig. 23.9: Courtesy Lennart Nilsson, *Behold Man*, Little, Brown and Co., Boston. Fig. 23.10b: Alan & Sandy Carey. Human Perspective: Science Vu/Visuals Unlimited. Fig. 23.14: Focus on Sports. Fig. 23.15: The Image Works. Fig. 23.20a: Doug Wechsler/Animals Animals. Fig. 23.20b: Robert F. Sisson/National Geographic Society. **Chapter 24** Fig. 24.1a: Anthony Bannister/NHPA. Fig. 24.1b: Giddings/The Image Bank. Fig. 24.1c: Michael Fogden/Bruce Coleman, Inc. Fig. 24.4: (left) Omikron/Photo Researchers. Fig. 24.4: (inset) Don Fawcett/K. Saito/K. Hama/Photo Researchers. Fig. 24.5a: (top) Courtesy Lennart Nilsson, *Behold Man*, Little Brown & Co., Boston; (center) Don Fawcett/K. Saito/K. Hama/Photo Researchers. Fig. 24.5b: Star File. Fig. 24.6: Philippe Petit/Sipa Press. Fig. 24.7b: Jerome Shaw. Fig. 24.10a: Oxford Scientific Films/Animals Animals. Fig. 24.10b: Kjell Sandved/Photo Researchers. Fig. 24.10c: George Shelley; (inset) Raymond A. Mendez/Animals Animals. **Chapter 25** Fig. 25.1: Robert & Linda Mitchell. Fig. 25.7: Courtesy Circus World Museum. Fig. 25.8: Courtesy A.I. Mindelhoff and D.E.. Smith, *American Journal of Medicine*, 20: 133 (1956). Fig. 25.13: Graphics by T. Hynes and A.M. de Vos using University of California at San Francisco MIDAS-plus software; photo courtesy of Genentech, Inc. **Chapter 26** Fig. 26.1: Ed Reschke/Peter Arnold. Fig. 26.2a: Joe Devenney/The Image Bank. Fig. 26.2b: John Cancalosi/DRK Photo. Fig. 26.3: Wallin/Taurus Photos. Fig. 26.4a: E.S. Ross/Phototake. Fig. 26.5a: D. Holden Bailey/Tom Stack & Associates. Fig. 26.5b: Jany Sauvanet/Photo Researchers. Fig. 26.6a: (top) Courtesy Lennart Nilsson, *Behold Man*, Little Brown & Co., Boston; (center) Michael Abbey/Science Source/Photo Researchers. Fig. 26.6b: F. & A. Michler/Peter Arnold. Human Perspective: Courtesy Professor Philip Osdoby, Department of Biology, Washington University, St. Louis. Fig. 26.8: Duomo Photography, Inc. Fig. 26.12: A. M. Siegelman/FPG International. Fig. 26.19a: Alese & Mort Pechter/The Stock Market. Fig. 26.19b-c: Stephen Dalton/NHPA. **Chapter 27** Bioline, Fig. 27.3, and Fig. 27.5: Courtesy Lennart Nilsson, *Behold Man*, Little Brown & Co., Boston, Fig. 27.6: Micrograph by S.L. Palay, courtesy D.W. Fawcett, from *The Cell*, © W..B. Saunders Co. Fig. 27.8a: Courtesy Gregory Antipa, San Francisco State University. Fig. 27.8b: Gregory Ochocki/Photo Researchers. Human Perspective: Derik Muray Photography, Inc. Fig. 27.9a: Warren Garst/Tom Stack & Associates. Fig. 27.9b: Hervé Chaumeton/Jacana. Fig. 27.9c: D. Wrobel/Biological Photo Service. Fig. 27.10: Sari Levin. **Chapter 28** Fig. 28.2: Biophoto Associates/Photo Researchers. Fig. 28.4: Courtesy Lennart Nilsson, *Behold Man*, LIttle Brown & Co., Boston. Fig. 28.10: CNRI/Science Photo Library/Photo Researchers. Fig. 28.11a-b: Howard Sochurek. Fig. 28.11c: Dan McCoy/Rainbow. Fig. 28.12: Jean-Claude Revy/Phototake. Fig. 28.15: Dr. Tony Brain/Science Photo Library/Photo Researchers. Fig. 28.16: Alan Kearney/FPG International. Fig. 28.17a: Ed Reschke/Peter Arnold. Fig. 28.18: Manfred Kage/Peter Arnold. Fig. 28.19b: Runk/Schoenberger/Grant Heilman Photography. **Chapter 29** Fig. 29.1: Richard Kane/Sports Chrome, Inc. Fig. 29.2: Courtesy Ewald R. Weibel, Anatomisches Institut der Universität Bern, Switzerland, from "Morphological Basis of Alveolar-Capillary Gas Exchange", *Physiological Review*, Vol. 53, No. 2, April 1973, p. 425. Fig. 29.3: (top left and right insets) Courtesy Lennart Nilsson, *Behold Man*, Little Brown & Co., Boston; (center) CNRI/Science Photo Library/Photo Researchers. Fig. 29.7: Four By Five/SUPERSTOCK. Human Perspective: (Fig. 2) Vu/O. Auerbach/Visusals Unlimited; (Fig. 3) Photofest. Fig. 29.8: Kjell Sandved/Bruce Coleman, Inc. Fig. 29.10: Robert F. Sisson/National Geographic Society. Fig. 29.11a: Robert & Linda Mitchell. **Chapter 30** Fig. 30.1 and Fig. 30.5: Courtesy Lennart Nilsson, Boehringer Ingelheim International GmbH. Bioline: Courtesy Steven Rosenberg, National Cancer Institute. Fig. 30.6: G. Robert Bishop/AllStock, Inc. Fig. 30.7b: Courtesy A.J. Olson, TSRI, Scripps Institution of Oceanography. Fig. 30.10: Alan S. Rosenthal, from "Regulation of the Immune Response—Role of the Macrophage', *New England Journal of Medicine*, November 1980, Vol. 303, #20, p. 1154, courtesy Merck Institute for Therapeutic Research. Fig. 30.11: Nancy Kedersha. Human Perspective: (Fig. 1a) David Scharf/Peter Arnold; (Fig. 1b) Courtesy Acarology Laboratory, Museum of Biological Diversity, Ohio State University, Columbus; (Fig. 1c) Scott Camazine/

Photo Researchers; (Fig. 2) Courtesy Lennart Nilsson, Boehringer Ingelheim International GmbH. **Chapter 31** Fig. 31.1a: R. La Salle/Valan Photos. Fig. 31.1b: Richard Campbell/Biological Photo Service. Fig. 31.1c: Robert Harding Picture Library. Fig. 31.2a: Doug Perrine/DRK Photo. Fig. 31.2b: George Grall. Fig. 31.2c: Biological Photo Service. Fig. 31.2d: Densey Clyne/Ocford Scientific Films/Animals Animals. Fig. 31.3a: Chuck Nicklin. Fig. 31.3b: Robert & Linda Mitchell. Bioline: Hans Pfletschinger/Peter Arnold. Fig. 31.5: Courtesy Richard Kessel and Randy Kandon, from *Tissues and Organs,* W.H. Freeman. Fig. 31.9: C. Edelmann/Photo Researchers. Fig. 31.13: From A.P. McCauley and J.S. Geller, "Guide to Norplant Counseling", *Population Reports,* Sept. '92, Series K, No. 4, p. 4; photo courtesy Johns Hopkins University Population Information Program. **Chapter 32** Fig. 32.1a: Courtesy Jonathan Van Blerkom, University of Colorado, Boulder. Fig. 32.1b: Doug Perrine/DRK Photo. Fig. 32.2a: G. Shih and R. Kessel/Visuals Unlimited. Fig. 32.2b: Courtesy A.L. Colwin and L.H. Colwin. Fig. 32.3: Courtesy E.M. Eddy and B.M. Shapiro. Fig. 32.4: Courtesy Richard Kessel and Gene Shih. Fig. 32.10b: Courtesy L. Saxen and S. Toivonen. Fig. 32.13b: Courtesy K. Tosney, from *Tissue Interactions and Development* by Norman Wessels. Fig. 32.14: Dwight Kuhn. Fig. 32.17: Courtesy Lennart Nilsson, from *A Child Is Born.* **Part 6 Opener:** David Doubilet/National Geographic Society. **Chapter 33** Fig. 33.1: Ivan Polunin/NHPA. Fig. 33.3: Courtesy of Victor A. McKusick, Medical Genetics Department, Johns Hopkins University. Fig. 33.4a: Sharon Cummings/Dembinsky Photo Associates. Fig. 33.4b: © Ron Kimball Studios. Fig. 33.5b-c: Courtesy Professor Lawrence Cook, University of Manchester. Fig. 33.8a: COMSTOCK, Inc. Fig. 33.8b: Tom McHugh/AllStock, Inc. Fig. 33.8c: Kevin Schafer & Martha Hill/Tom Stack & Associates; (inset) Courtesy K.W. Barthel, Museum beim Solenhofer Aktienverein, Germany. Bioline: (Fig. a) Robert Shallenberger; (Fig. b) Scott Camazine/Photo Researchers; (Fig. c) Jane Burton/Bruce Coleman, Inc.; (Fig. d) R. Konig/Jacana/The Image Bank; (Fig. e) Nancy Sefton/Photo Researchers; (Fig. f) Seaphoto Limited/Planet Earth Pictures. Fig. 33.10a: John Garrett/Tony Stone World Wide. Fig. 33.10b: Heather Angel. Fig. 33.11a: Frans Lanting/Minden Pictures, Inc. Fig. 33.11b: S. Nielsen/DRK Photo. Fig. 33.12: Zeisler/AllStock, Inc. Fig. 33.13a: Zig Leszczynski/Animals Animals. Fig. 33.13b: Art Wolfe/AllStock, Inc. Fig. 33.16a: Gary Milburn/Tom Stack & Associates. Fig. 33.16b: Tom McHugh/Photo Researchers. Fig. 33.16c: Nick Bergkessel/Photo Researchers. Fig. 33.16d: C.S. Pollitt/Australasian Nature Transparencies. **Chapter 34** Fig. 34.1a: Leonard Lee Rue III/Earth Scenes/Animals Animals. Fig. 34.1b: Bradley Smith/Earth Scenes/Animals Animals. Fig. 34.4a: Courtesy Professor George Poinar, University of California, Berkeley. Fig. 34.4b: David Muench Photography. Fig. 34.5a: William E. Ferguson. Fig. 34.7a: Courtesy Merlin Tuttle. Fig. 34.7b: Stephen Dalton/Photo Researchers. Fig. 34.9: David Brill. Fig. 34.10: Courtesy Institute of Human Origins. Fig. 34.12: John Reader/Science Photo Library/Photo Researchers. **Chapter 35** Fig. 35.2a: Courtesy Dr. S.M. Awramik, University of California, Santa Barbara. Fig. 35.2b: William E. Ferguson. Fig. 35.4: Kim Taylor/Bruce Coleman, Inc. Fig. 35.5: Courtesy Professor Seilacher. Biuoline: Louie Psihoyos/Matrix. Fig. 35.11 and 35.13 Carl Buell. **Part 7 Opener:** Y. Arthus/Peter Arnold. **Chapter 36** Fig. 36.1: Courtesy Zoological Society of San Diego. Fig. 36.2a-b: David M. Phillips/Visuals Unlimited. Fig. 36.2c: Omikron/Science Source/Photo Researchers. Fig. 36.3a: Courtesy Wellcom Institute for the History of Medicine. Fig. 36.3b: Courtesy Searle Corporation. Fig. 36.5a: A.M. Siegelman/Visuals Unlimited. Fig. 36.5b: Science Vu/Visuals Unlimited. Fig. 36.5c: John D. Cunningham/Visuals Unlimited. Fig. 36.6: (top) Courtesy Dr. Edward J. Bottone, Mount Sinai Hospital, New York; (bottom) Courtesy R.S. Wolfe and J.C. Ensign. Fig. 36.7: CNRI/Science Source/Photo Researchers. Fig. 36.8: Courtesy C.C. Remsen, S.W. Watson, J.N. Waterbury and H.S. Tuper, from *J. Bacteriology,* vol 95, p. 2374, 1968. Human Perspective: Rick Rickman/Duomo Photography, Inc. Fig. 36.9: Sinclair Stammers/Science Source/Photo Researchers. Fig. 36.10: Courtesy R.P. Blakemore and Nancy Blakemore, University of New Hampshire. Bioline: (Fig. 1) Courtesy B. Ben Bohlool, NiftAL; (Fig. 2) Courtesy Communication Arts. Fig. 36.11: (left) Courtesy Dr. Russell, Steere, Advanced Biotechnologies, Inc.; (inset) CNRI/Science Source/Photo Researchers. 36.12b: Richard Feldman/Phototake. **Chapter 37** Fig. 37.1a: Jerome Paulin/Visuals Unlimited. Fig. 37.1b: Stanley Flegler/Visuals Unlimited. Fig. 37.1c: Victor Duran/Sharnoff Photos. Fig. 37.1d: Michael Fogden/DRK Photo. Figure 37.2: R. Kessel and G. Shih/Visuals Unlimited. Fig. 37.3: Courtesy Romano Dallai, Department of Biology, Universitá de Siena. Fig. 37.5a: M. Abbey/Visuals Unlimited. Fig. 37.5b: James Dennis/CNRI/Phototake. Fig. 37.8a: Manfred Kage/Peter Arnold. Fig. 37.8b: Eric Grave/Science Source/Photo Researchers. Fig. 37.9: Mark Conlin. Fig. 37.10: John D. Cunningham/Visuals Unlimited. Fig. 37.11a: David M. Phillips/Visuals Unlimited. Fig. 37.11b: Kevin Schafer/Tom Stack & Associates. Fig. 37.12a: E.R. Degginger. Fig. 37.12b: Waaland/Biological Photo Service. Fig. 37.13: Dr. Jeremy Burgess/Science Source/Photo Researchers. Fig. 37.14: Sylvia Duran Sharnoff. Fig. 37.15: Herb Charles Ohlmeyer/Fran Heyl Associates. Fig. 37.16 and 37.17a: Michael Fogden/Earth Scenes/Animals Animals. Fig. 37.17b: Stephen Dalton/NHPA. **Chapter 38** Fig. 38.1a: (left) Courtesy Dr. C.H. Muller, University of California, Santa Barbara; (inset) William E. Ferguson. Fig. 38.1b: Doug Wechsler/Earth Scenes. Fig. 38.2: T. Kitchin/Tom Stack & Associates. Fig. 38.3a: Michael P. Gadomski/Earth Scenes/Animals Animals. Fig. 38.3b: Courtesy Edward S. Ross, California Academy of Sciences. Fig. 38.3c: Rod Planck/ Photo Researchers. Fig. 38.3d: Kurt Coste/Tony Stone World Wide. Bioline: (Fig. a) Michael P. Gadomski/Photo Researchers; (Fig. b-c) E.R. Degginger; (Fig. d) COMSTOCK, Inc. Fig. 38.6a: Kim Taylor/Bruce Coleman, Inc. Fig. 38.6b: Breck Kent. Fig. 38.6c: Flip Nicklin/Minden Pictures, Inc. Fig. 38.6d: D.P. Wilson/Eric & David Hosking/Photo Researchers. Fig. 38.7: J.R. Page/Valan Photos; (inset) Stan E. Elems/Visuals Unlimited. Fig. 38.8a: Runk/Schoenberger/Grant Heilman Photography. Fig. 38.8b: John Gerlach/Tom Stack & Associates. Fig. 38.8c: Robert A. Ross/R.A.R.E. Photography. Fig. 38.11: John H. Trager/Visuals Unlimited. Fig. 38.13: John D. Cunningham/Visuals Unlimited. Fig. 38.14: Milton Rand/Tom Stack & Associates. Fig. 38.17: Cliff B. Frith/Bruce Coleman, Inc. Fig. 38.18: Leonard Lee Rue/Photo Researchers. Fig. 38.19: Anthony Bannister/Earth Scenes. Fig. 38.21a: J. Carmichael/The Image Bank. Fig. 38.21b: Sebastio Barbosa/The Image Bank. Fig. 38.21c: Holt Studios/Earth Scenes. Fig. 38.21d: Gwen Fidler/COMSTOCK, Inc. **Chapter 39** Fig. 39.1a: Courtesy Karl G. Grell, Universität Tübingen Institut für Biologie. Fig. 39.1b: François Gohier/Photo Researchers. Fig. 39.2a: Larry Ulrich/DRK Photo. Fig. 39.2b: Christopher Newbert/Four by Five/SUPERSTOCK. Fig. 39.6a: Chuck Davis. Fig. 39.7a: Robert & Linda Mitchell. Fig. 39.7b: Runk/Schoenberger/Grant Heilman Photography. Fig. 39.7c: Fred Bavendam/Peter Arnold. Fig. 39.10a: Carl Roessler/Tom Stack & Associates. Fig. 39.10b: Goivaux Communication/Phototake. Fig. 39.10c: Biomedia Associates. Fig. 39.14a: Gary Milburn/Tom Stack & Associates. Fig. 39.14b: E.R. Degginger. Fig. 39.14c: Christopher Newbert/Four by Five/SUPERSTOCK. Fig. 39.15: Courtesy D. Phillips. Fig. 39.16a: Marty Snyderman/Visuals Unlimited. Fig. 39.16b: Robert & Linda Mitchell. Fig. 39.18a: John Cancalosi/Tom Stack & Associates. Fig. 39.18b: John Shaw/Tom Stack & Associates. Fig. 39.18c: Patrick Landman/Gamma Liaison, Fig. 39.20a: Robert & Linda Mitchell. Fig. 39.20b: Edward S. Ross, National Geographic, *The Praying of Predators,* February 1984, p. 277. Fig. 39.20c: Maria Zorn/Animals Animals. Fig. 39.21: Courtesy Philip Callahan. Fig. 39.22a: Biological Photo Service. Fig. 39.22b: Kim Taylor/Bruce Coleman, Inc. Fig. 39.22c: Tom McHugh/Photo Researchers. Fig. 39.23a: E.R. Degginger. Fig. 39.23b: Christopher Newbert/Four by Five/SUPERSTOCK. Fig. 39.26b: Dave Woodward/Tom Stack & Associates. Fig. 39.26c: Terry Ashley/Tom Stack & Associates. Fig. 39.28a: Russ Kinne/COMSTOCK, Inc. Fig. 39.28b: Ken Lucas/Biological Photo Service. Fig. 39.30: Marc Chamberlain/Tony Stone World Wide. Fig. 39.31a: W. Gregory Brown/Animals Animals. Fig. 39.31b: Chris Newbert/Four by Five/SUPERSTOCK. Fig. 39.32a: Zig Leszcynski/Animals Animals. Fig. 39.32b: Chris Mattison/Natural Science Photos. Fig. 39.33a: Michael Fogden/Animals Animals. Fig. 39.33b: Jonathan Blair/Woodfin Camp & Associates. Fig. 39.34a: Bob and Clara Calhoun/Bruce Coleman, Inc. Fig.

39.35a-b: Dave Watts /Tom Stack & Associates. Fig. 39.35c: Boyd Norton. **Part 8 Opener:** Carr Clifton. **Chapter 40** Fig. 40.1: Courtesy Earth Satellite Corporation. Fig. 40.4: Norbert Wu/Tony Stone World Wide. Fig. 40.5: Stephen Frink/AllStock, Inc. Fig. 40.6a-b: Nicholas De Vore III/Bruce Coleman, Inc. Fig. 40.6c: Juergen Schmitt/The Image Bank. Fig. 40.7a: Frans Lanting/Minden Pictures, Inc. Fig. 40.7b: D. Cavagnaro/DRK Photo. Fig. 40.7c: Jeff Gnass Photography. Fig. 40.7d: Larry Ulrich/DRK Photo. Fig. 40.9: Joel Arrington/Visuals Unlimited. Fig. 40.11a: K. Gunnar/Bruce Coleman, Inc. Fig. 40.11b: Claude Rives/Agence C.E.D.R.I. Fig. 40.11c: Maurice Mauritius/Rapho/Gamma Liaison. Human Perspective: Will McIntyre/AllStock, Inc. Fig. 40.13: Jim Zuckerman/West Light. Fig. 40.14: James P. Jackson/Photo Researchers. Fig. 40.15: Steve McCutcheon/Alaska Pictorial Service. Fig. 40.16: Linda Mellman/Bill Ruth/Bruce Coleman, Inc. Fig. 40.17: Bill Bachman/Photo Researchers. Fig. 40.18: Nigel Dennis/NHPA. Fig. 40.19: Bob Daemmrich/The Image Works. Fig. 40.20: B.G. Murray, Jr./Animals Animals. Fig. 40.21a: Breck Kent. Fig. 40.21b: Carr Clifton. **Chapter 41** Fig 41.1a: Gary Braasch/AllStock, Inc. Fig. 41.1b: George Bermard/NHPA. Fig. 41.1c: Stephen Krasemann/NHPA. Fig. 41.2: Robert & Linda stures, Inc. Fig. 41.7a: E.R. Degginger/Photo Researchers. Fig. 41.7b: Anthony Mercieca/Natural Selection. Fig. 41.9: Walter H. Hodge/Peter Arnold. Fig. 41.16: Larry Ulrich/DRK Photo. Fig. 41.18: © Gary Braasch. **Chapter 42** Fig. 42.1a: Andy Callow/NHPA. Fig. 42.1b: Al Grotell. Fig. 42.5a: Cosmos Blank/Photo Researchers. Fig. 42.5b: Karl and Steve Maslowski/Photo Researchers. Fig. 42.5c: Michael Fogden/Bruce Coleman, Inc. Fig. 42.5d: Gregory G. Dimijian, M.D./Photo Researchers. Fig. 42.6: B&C Calhoun/Bruce Coleman , Inc. Fig. 42.7: Kjell Sandved/Photo Researchers. Fig. 42.8a: Wolfgang Bayer/Bruce Coleman, Inc. Fig. 42.8b: J. Carmichael/The Image Bank. Fig. 42.9: Anthony Bannister/NHPA. Fig. 42.10a: Adrian Warren/Ardea London. Fig. 42.10b: Peter Ward/Bruce Coleman, Inc. Fig. 42.10c: Michael Fogden. Fig. 42.11: P.H. & S.L. Ward/Natural Science Photos. Fig 42.12: Tom McHugh/Photo Researchers. Fig. 42.13: E.R. Degginger. Fig. 42.14: William H. Amos. Fig. 42.16: Dr. J.A.L. Cooke/Oxford Scientific Films/Animals Animals. Fig. 42.17: David Cavagnaro/Visuals Unlimited. Fig. 42.19: Stephen G. Maka. Fig. 42.20: Raymond A. Mendez/Animals Animals; (inset): Eric Grave/Phototake. **Chapter 43** Fig. 43.1a: Steve Krasemann/DRK Photo. Fig. 43.1b: Carr Clifton. Fig. 43.1c: Willard Clay. Fig. 43.3: Robert & Linda Mitchell. **Chapter 44** Fig. 44.1: Johnny Johnson/DRK Photo. Fig. 44.2: Dwight R. Kuhn. Fig. 44.3: Mike Severns/Tom Stack & Associates. Fig. 44.4: Bettmann Archive. Fig. 44.5: Robert Maier/Animals Animals. Fig. 44.6: Courtesy Masao Kawai, Primate Research Institute, Kyoto University. Fig. 44.7: Glenn Oliver/Visuals Unlimited. Fig. 44.8: Konrad Wothe/Bruce Coleman, Ltd. Fig. 44.9: Gordon Wiltsie/Bruce Coleman, Inc. Fig. 44.10: Hans Reinhard/Bruce Coleman, Ltd. Fig. 44.11: Oxford Scientific Films/Animals Animals. Fig. 44.12: Patricia Caulfield. Fig. 44.13: Roy P. Fontaine/Photo Researchers. Fig. 44.14: Anup and Manoj Shah/Animals Animals. **Appendix B** Fig. b: Bob Thomason/Tony Stone World Wide. Fig. c: Dr. Gennavo/Science Photo Library/Photo Researchers. Fig. d: F.R. Turner, Indiana University/Biological Photo Service.

Index

Note: A t following a page number denotes a table, f denotes a figure, and n denotes a footnote.

A

Abiotic environment, 934–936, 936–937
ABO blood group antigens, 240
"Abortion" pill, 679
Abscisic acid, 427t, 433
Abscission, 433
Absorption of nutrients, 568, 574
Accidental scientific discovery, 35–37
Accommodation, 503
Acetylcholine, 475t
Acetyl coenzyme A (acetyl CoA), 181
Acid rain, 392, 918
Acids, 61–62
Acoelomate, 870
Acquired immune deficiency syndrome (AIDS), 657, 794, 796–797
Acromegaly, 526, 527f
Acrosome, 670
ACTH (adrenocorticotropic hormone), 522t, 526
Actin, 555
Action potential, 471–473
Activation energy, 123, 124f
Active immunity, 653
Active site of enzyme, 124f, 125
Active transport, 142–144
"Adam's apple," 625
Adaptation, 4–5, 728–729
  and evolution, 4–5, 18
  to food type, 577, 581
  and nitrogenous wastes, 613
  and osmoregulation, 611–613
  and photosynthesis, 164–166
  of respiratory system, 632–635
Adaptive radiation, 734
Addison's disease, 528–529
Adenine, 269, 270f
  base pairing
    in DNA, 271f, 272
    in RNA, 278
Adenosine triphosphate, *see* ATP
Adenylate cyclase, 534, 535f
ADH, *see* Antidiuretic hormone
Adipocytes, 76
Adrenal gland, 522t, 526–529
  cortex, 522t, 526, 528–529
  medulla, 522t, 529
Adrenaline (epinephrine), 522t, 529
Adrenocorticotropic hormone (ACTH), 522t, 526
Adventitious root system, 373, 374f
Aerobes, appearance on earth, 174
Aerobic respiration, 180–184
  versus fermentation, 175, 178
Aestivation, 927
Agar, 841
Age of Mammals, 774
Age of Reptiles, 771
Agent Orange, 431
Age–sex structure of population, 988
  for humans, 999, 1000f
Aggregate fruits, 412, 413f
Agnatha, 886–887
AIDS (acquired immune deficiency syndrome), 642–643, 657, 794, 796–797
AIDS virus, *see* Human immunodeficiency virus
"Air breathers," 633, 635
Alarm calls, 1023
Albinism, 240
Alcohol, in pregnancy, 704
Alcoholic fermentation, 178, 179
Aldosterone, 522t, 528, 610–611
Algae, 816
  as bacteria, 792
  as lichens, 821
  as plants, 834, 835–842
  as protists, 816–819
Algin, 840
Alkalinity, 61
Allantois, 703
Allele frequency, 717–718
Alleles, 234–235, 240, 241
  linkage groups and, 249, 250
Allelochemicals, 972
Allelopathy, 974–975
Allergies, 656
Alligators, 889
Allopatric speciation, 732
Allosteric site, 130f
Alpha-helix, 67, 80
Alpine tundra, 922
Alternation of generations, 833
Altruism, 1022–1026
Alveolar-capillary gas exchange, 625–626, 627
Alveoli, 625–626
Alzheimer's disease, 101, 483, 484
Ames test, 286–287
Amino acids, 79–80
  assembly into proteins, 78f, 277, 278, 284–285, 286f
  and diet, 578
  and genetic code, 278–279, 282f
  sequencing, 66–67, 84, 750
Amino group, 78f, 79
Ammonia, excretion, 607, 613
Amniocentesis, 347f, 348
Amnion, 703
Amniotic fluid, 703
Amoebas, 813, 816
Amphibians, 888–889
  heart in, 605f
Anabolic pathways, 128
Anabolic reactions, electron transfer in, 128–129
Anaerobes, 174, 791
Anaerobic respiration, 179n
Analogous, versus homologous, 458, 744–746
Analogous features, 744
Anaphase, of mitosis, 203f, 206
Anaphase I, of meiosis, 217f, 218–219
Anaphase II, of meiosis, 217f, 221
Anaphylaxis, 656
Anatomy, and evolution, 749–750
Androgen-insensitivity syndrome, 305
Anecdotal evidence, 31
Angiosperms, 849, 852–853
Angiotensin, 591
Animal behavior, 1006–1030
Animal kingdom, 858–897
  phyla, 863, 864t
  phylogenetic relations, 865f
  phylum Annelida, 864t, 875–876
  phylum Arthropoda, 864t, 876–882
  phylum Chordata, 864t, 884–886
  phylum Cnidaria, 864t, 867–869
  phylum Echinodermata, 864t, 882–884
  phylum Mollusca, 864t, 874–875
  phylum Nematoda, 864t, 871–873
  phylum Platyhelminthes, 864t, 869–871
  phylum Porifera, 864t, 864–867
Animal models, 320
Animals:
  basic characteristics, 860–861
  body plan, 861–863
  cognition in, 1024
  desert adaptations, 926–927
  evolution, 861, 865f, 873
  form and function, 448–464
  growth and development, 686–709
Annelida, 864t, 875–876, 877f
Annual plants, 361
Annulus of fern, 847, 848f
Antarctic ozone hole, 902–903
Anteater, 893
Antennae of insects, 879
Antenna pigments, 158f, 159
Anther, 403
Antheridium, 842
Anthophyta, 839t, 852–853
Anthropoids, 894f
Antibiotics, discovery of, 37
Antibodies, 646, 647, 650–654
  formation of, 650–652
    DNA rearrangement, 651, 652f
  monoclonal, 654
  to oneself (autoantibodies), 654
  production of, 652–653
  specificity, 650
  structure, 650, 651f
Anticodon, 279–280, 283f
Antidiuretic hormone (ADH; vasopressin), 522t, 525, 610, 611
Antigen-binding site, 650, 651f
Antigens, 650
  and antibody production, 652
  defined, 646
  presentation of, 652, 654f
Anti-oncogenes, 260
Antioxidants, and free radicals, 56
Aorta, 595
Apes, 893
  genetic relation to human, 259, 329
Aphids:
  and ants, 962f
  and phloem research, 394
Apical dominance, 430, 431f
Apical meristem, 362, 372f, 373, 375f
Aposematic coloring, 970
Appendicitis, 575
Appendicular skeleton, 548–550
Appendix, 574–575, 576f
Aquatic animals, respiration in, 620–621, 632–633
Aquatic ecosystems, 908–916
Archaebacteria, 791, 796–798
Archaeopteryx, 746–747, 748f
Archegonium, 842
Archenteron, 693f
Arctic tundra, 922
Arteries, 589–590, 595
Arterioles, 590
Arthropoda, 864t, 876–882
Artificial selection, 721
Asbestosis, 101

Ascaris, 873
Ascocarp, 823
Ascomycetes, 822t, 823–824
Ascorbic acid (vitamin C), 540–541, 579t
  and cancer, 31, 39–40
  and colds, 30–31
Ascospores, 823
Ascus, 823
Asexual reproduction, 199, 664–665
  in flowering plants, 417–418
Association, defined, 933
Asthma, 656
Atherosclerosis, 75, 146, 590
Athletes and steroids, 528
Atmosphere, 904
Atolls, 912
Atomic basis of life, 48–64
Atomic mass, 52
Atomic number, 52
Atoms, structure of, 51–55
ATP (adenosine triphosphate), 121–123
  in active transport, 142–143, 144f
  chemical structure, 121f
  formation of
    by chemiosmosis, 161, 184
    by electron transport, 182–184
    by glycolysis, 175–178
    by Krebs cycle, 180f, 181, 182f
    energy yield, 178, 179, 180, 182, 185, 188
    proton gradient and, 161, 184
    site of, 172–173, 184f
  and glucose formation, 163–164
  hydrolysis of, 121f, 122–123
  and muscle, 186–187, 552, 556
  photosynthesis and, 160f, 161–163
ATPase, 142
ATP synthase, 160f, 161, 173, 184
Atrioventricular (AV) node, 599
Atrioventricular (AV) valves, 597
Atrium of heart, 593f, 594
Australopithecines, 752, 754, 755f
Australopithecus species, 743, 752, 754, 755f
Autoantibodies, 654
Autoimmune diseases, 656–657
Autonomic nervous system, 490–491
Autosomal dominant disorders, 343–345
Autosomal recessive disorders, 343
Autosomes, 253
Autotoxicity, 974
Autotrophs, 154–155, 764
Auxins, 425, 427, 430–431
Aves, 889–892
AV (atrioventricular) node, 599
AV (atrioventricular) valves, 597
Axial skeleton, 548, 549f
Axillary bud, 377
Axon, 468, 469f
AZT (azidothymidine), 657

◄ B ►

Backbone, 886
  origin of, 694
Bacteria, 786–800. *See also* Prokaryotes
  beneficial, 799
  chemosynthetic, 167–168, 798
  in digestive tract, 575, 581
  disease-causing, 792
  enzyme inhibition in, 126
  in fermentation, 179
  in genetic engineering, 314, 316, 324
  in genetics research, 266–267
  gene transfer between, 804
  growth rate, 788
  metabolic diversity, 791
  motility, 792
  and nitrogen fixation, 390, 799, 949–951
  structure of, 793f
  taxonomic criteria, 788–791
  ubiquity of, 799
Bacterial chlorophyll, 791–792
Bacteriophage, 268, 269f, 801
Balance, 506, 508–509
Ball-and-socket joint, 550
Bark, 369
Barnacles, 876
Barr body, 256f
Base pairing:
  in DNA, 271f, 272
  in RNA, 278
Bases, acids and, 61–62
Basidia, 824
Basidiocarp, 824–825
Basidiomycetes, 822t, 824–825
Basidiospores, 824
Basophils, 601–602
Bats, 498–499, 750, 751f
B cells, 647, 652–653
Beagle (HMS), voyage of, 20f, 21
Bean seedling development, 416, 417f
Beards, and gene activation, 305
Bees:
  dances, 1019, 1020f
  haplodiploidy in, 1025–1026
Behavioral adaptation, 729
Behavior of animals, 1006–1030
  development of, 1014–1015
  genes and, 1008–1009, 1010–1011
  mechanisms, 1008–1011
Beri-beri, 566–567
Beta-pleated sheet, 80
Bicarbonate ion, in blood, 629
Biennial plants, 361
Bile salts, 574
Binomial system of nomenclature, 13
Biochemicals, defined, 68
Biochemistry, 66–86, 750
Bioconcentration, 946
Biodiversity Treaty, 997
Bioethics, 25, 40–41. *See also* Ethical issues
Biogeochemical cycles, 947–952
Biogeography, 750–752
Biological classification, 11–15
Biological clock, of plants, 434, 438
Biological magnification, 946
Biological organization, 7, 11, 12f
Biological species concept, 732, 830
Biology, defined, 6
Biomass, 943
  pyramid of, 944
Biomes, 907f, 916–927, 934
Biosphere, 11, 902–930
  boundaries, 904
Biotechnology:
  applications, 314–321
  DNA "fingerprints," 328, 329
  genetic engineering techniques, 321–329
  opposition to, 321
  patenting genetic information, 315
  recombinant DNA techniques, 321–325. *See also* Recombinant DNA
  without recombinant DNA, 325–329
Biotic environment, 934–936, 936–937
Biotic (reproductive) potential, 986, 989–990
Biotin, 579t
Birds, 889–892
  fossil, 746–747, 748f
Birth:
  canal, 673
  control, 677–681
  control pills, 678–679
  in human, 705
Bivalves, 874
Blade of leaf, 377
Blastocoel, 690
Blastocyst, 700
Blastodisk, 691f
Blastomere, 690, 692
Blastopore, 693f, 873
Blastula, 690, 691f
Blastulation, 690
"Blind" test, 39
Blood:
  clotting, 450, 602–603
  composition, 600–603
  flow
    through circulatory system, 594f
    through heart, 593f
  gas exchange in, 627–629
Blood-brain barrier, 110
Blood:
  pressure, 589–590. *See also* Hypertension
  type, 94, 240
  vessels, 588–593
Blooms of algae, 818
Blue-green algae, *see* Cyanobacteria
Body:
  cavities, 861
  defense mechanisms
    nonspecific, 644–645
    specific, 646–650
  fluids:
    compartments, 588
    regulation of, 606–613
  mechanics, 558–559
  organization, 696–698
    genetic control of, 686–687
  plan, 861–863
  size, 460
  surface area, 460–462
    and respiration, 622–623, 626
  symmetry, 861, 862f
  temperature, 449, 450, 451f, 588, 614
  weight, 76, 460
Body contact and social bonds, 1021
Body planes, 862f
Body segmentation, 863, 875–876
Bohr effect, 629
Bonding, 1021–1022
Bones, 546–548
  human skeleton, 548, 549f
  repair and remodeling, 550
Bony fishes, 887, 888f
Bottleneck, genetic, 720
Botulism, 478
Bowman's capsule, 608–609
Brain, 481–487
  evolution, 492
  left versus right, 504
  and sensory stimuli, 510–511
Brainstem, 481f, 486
Breast-feeding, and immunity, 655
Breathing:
  mechanics of, 626–627
  regulation of, 629, 632
Bristleworms, 875–876
Bronchi, 625
Bronchioles, 625
Brown algae, 838–840
Bryophyta, 839t
Bryophytes, 835, 842–843
Budding, in animals, 664
Buffers, 62
Bundle-sheath cells, 165f, 166
Bursitis, 551

C

C_3 synthesis, 163–164
C_4 synthesis, 163t, 164–166
Caecilians, 888–889
Calciferol (vitamin D), 579t
Calcitonin, 522t, 531
Calcium:
  blood levels, 520, 530f, 531
  in diet, 580, 580t
  in muscle contraction, 557
Calorie, 60
Calvin–Benson cycle, 164
Calyx of flower, 403
Cambia, 362, 372–373
Camel's hump, 927
Camouflage, 967–968
cAMP (cyclic AMP), 533–534, 535f, 819
CAM synthesis, 163t, 166
Cancer, *see also* Carcinogens
  and cell cycle, 202
  and chromosomal aberrations, 259, 260
  genetic predisposition to, 346
  immunotherapy for, 648
  and lymphatic system, 606
  of skin, 280–281
  and smoking, 630
  viruses causing, 803
Capillaries, 587, 590–593
  lymphatic, 606
Capillary action, 60
Capsid, viral, 801
Capsule, bacterial, 792
Carbohydrases, 573
Carbohydrates, 70–74
  as energy source, 188
  in human diet, 578
  of plasma membrane, 94
  synthesis in plants, 163–166
Carbon:
  biological importance, 68–69
  in oxidation and reduction, 128
  properties, 68–69
Carbon cycle, 948–949
Carbon dating of fossils, 54
Carbon dioxide:
  and bicarbonate ions, 629
  in respiration, 628f, 629, 632
Carbon dioxide fixation, 163–164
  and leaf structure, 165–166
Carbonic acid, 629
Carbonic anhydrase, 629
Carbon monoxide poisoning, 627
Carboxyl group, 78f, 79
Carcinogens, 286–287
  testing for, 37–39, 286–287
Cardiac muscle, 558, 559f
Cardiac sphincter, 569
Cardiovascular system, *see* Circulatory system
Careers in biology, Appendix D
Carnivore, 966
Carotenoids, 157
Carpels, 403
Carrier of genetic trait, 254–255, 343
Carrier proteins of plasma membrane, 141, 143f
Carrying capacity (*K*) of environment, 992, 999–1001
Carson, Rachel, 714–715
Cartilage, 548
Cartilaginous fishes, 887
Casparian strips, 374, 376f, 387, 389f
Catabolic pathways, 128
Catalysts, enzymes as, 123
Cavemen, 742
Cecum, 574–575, 576f, 581
Cell body of neuron, 468, 469f
Cell cycle, 199–202
Cell death, and development, 697
Cell differentiation, 302–303, 698–699
Cell division, 194–210, 212–225. *See also* Meiosis; Mitosis
  and cancer, 202
  defined, 196
  in eukaryotes, 197–199
  in prokaryotes, 197
  regulation of, 201–202
  triggering factor, 194–195
  types, 196–197
Cell fate, 690–691
  induction and, 695
Cell fusion, 89
Cell-mediated immunity, 647
Cell plate, in plant cytokinesis, 206
Cells, 88–114. *See also* Eukaryotic cells; Prokaryotic cells
  discovery of, 90–91
  exterior structures, 108–110
  as fundamental unit of life, 91
  genetically engineered, 314
  how large particles enter, 136–137, 144–147
  how molecules enter, 138–144
  junctions between, 109–110
  locomotion, 106–107
  membrane, *see* Plasma membrane
  as property of living organism, 7, 11
  size limits, 92–93
  specialization, 302–303
  totipotent, 297
Cell surface receptor, 94, 147–148, 533–534, 535f
Cell theory, 90–91
Cellular defenses
  nonspecific, 645
  specific (T cells), 647, 649–650
Cellular slime molds, 819
Cellulose, 73, 119f
  in diet, 578, 581
Cell wall, 108
  of bacteria, 787
  of plants, 108–109
Cenozoic Era, 774
Centipedes, 876, 881–882
Central nervous system (CNS), 479, 481–489
  brain, *see* Brain
  spinal cord, 488–489
Centromere, 198, 204f, 218f
  versus kinetochore, 205n
Cephalization, 492, 861
Cephalopods, 874–875
Cerebellum, 481f, 486
Cerebral cortex, 481f, 482
Cerebrospinal fluid, 482
Cerebrum, 482–485
Certainty and science, 40–41
Cervix, 673
Cestodes, 869, 870f, 871
Chain (polypeptide) elongation and termination, 285
Chain of reactions, 1009
Chaparral, 924, 925f
Character displacement, 965
Chargaff's rules, 272
Charging enzymes, 280, 286f
Cheetah, 730
Chelicerates, 876, 878f, 881
Chemical bonds, 55–58
Chemical defenses, 972, 974–975
Chemical energy, from light energy, 156–163
Chemical evolution, 760–761
Chemical reactions:
  electron transfer in, 128–129
  endergonic, 122–123
  energy in, 121–123
  exergonic, 122
  favored versus unfavorable, 122
Chemiosmosis, 161
Chemoreception, 509–510, 512
Chemoreceptors, 500, 501t
  and breathing control, 632
Chemosynthesis, 167–168
Chemosynthetic bacteria, 167–168, 798
Chest cavity, 626
Chiasmata (of chromosomes), 218f
Chimpanzee, 894f
Chitin, 73
Chlamydia infections, 795
Chlorophylls, 157
  bacterial, 791–792
Chlorophyta, 835, 837
Chloroplasts, 104
  forerunners of, 792, 793f
  and origin of eukaryote cell, 110, 765, 962
  in photosynthesis, 155, 156, 158f
Cholecystokinin, 522t, 573
Cholera, 573
Cholesterol, 75, 77f, 136–137
  and atherosclerosis, 75, 146
Chondrichthyes, 888f
Chondrocytes, 548
Chordamesoderm, 694–695
Chordata, 864t, 884–886
Chorion, 700, 703
Chorionic villus sampling, 347f, 348
Chromatids, 198, 204f, 216, 218
Chromatin, 97
  and DNA packing, 274f
Chromosomal exchange, 250, 292
Chromosomal puffs, 305
Chromosomes, 97, 246–264, 334–353. *See also* Sex chromosomes; Homologous chromosomes; Meiosis; Mitosis
  abnormalities, 221, 257–259
    deletions, 258
    duplications, 258
    inversions, 259
    translocations, 259
    wrong number, 221, 336–337
      detection in fetus, 349
      nondisjunction, 236
      polyploidy, 260–261
  banding patterns, 198f, 253f
  condensation, 203–204
  discovery of functions, 246–247
  DNA in, 273, 274f, 275f
  in eukaryotes, 273
  giant (polytene), 252–253, 305
  mapping, 252, 338
    Human Genome Project, 340
  mitotic, 198–199, 204
  number, 199
    in humans, 212–213
  in prokaryotes, 273
  structure, 273, 274f, 275f
Chrysophytes, 817–818
Chyme, 570
Chymotrypsin, 573
Cigarette smoking, 630–631, 704
Cilia:
  of cell, 106–107
  of protozoa, 814–816
Ciliata, 815t
Circadian plant activities, 438
Circulatory (cardiovascular) systems, 455, 588–605
  discovery of, 586–587
  evolution of, 603–605
  in humans, 456f, 588–600
  open versus closed, 605
Circumcision, 669

Citric acid cycle, 181
Clams, 864t, 874
Class, in taxonomic hierarchy, 15
Classical conditioning, 1011–1012
Cleavage, 690
Cleavage furrow, in cell division, 206
Climate, 905–907, 916
Climax community, 952
Clitoris, 673
Clock, biological, 434, 438
Clonal deletion, 654
Clonal selection theory, 651–653
Clones, 296–297
Clotting factors, 602
Clotting of blood, 450, 602–603
Club fungi, 824
Club mosses, 845–846
Clumped distribution patterns, 987
Cnidaria, 864t, 867–869
CNS, *see* Central nervous system; Brain
Coagulation (clotting) of blood, 450, 602–603
Coastal water habitats, 908, 910–912
Cobalamin (vitamin B_{12}), 579t
Cocaine, 487
Cochlea, 507f, 508
Codominance, 239–240
Codons, 278
  role in mutations, 286
  role in translation, 279–281, 284–285
Coelacanth fish, 859
Coelom, 861, 863f, 871–872, 875–876
Coenzyme, 127
Coenzyme A (CoA), 181
Coevolution, 734f, 735
Cofactors, 127
Cognition in animals, 1024
Coherence theory of truth, 41
Cohesion of water, 60
Coitus interruptus, 681
Cold (disease), 803
Coleoptile, 417
Coleoptile tip, 424–425
Collagen, 453–454, 541
Collecting duct, 608–609
Collenchyma cells, 364
Colon, 576f
Color adaptations, 968, 970
Colorblindness, 255f, 256f
Combat, 1017, 1018–1019
Commensalism, 976
Communication, 1018–1022
  functions, 1018–1019
  types of signals, 1019–1021
Communities, 11, 932–933, 960–982
  changes in, 952
  climax, 952
  pioneer, 952
Compact bone, 546, 547f
Companion cell of plant, 372
"Compass" bacterium, 798
Competition, 960–961, 963–965
Competitive exclusion, 941, 961, 964
Complement, 645
Complementarity in DNA structure, 272
Complete flower, 407
Composite flower, 406f
Compound (chemical), 55
Compound leaf, 377
Concentration gradient, 138, 142, 144f
  and energy storage, 143–144, 145f
  of protons, 160f
Condensation of molecules, 69, 70f
Conditioned response, 1011–1012
Conditioning, 1011–1012
Condoms, 681
Cones of gymnosperms, 849
Cones of retina, 503, 505f, 506
Confirmation in scientific method, 33
Conformational changes in proteins, 81
Coniferophyta, 839t, 851–852
Coniferous forests, 917, 920, 921f
Conjugation in protists, 222
Connective tissue, 453–454
  and vitamin C, 541
Consumers, 937, 942
Continental drift, 767–770
  and natal homing, 312–313
Continuous variation, 240
Contraception, 677–681
Control group, 33, 37
Controlled experiments, 37–39
Convergent evolution, 734f, 735, 736f
Conversion charts, Appendix A
Cooperation between animals, 1023–1024
Copepods, 881
Coral reefs, 911–912
Corals, 864t, 867–869
Corepressor of operon, 300f
Cork cambium, 362
Corn seedling development, 417, 418f
Corolla of flower, 403
Coronary arteries, 595, 596f
Coronary bypass surgery, 597
Corpolites, 746
Corpus callosum, 481f, 482, 504
Corpus luteum, 676f, 677
Cortex of plants, 372
Cortisol (hydrocortisone), 522t, 526, 528
Cotyledons, 411
Countercurrent flow, 633
Coupling, in chemical reactions, 122
Courtship behavior, 668, 1018
Covalent bonds, 55
  in oxidation and reduction, 128
Covalent modification of enzyme activity, 129
Cowper's gland, 672
Crabs, 864t, 876, 881
Cranial nerves, 489–490
Cranium, skull, 482
Crassulacean acid metabolism, 163t, 166
Cretinism, 531
Crick, Francis, 269–272
Cri du chat syndrome, 258
Crinoids, 864t
Crocodiles, 889
Crops, biotechnology and, 316–319, 418–419
Crosses, genetic
  dihybrid, 238
  monohybrid, 233, 234f
  test cross, 237
Crossing over, 216, 218f, 250, 251f, 292
  and genetic mapping, 252
Crustaceans, 876, 878f, 881
Cryptic coloration, 968
Curare, 478
Cutaneous respiration, 622
Cuticle of arthropods, 545
Cuticle of plants, 366
  and transpiration, 392
Cyanide, 184
Cyanobacteria, 791, 792, 796
  role in evolution, 174, 764
Cycadophyta (cycads), 839t, 849
Cyclic AMP (cAMP), 533–534, 535f, 819
Cyclic photophosphorylation, 161–163
Cystic fibrosis, 240, 337t, 338, 341f
  and evolution, 342
  gene for, 338, 341, 342f
  gene therapy for, 349
  screening test for, 347
Cysts, protozoal, 814
Cytochrome oxidase, 184
Cytokinesis:
  in meiosis, 217f, 219
  in mitosis, 206
Cytokinins, 427t, 432
Cytoplasm:
  components of, 97–105
  receptors in, 534, 535f
Cytosine, 269, 270f
  base pairing, 271f, 272
Cytoskeleton, 104–105
Cytotoxic (killer) T cells, 648

D

Dancing bees, 1019, 1020f
Dark (light-independent) photosynthetic reactions, 155f, 156, 158f, 163–166
Darwin, Charles, 19–24, 721–722, 750–752
  and age of earth, 746
  and childhood sexuality, 662–663
  and plant growth, 424–425
Darwin's finches, 21–23
Daughter cells, 196
Davson-Danielli model of membrane structure, 89
Day-neutral plants, 437
DDT, 714–715, 946
Deciduous forests, 917, 919–920, 921f
Decomposers, 937
Deep-sea-diving mammals, 620–621
Defecation reflex, 574
Defense adaptations, 966t, 967–972
Defense mechanisms of body
  nonspecific, 644–645
  specific, 646–650
Denaturing of protein, 81, 127
Dendrites, 468, 469f
Denitrifying bacteria, 950
Density-dependent and -independent population-control factors, 996
Deoxyribonucleic acid, *see* DNA
Deoxyribose, 278
Derived homologies, 746
Dermal bone, 543
Dermal tissue system of plants, 365, 366–369, 367f
Dermis, 542
Deserts, 926
  and mycorrhizae, 388
  and osmoregulation, 613
  plant adaptations to, 164–166, 926
Detritovores, 937, 942–943
Deuteromycetes, 822t, 825–826
Deuterostomes, 873
Development, 10, 686–709
  embryonic, *see* Embryo, development
  postembryonic, 699–700
Developmental defects, 686
Developmental "fate," 690–691, 695
Diabetes, 531–532
Dialysis, 611
Diaphragm
  anatomic, 626
  contraceptive, 681
Diatoms, 817–818
Dichloro-diphenyl-trichloro-ethane (DDT), 714–715, 946
Dicotyledons (dicots), 361, 411, 853
  seedling development, 416, 417f
  versus monocots, 361t, 372
Dieback of population, 992
Diet, *see* Food; Nutrients
Dietary deficiencies, 566–567
Diffusion, 138–139
  facilitated, 141, 143f
DiGeorge's syndrome, 687
Digestion, 568
  intracellular versus extracellular, 576

Digestive systems, 455, 566–584
evolution of, 576–577, 581
incomplete versus complete, 576
in human, 456f, 568–575
hormones of, 522t
Digestive tract, 568, 861
Dihybrid cross, 238
Dimer of thymine bases, 280
Dimorphism, 727
sexual, 727–728
Dinoflagellates, 818
Dinosaurs, 771, 772–773
Dioecious plants, 407, 849
Diphosphoglycerate (DPG), 129
Diploid (2N) cells, 199, 214, 833
Directional selection, 725–726
Disaccharides, 70
Discontinuous variation, 240
Disruptive coloration, 968
Disruptive selection, 726–727
Distal convoluted tubule, 608–609, 610
Divergent evolution, 734
Division, in taxonomic hierarchy, 15
DNA, 82, 266–290, 312–332. *See also* Nucleic acids; Recombinant DNA
amplification, 326–327, 329
base pairs in, 271f, 272
cloning, 324
damage to, 280–281, 286–287
functions, 272–285
discovery of, 266–267
gene expression, 277–285
information storage, 272–275
inheritance, 275–277
protein synthesis, 277–285, 286f. *See also* Transcription
as template, 275
in mitosis, 203–204
noncovalent bonds in, 57f, 58
packaging in chromosomes, 273, 274f
rearrangement, 651, 652f
regulatory sites, 304
repeating sequences, 338, 339f
replication, 275–277
sequencing, 325–326
and evolution, 329, 750
legal aspects, 315
and turtle migration, 313
splicing, 321
structure, 268–272
double helix, 271f
Watson-Crick model, 271–272
unexpressed, 303
DNA "fingerprints," 328
DNA ligase, in recombinant DNA, 323
DNA polymerase, 275–277
DNA probe, 326
DNA repair enzymes, 287
DNA restriction fragments, 325
Domain, in taxonomic hierarchy, 15n
Dominance, 234
incomplete (partial), 239–240
Dominant allele, characteristic, trait, 234–235, 236–237
and genetic disorders, 343–345
Dopamine, 475t, 487
Dormancy in plants, 434
Dorsal hollow nerve cord, 884
Double-blind test, 39
Double bonds, 56, 75
"Double fertilization," 410
Double helix of DNA, 271f
Doves and hawks, 1017
Down syndrome (trisomy 21), 213, 221, 336, 337t
DPG (diphosphoglycerate), 129
Drosophila melanogaster, use in genetics, 249–253
Drugs, *see also* Medicines
mood-altering, 487
and pregnancy, 704
Duchenne muscular dystrophy, 337t, 345
Duodenum, 573
Dystrophin, 345

E

Ear, 506–509
bones of, evolution, 459–460
Earth's formation, 762–763
Earthworms, 875–876, 877f
Eating, 568
Ecdysone, 305–306
Echinodermata, 864t, 882–884
Echolocation (sonar), 498–499
Eclipse phase in viruses, 800
Ecological equivalents, 942
Ecological niches, 939–941
Ecological pyramids, 943–946
Ecology, 902–930, 932–958, 960–982, 984–1004
defined, 905
ethical issues, 909
and evolution, 905
Ecosystems, 11, 932–958
aquatic, 908–916
energy flow in, 942–946
linkage between, 934
nutrient recycling in, 947–952
structure, 934–939
succession in, 952–955
Ecotype, 831, 938
Ectoderm, 693
Ectotherms, 614
Edema, 593
Egg (ovum), 669
human, 673–675
Egg-laying mammals, 892
Ehlers-Danlos syndrome, 337t, 541
Ejaculation, 672
Ejaculatory fluid, 672
Elaters, 846
Electrocardiogram (EKG), 599f
Electron acceptor, 158f, 159
Electron carriers:
and NAD + conversion to NADH, 177f
and NADP + conversion to NADPH, 160–161
Electron donor, 129
Electrons, 51, 52–55
in oxidation, 128–129
"photoexcited," 156, 159
in reduction, 128–129
Electron transfer in chemical reactions, 128–129
Electron transport system, 160–161
and ATP formation, 182–184
in glycolysis, 177
in photosynthesis, 160–161, 162
in respiration, 182–185
Elements, 50–51
Elephantiasis, 873
Ellis-van Creveld syndrome, 720f, 721
Embryo, 686–709
defined, 688
development, 688–699
blastulation, 690
cleavage, 690
energy source for, 688
and evolution, 705
fertilization, 689
gastrulation, 692–693
genetic control of, 686–687
in humans, 700–705
implantation, 677
induction of, 695
metamorphosis, 699–700
morphogenesis, 696–697
neurulation, 694–695
organogenesis, 696
in flowering plants, 411
frozen, 679, 680
risks to, 704
Embryology, and evolution, 750
Embryonic disk, 703
Embryo sacs of flowering plants, 408
Emigration, 988
Endangered species, 995–996
Endergonic reactions, 122–123
Endocrine glands, 521f, 522t
Endocrine systems, 457, 518–538
digestion and, 572f
evolution of, 534, 536
functions, 520–521
in humans, 456f, 523–534
nervous system and, 523–524
properties, 521–523
Endocytosis, 144–148
Endoderm, 693
Endodermis, 373–374, 376f
Endogenous, defined, 438
Endoplasmic reticulum (ER), 98
Endorphins, 487
Endoskeletons, 546–548
Endosperm, 408, 411
Endosperm mother cell, 408
Endospores, 792
Endosymbiosis hypothesis (endosymbiont theory), 110, 111f, 962
Endothelin, 591
Endotherms, 614
Energy:
acquisition and use, 10, 18, 118–123
biological transfers of, 119f
in chemical reactions, 121–123
conversions of, 119, 156–163
genetic engineering and, 316
defined, 118
in electrons, 53–55
expenditure by organ systems, 457
flow through ecosystems, 942–946
forms of, 118
in human diet, 578
and membrane transport, 142–144, 145f
pyramid of, 943–944, 946
storage of, 118
and ATP, 121–122, 143–144, 161
and ionic gradients, 143–144, 145f
yield:
from aerobic respiration, 185, 188
from fermentation, 179
from glycolysis, 178, 185
from substrate-level phosphorylation, 185
Energy conservation, law of, 119
Enkephalins, 487
Entropy, 120–121
Environment:
adaptation to, 4–5
photosynthesis and, 164–166
influence on genetics, 241
and plant growth, 427, 434–438
Environmental resistance to population growth, 991–992
Environmental science, 909
Enzymes, 10, 123–127
discovery of, 116–117
effect of heat on, 127
effect of pH on, 127
and evolution, 131
how they work, 124f, 125
inhibition of, 126
interaction with substrate, 124f, 125
in lysosomes, 100–102

Enzymes (Cont'd)
of pancreas, 573
regulation of, 129–130
of stomach, 570
synthetic modifications of, 131
Eosinophils, 601–602
Epicotyl, 411
Epidermal cells of plants, 369
Epidermal hairs of plants, 393f
Epidermis of skin, 542
Epididymis, 670
Epiglottis, 569f, 624
Epinephrine (adrenaline), 522t, 529
Epiphyseal plates, 547f
Epiphytes, 845
Epistasis, 240
Epithelial tissue, 452–453
Epithelium, 452
ER (endoplasmic reticulum), 98
Erectile tissue, 669, 670f
Erection, 669
Erythrocytes (red blood cells), 600–601, 627
Erythropoietin, 522t
Escape adaptations, 966t, 969–970
Esophagus, 569
Essential amino acids, 578
Essential nutrients, in plants, 388, 390t
Estivation, 927
Estrogen, 75, 77f, 522t, 532, 675, 676f
Estrus, 668
Estuary habitats, 912–914, 915f
Ethical issues:
anencephalic organ donors, 489
facts versus values, 909
fertility practices, 680
gene therapy, 349
genetic screening tests, 350
patenting genetic sequences, 315
Ethics and biology, 25, 40–41
Ethylene gas, 427t, 432
Etiolation, 438
Eubacteria, 791–796
Euglenophyta, 818–819
Eukaryota (proposed kingdom), 809
Eukaryotes:
chromosomes in, 273
evolution of, 764–765
gene regulation in, 302–308
versus prokaryotes, 91–92, 787
Eukaryotic cells, 91–93, 95f
cell cycle in, 201f
cell division in, 197–199
components, 91, 94t, 95–107
evolutionary origin, 110, 111f, 764–765, 787–788, 789f, 962
glucose oxidation in, 183f
respiration in, 180f
Eusociality, 1025–1026
Eutrophic lakes, 916, 917f
Evergreens, 851
Evolution, 4–5, 18–25, 716–740, 742–758, 760–779
amino acid sequencing and, 84, 750
of Animal kingdom, 861, 865f, 873
basis of, 717–731
chemical, 760–761
Darwin's principles, 24–25
DNA sequencing and, 329, 750
of eukaryotic cell, 110, 111f, 764–766, 787–788, 789f, 962
evidence for, 742–758
of fungi, 826
genetic changes and, 258, 259, 287, 293, 303
of humans, 742–743, 752–756, 755f
of mammals, 892
of organ systems, 457–460
circulatory, 603–605
digestive, 576–577, 581
endocrine, 534, 536
excretory, 611–612
immune, 655
integument, 543–544
nervous, 491–493
respiratory, 636
skeletomuscular, 560–561
of osmoregulation, 611–613
patterns of, 734–735
of plants, 833–834, 835f
of primates, 894f
prokaryotes and, 787, 789f
protists and, 811–812
viruses and, 803–804
Evolutionary relationships, 744–746
Evolutionary stable strategy, 1017
Excitatory neurons, 468
Excretion, 607
Excretory systems, 455, 606–613
evolution of, 611–612
in humans, 456f, 607–611
Exercise, muscle metabolism in, 186–187
Exergonic reactions, 122
Exocytosis, 99
Exogenous, defined, 438
Exons, 302f, 303, 307
Exoskeletons, 544–546, 877
Experimental group, 33, 37
Experimentation, 32
controlled, 37–39
Exploitative competition, 963–964
Exponential growth of population, 990–991
Extensor muscles, 552–553
External fertilization, 665
Extinction, 736, 995–996
acceleration of, 997
Permian, 770–771
Extracellular digestion, 576
Extracellular matrix, 108, 546
Eye, 503, 505f, 506
Eyeball, 503

F

Facilitated diffusion, 141, 143f
Facultative anaerobes, 791
FAD, $FADH_2$ (flavin adenine dinucleotide), 181, 182f, 185
Fallopian tube (oviduct), 673
ligation (tieing) of, 681
Familial hypercholesterolemia, 146, 337t
Family, in taxonomic hierarchy, 14
Fat cells, 76
Fats, 75, 188
Fatty acids, 75, 578
Favored chemical reaction, 122
Feathers, 889–890
Feces, 574
Feedback mechanisms, 450
and enzyme activity, 129–130
and hormone regulation, 523
Female reproductive system, 673–677
Female sex hormones, 675, 676f, 677
Fermentation, 178–179
and exercise, 187
versus aerobic respiration, 175, 178
Ferns, 847, 848f
Fertility rate, 999
Fertilization, 689
in flowering plants, 408–410
in humans, 677
Fertilization membrane, 689
Fetal alcohol syndrome, 704
Fetus, 700
genetic screening tests, 347–350
risks to, 704
Fever blisters, 796, 803
Fibrin, 602
Fibrinogen, 602
Fibrous root system, 373, 374f
Fighting, 1017, 1018–1019
Filament of flower, 403
Filamentous fungi, 820–821, 822
Filial imprinting, 1013–1014
Filter feeders, 577, 581
and evolution of gills, 636
Fire algae, 818
First Family of humans, 754, 755f
Fishes, 886–887
air-breathing, 636
heart of, 605f
sense organs of, 512
Fission, 664
Fitness, 1022
Fixed action patterns, 1009
Flagella:
of bacteria, 792, 793f
of cell, 106–107
of protozoa, 814–816
Flatworms, 864t, 869–871, 872
Flavin adenine dinucleotide (FAD, $FADH_2$), 181, 182f, 185
Fleming, Sir Alexander, 35–37
Flexor muscles, 552–553
Flight, adaptations for, 890, 891f
Florigen, 434
Flowering, 400–401, 437
Flowering plants, 852–853
asexual (vegetative) reproduction, 417–418
pollination, 402–407
sexual reproduction, 402–417
Flowers, 378
color, 239
structure, 403
types, 407
why they wilt, 391
Fluid compartments of body, 588
Fluid-mosaic model of membrane structure, 89
Flukes, 869, 870f, 871, 872
Folic acid, 579t
Follicle of ovary, 673–674
Follicle-stimulating hormone (FSH), 522t, 526
in females, 675, 676f
in males, 672
Food, *see also* Nutrients
adaptations to, 577, 581
conversion to nutrients, 568
sharing of, 1024–1025
transport in plants, 393–396
Food chains, 942
Food crops, biotechnology and, 316–319, 418–419
Food webs, 942–943, 944f
"Foreskin," 669
Forests, 917–920
Form, relation to function, 15
Fossil dating, 54, 746
Fossil fuels, 949, 950
Fossil record, 746, 747, 749
Founder effect, 720–721
Frameshift mutations, 286
Free radicals, 56, 174
Freeze-fracturing, 89
Freshwater habitats, 914–916
and osmoregulation, 612–613
Freud, Sigmund, 663
Frogs, 888–889
Fronds, 847
Fructose, 70, 71f
Fruit:
development of, 412
dispersal of, 412, 415f

and human civilization, 414
protective properties, 833
ripening, 432
types, 412, 413f
Fruit fly, in genetics, 249–253
Fruiting body of slime mold, 819
FSH, *see* Follicle-stimulating hormone
Function, relation to form, 15
Functional groups, in molecules, 69
Fungi, 820–826
diseases from, 822, 823–824, 825–826
evolution of, 826
imperfect, 825
nutrition of, 822
reproduction in, 822
uses of, 820
Fungus kingdom, 820–826
Fur, 542

G

GABA (gamma-aminobutyric acid), 475t
Galactosemia, 337t
Galen, 586–587
Gallbladder secretions, 574
Gametangia, 833
Gametes, 199
in flowering plants, 407–408, 409f
in human female, 674–675
in life cycle, 214, 215f
in meiosis, 217f
Gametophytes, 408, 409f, 833
Gamma-aminobutyric acid (GABA), 475t
Gaseous nutrient cycles, 947–951
Gas exchange, 622–623
alveolar-capillary, 625–626
in lung, 627
in tissues, 627–629
Gastric juice, 570, 572
Gastrin, 522t
Gastrointestinal tract, 568. *See also* Digestive systems; Digestive tract
hormones of, 522t
Gastropods, 874
Gastrula, 692
Gastrulation, 692–693
in humans, 703
Gated ion channels, 471
Gause's principle of competitive exclusion, 961, 964
Gel electrophoresis, 325, 326f
Gender, *see* Sex
Gene dosage, 256–257, 337
Gene flow, 719
Gene frequency, 717–718
Genentech, 314
Gene pool, 717
Gene products, 239, 338
Generative cell, 408, 409f
Gene regulatory proteins, 295, 298
in eukaryotes, 304
in prokaryotes, 301
Genes, 246–264, 266–290, 292–310. *See also* Alleles; Mutations
and behavior, 1008–1009, 1010–1011
duplication, 258
exchange of, *see* Crossing over
expression of, 277–285
visualization of, 305–306
homeotic, 686–687
and hormone action, 534, 535f
mapping, 252, 338, 341, 342f
Human Genome Project, 340
rearrangements, 651, 652f
regulation of expression, 292–310
in eukaryotes, 302–308
need for, 294
in prokaryotes, 298–301
selective expression, 294, 295f
split, 302f, 303
structural, 299, 300f-301f
transfer between bacteria, 804
transposition, 293
Gene therapy, 348–349
Genetic code, 272, 278–279, 282f
Genetic disorders, 334–350, 337t. *See also* Chromosomes, abnormalities
gene therapy for, 348–349
tests for, 346–350
in fetus, 347–350
X-linked, 254
Genetic drift, 719, 720–721
Genetic engineering, *see* Biotechnology; Recombinant DNA
Genetic equilibrium, 718
Genetic information
and living organism, 10
storage in DNA, 272–275
universality of, 18
Genetic markers, 249
in gene mapping, 338, 342f
Genetic mosaic, 256f, 257
Genetic recombination, 216, 218f, 219f
Genetics, 230–244, 246–264, 266–290, 292–310, 312–332, 334–353. *See also* Inheritance
Mendelian, 232–239
Genetic variability, 5
and meiosis, 214–216, 219
and sexual reproduction, 665
Genital herpes, 796
Genitalia, in human:
female, 673
male, 669
Genome, defined, 340
Genotype, 235
Genus:
in binomial system, 13
in taxonomic hierarchy, 14
Geography, and evolution, 750–752
Geologic time scale, 764, 765f
German measles, and embryo, 704
Germ cells, 214
Germination, 416
Germ layers, 693
Germ theory of disease, 785
Giant (polytene) chromosomes, 252–253, 305
Gibberellins, 427t, 431–432
Gibbon, 894f
Gill filaments, 633
Gill(s), 632–633, 634f
evolution of, 636
slits, 706, 884
Ginkgophyta, 839t, 849
Girdling of plants, 384–385
Gizzard, 570
Glans penis, 669
Global warming, 950
Glomerular filtration, 609–610
Glomerulus, 608
Glottis, 569f, 624
Glucagon, 522t, 531, 532f, 533–534
Glucocorticoids, 526, 528
Glucoreceptors, 572
Glucose, 70, 71f
blood levels, 531, 532f
energy yield from, 175, 185, 188
oxidation, 174–188
in photosynthesis, 163–164
Glycerol, 75
Glycine, 475t
Glycogen, 72–73
Glycolysis, 175–178, 176f
energy yield from, 185
Glycoproteins of plasma membrane, 94
Gnetophyta, 839t, 849
Goiter, 529, 531
Golgi complex, 98–99, 100
Gonadotropic hormones (gonadotropins), 522t, 526
in females, 675, 676f
in males, 672
Gonadotropin-releasing hormone (GnRH):
in females, 675, 676f
in males, 672
Gonads, 532
Gondwanaland, 769
Gonorrhea, 794, 795
Gorilla, 893f, 894f
Gradualism, 737
Graft rejection, 650
Grain, and human civilization, 414
Granulocytes, 601–602
Grasshopper, 878, 879f
Grasslands, 922–923
Graves' disease, 531
Gravitropism, 438
Gray crescent, 692
Gray matter:
of brain, 481–482, 481f
of spinal cord, 488f, 489
Green algae, 835, 837–838
Greenhouse effect, 798, 950
Grooming of others, 1022
Ground tissue system of plants, 365, 367f, 372
Group courtship, 1018
Group defense responses, 970
Group living, 1018
Growth, 10
in animals, 686–709
in plants, 362, 424–440
of populations, 986, 988–1001
Growth curves, 991, 993–994
Growth factor, 202
Growth hormone (GH; somatotropin), 522t, 525, 527f, 534f
athletes and, 528
Growth regulators, 432
Growth ring, 370
Guanine, 269, 270f
base pairing, 271f, 272
Guard cells, 368
Guilds, 941
Gut, 568, 861
Guthrie test for phenylketonuria (PKU), 335
Guttation, 391
Gymnosperms, 847, 849

H

Habituation, 483, 1011
Hagfish, 886–887
Hair, 542, 892
Hair cells of ear, 506, 507f, 508
Half-life of radioisotopes, 54
Halophilic bacteria, 796
Haplodiploidy, 1025–1026
Haploid (1N) cells, 199, 214, 216, 833
"Hardening" (narrowing) of the arteries, 75, 146
Hardy-Weinberg law, principle, 718, Appendix C
Harvey, William, 586–587
Haversian canals, 547f
Hawks and doves, 1017
Hearing, 506–508
in fishes, 512
Heart, 593–600
blood flow in, 593f
changes at birth, 705
evolution of, 605f
excitation and contraction, 598–599
pulse rate, 599–600

Heart (Cont'd)
  valves of, 597
Heart attack, 590, 597, 602–603
Heartbeat, 597–600
"Heartburn," 569
Heat, plant adaptations to, 164–166
Helper T cells, 649
Helping behavior, 1023–1024
Hemocoel, 861, 878
Hemoglobin:
  and gene duplication, 258, 302f
  as oxygen carrier, 627
  shape, 80f
  in sickle cell anemia, 81, 82f
Hemophilia, 337t, 602
  gene therapy for, 349
  inheritance of, 254–255
Herbaceous plants, 362
Herbivore, 966
Herbivory, 966
Heredity, defined, 232. *See also* Inheritance
Hermaphrodite, 667
Heroin, 487
Herpes virus, 796, 803
Heterocyst, 796
Heterosporous plants, 845
Heterotrophs, 155
Heterozygous individual, 235
Hierarchy, taxonomic, 14–15
Hierarchy of life, 11
Hinge joint, 550
Hippocampus, 483, 486
Hirudineans, 875–876
Histocompatibility antigens, 650
Histones, 273, 274f
HIV *see* Human immunodeficiency virus
Homeobox, 687
Homeostasis, 10, 15, 448–449
  and circulatory system, 588
  effect of environment, 455
  and energy expenditure, 457
  mechanisms for, 450
Homeotic genes, 686–687
Hominids, 742–743, 752–756, 755f
Homo erectus, 743, 752, 755f
Homo habilis, 753, 755f
Homologous, versus analogous, 458, 744–746
Homologous chromosomes, 199
  in genetic recombination, 216, 218f
  independent assortment, 219, 220f
Homology, 705, 744–746
Homoplasy, 744
Homo sapiens, 755f
Homo sapiens neanderthalensis, 755f
Homo sapiens sapiens, 752f, 755f
Homosexuality, hypothalamus and, 663
Homosporous plants, 843
Homozygous individual, 235
Hookworms, 873
Hormones, 518–538
  for birth control, 678–679
  discovery of, 521–522
  in labor, 705
  list of, 522t
  of plants, *see* Plants, hormones
  receptors for, 533–534, 535f
  regulation of, 523
  in reproduction, 669
    female, 675, 676f
    male, 672–673
  types, 522
  and urine formation, 610–611
Horsetails, 846–847
Hot dry environment, 164–166. *See also* Desert
Human chorionic gonadotropin (HCG), 677
Human Genome Project, 340
Human immunodeficiency virus (HIV; AIDS virus), 643, 794, 797, 803
  discovery of, 642–643
  and helper T cell, 649
  in pregnancy, 704
Humans:
  body cavities, 863f
  body systems, 456f
    circulatory, 456f, 588–600
    digestive, 456f, 568–575
    endocrine, 456f, 523–534
    excretory, 456f, 607–611
    immune, 456f, 642–660
    integumentary, 456f
    lymphoid (lymphatic), 456f, 646f
    nervous, 456f, 481–491
    reproductive, 456f, 667–677
    respiratory, 456f, 623–627
  chromosome number, 212–213
  embryonic development, 700–705
  evolution, 742–743, 752–756, 755f
  First Family, 754, 755f
  genetically engineered proteins, 314, 315t
  genetic disorders, 334–350
  nutrition, 578–580
  population growth, 997–1001
    doubling time, 998
    earth's carrying capacity and, 999–1001
    fertility rate and, 999
  as research subjects, 39–40
  skeleton, 548–551
  vestigial structures, 749
Hunger, 572
Huntington's disease, 337t, 343–345
  screening test for, 347
Hybridization, 733–734
Hybridomas, 654
Hybrids, 233
Hydras, 864t, 867–869
Hydrocortisone (cortisol), 522t, 526, 528
Hydrogen bonds, 58
  and properties of water, 59
Hydrogen ion, and acidity, 61–62
Hydrologic cycle, 947–948
Hydrolysis, 69, 70f
Hydrophilic molecules, 59
Hydrophobic molecules, 57f, 58
Hydroponics, 387–388
Hydrosphere, 904
Hydrostatic skeletons, 544
Hydrothermal vent communities, 167
Hypercholesterolemia, 146, 337t
  and membrane transport, 136–137
Hypertension, 590, 591, 610
Hypertonic solution, 140, 141f
Hyperventilation, 632
Hyphae, 820
Hypocotyl, 411
Hypoglycemia, 578
Hypothalamus, 481f, 486
  and body temperature, 449
  in human sexuality, 663
  and pituitary, 524, 525f
  and urine volume, 611
Hypothesis, scientific, 32, 33
Hypotonic solution, 140, 141f

I

Ice, properties of, 60–61
Ice-nucleating protein, 316
Icons for themes in this book, 15
Ileocecal valve, 576f
Immigration, 988
Immune response, 650
  secondary, 653
Immune system, 457, 642–660
  components, 647
  disorders of, 656–657
  evolution of, 655
  and fetus, 705
  mechanics of, 646–647, 649–650
  and viral diseases, 803
Immune tolerance, 654
Immunity:
  active, 653
  passive, 655
Immunization, 646, 654–655
Immunodeficiency disorders, 657, *See also* AIDS
Immunoglobulins, 650. *See also* Antibodies
Immunological memory, 652–653
Immunological recall, 653
Immunotherapy for cancer, 648
Imperfect flower, 407
Imperfect fungi, 825
Implantation of embryo, 677
Impotence, 669
Imprinting, 1013–1014
Inbreeding, 730–731
Inclusive fitness, 1022
Incomplete (partial) dominance, 239–240
Incomplete flower, 407
Incubation period, 643
Independent assortment:
  of homologous chromosomes, 219, 220f
  Mendel's law of, 238–239
Indoleacetic acid, 430
Inducer of operon, 300f
Inducible operon, 300f, 301
Inductive reasoning, 40
Indusium, 847
Industrial products, genetic engineering and, 316
Infertility, 677, 679, 680
Inflammation, 645
Inheritance:
  autosomal dominant, 343–345
  autosomal recessive, 343
  and genetic disorders, 343–346
  polygenic, 240–241
  predicting, 235–237
  sex and, 253–257
  X-linked, 345–346
Inhibiting factors of hypothalamus, 524
Inhibitory neurons, 468
Initiation (start) codon, 281, 284
Initiator tRNA, 284
Innate behavior, 1009–1010
Insecticides, *see also* Pesticides
  genetically engineered, 317
  neurotoxic, 478
  plant hormones as, 434, 975
Insects, 864t, 876, 878–881
  metamorphosis in, 305–306
  respiration in, 633, 635f
Insight learning, 1012–1013
Insulin, 522t, 531, 532f
  discovery of, 518–519
  genetically engineered, 314
Integument, 542–544
Integumentary systems, 456f, 457
Interactions between animals, types and outcomes, 963t
Intercellular junctions, 109–110
Intercostal muscles, 627
Interference competition, 963–964
Interferon, 645
Intergene spacers, 302f
Interkinesis, in meiosis, 219
Interleukin II, 649
Internal fertilization, 665
Interneurons, 470
Internode of plant stem, 377
Intertidal habitats, 912, 913f, 914f

Intestinal secretions, 572–573
Intestine:
large 574–575
small, 572–574
Intracellular digestion, 576
Intrauterine device (IUD), 679, 681
Intrinsic rate of increase (r_o), 990
Introns, 302f, 303, 306–307
Invagination hypothesis of eukaryote origin, 110, 111f, 765
Invertebrates, 863–886
In vitro fertilization, 679
Iodine, and thyroid, 529
Ion, defined, 58
Ion channels of plasma membrane, 139, 140f, 471
Ionic bonds, 57, 58
Ionic gradient, 143–144, 145f
Irish potato blight, 820
Iron in diet, 580, 580t
Islets of Langerhans, 518, 531, 532f
Isolating mechanism, 732, 733t
Isotonic solution, 140, 141f
Isotope, 52
IUD (intrauterine device), 679, 681

J

Jaundice, 574
Java Man, 743
Jaws, 887
Jellyfish, 864t, 867–869
Jenner, Edward, 646
"Jet lag," 533
Joints, 550–551
J-shaped curve, 991
"Jumping genes," 293
Junctions between cells, 109–110

K

K (carrying capacity) of environment, 992
K-selected reproductive strategies, 995–996
Kangaroo, 893
Kelp, commercial uses, 840
Kidney, 607–611
artificial, 611
function, 609–610
regulation of, 610–611
hormones of, 522t
Kidney failure, 611
Killer (cytotoxic) T cells, 648
Kinetic energy, 118
Kinetochore, 204f, 205–206, 217f, 219f
versus centromere, 205n
Kingdoms, 13f, 14f, 15
Animal, 858–897
Fungus, 820–826
Plant, 830–856
Monera (prokaryotes), 786–800
Protist (Protoctista), 810–819
taxonomy of, 808–809
Kin selection, 1022–1023
Kinsey, Alfred, 663
Klinefelter syndrome, 337, 337t
Knee joint, 551f
Koala, 892f, 893
Koch, Robert, 785
Krebs cycle, 180–182
Kwashiorkor, 579

L

Labia, 673
Labor, 705
Lac (lactose) operon, 300f, 301
Lactase, 574
Lacteals, 574
Lactic acid fermentation, 178, 187
Lactose, 70
Lake habitats, 915–916, 917f, 918
Lamellae:
of bone, 547f
of gills, 633
Lampreys, 886
Lancelets, 864t, 884–886
Language, 485
Lanugo, 703
Large intestine, 574–575
Larva, 688
Larynx, 624–625
Latent viral infection, 803
Lateral roots, 373, 374f
Laurasia, 769
Laws, principles, rules:
Chargaff's, 272
competitive exclusion, 941, 961
energy conservation, 119
Hardy-Weinberg, 718, Appendix C
independent assortment, 238–239
the minimum, 938–939
natural selection, 24–25
segregation, 235
stratigraphy, 746
thermodynamics, 119–120
LDLs (low-density lipoproteins), 136–137, 146
receptors for, 137, 146, 147f
Leaf, 377–378
and carbohydrate synthesis, 165–166
simple versus compound, 377
Leaflets, 377
Leaf nodules, 396
Leakey: Louis, Mary, Richard, 753
Learning, 483, 485, 1011–1014
Leeches, 875–876
Lek display, 1018
Lemur, 894f
Lens of eye, 503, 505f
Lenticels, 369
Lesch-Nyhan syndrome, 336, 337t
Leukocytes (white blood cells), 601–602
LH, *see* Luteinizing hormone
Lichens, 821
Liebig's law of the minimum, 938–939
Life:
origin of, 760–761, 763–764
properties of, 6–7, 8f-9f
Life cycle, 214, 215f
Life expectancy, and population density, 989
Ligaments, 551
Light:
energy from, 156–163
and plant growth, 435–438
Light-absorbing pigments:
of eye, 506
of plants, 156, 157–159
Light-dependent photosynthetic reactions, 155f, 156–163
Light-independent (dark) photosynthetic reactions, 155f, 156, 158f, 163–166
Lignin, 109
Limbic system, 486–487
Limiting factors, 938–939
Linkage groups, 248–249
Lipases, 573
Lipid bilayer of plasma membrane, 89, 138
Lipids, 71t, 75
in human diet, 578
Lithosphere, 904
Liver, 574
Living organisms
classification of, 11–15
properties of, 6–11
Lizards, 889
Lobes of cerebrum, 482
Lobsters, 881
Locomotion, 558–559
Logistic growth, 994
Long-day plants, 400, 437
Loop of Henle, 608–609, 610
Loris, 894f
Lotka-Volterra equation, 961
Low-density lipoproteins (LDLs), 136–137, 146
receptors for, 137, 146, 147f
Lucy (hominid), 753–754, 755f
Lungs, 623, 625–626, 635
changes at birth, 705
evolution of, 636
gas exchange in, 627
Lupus (systemic lupus erythematosus), 657
Luteinizing hormone (LH), 522t, 526
in females, 675, 676f
in males, 672
Lycophyta (lycopods), 839t, 845–846
Lyme disease, 792
Lymphatic systems, 456f, 606
Lymph nodes, 606
Lymphocytes, 647. *See also* B cells; T cells
Lymphoid tissues, 646f, 647
Lysosomes, 100–102

M

Macroevolution, 731
Macrofungi, 821
Macromolecules, 69, 70f, 71t
evolution of, 83–84
Macronutrients, in plants, 388, 390t
Macrophages, 602, 647
Magnesium, in diet, 580, 580t
Malaria, 813, 815f
and sickle cell anemia, 723f, 724, 812
Male reproduction, 669–673
hormonal control, 672–673
Male sex hormones, 672–673
Malignancy, *see* Cancer; Tumor
Mammals, 892–894
Age of, 774
egg-laying, 892
evolution of, 892
heart in, 605f
Mapping of genes, 252, 338, 341, 342f
Human Genome Project, 340
Marine habitats, 908–912
and osmoregulation, 612
Marrow of bones, 546, 547f
Marsupials, 892f, 893
Mastigophora, 815t
Mating, nonrandom, 719, 727–728, 730–731
Mating types of algae, 837
Matter, defined, 51
McClintock, Barbara, 292–293
Mechanoreceptor, 500, 501t
Medicines, genetically engineered, 314, 321
Medulla, 481f
Medusas, 868–869
Megaspores, 407, 408
Meiosis, 199, 212–225
chromosomal abnormalities and, 221
defined, 214
and genetic variability, 214–216, 219
in human female, 674–675
importance, 214–216
reduction division in, 216
stages, 216–221
versus mitosis, 200f, 216, 222
Meiosis I, 216, 217f
Meiosis II, 217f, 219, 221
Meissner's corpuscles, 502f
Melanoma, 280–281
Melatonin, 522t, 533
Membrane invagination hypothesis, 110, 111f, 765

Membrane potential, 470–471
Membranes, 136–150
of cell, *see* Plasma membrane
in cytoplasm, 97–98
of ear (eardrum), 507f, 508
of mitochondria, 102–104
of nucleus (nuclear envelope), 97
organization in, 160
transport across, 136–144
active, 142–144
passive diffusion, 138–139
Memory, 483
cells, 653
Mendel, Gregor, 232–235
Mendelian genetics, 232–239
exceptions to, 239–241
Mendel's law of independent assortment, 238–239
Mendel's law of segregation, 235
Meninges, 482
Menopause, 675
Menstrual (uterine) cycle, 675, 677
Menstruation, 676f, 677
Meristem, 362–363
of root, 373, 375f
of stem, 372
Mesoderm, 693
Mesophyll, 377
Mesophyll cells, and photosynthesis, 165f, 166
Mesotrophic lake, 916, 917f
Mesozoic Era, 771, 774
Messenger RNA (mRNA)
transcription, 278, 279f
primary transcript processing, 306–307
translation, 279–280, 284–285, 286f
and control of gene expression, 307–308
Metabolic intermediates, 127
Metabolic ledger, 185, 188
Metabolic pathways, 127–130
and glucose oxidation, 188–189
Metabolism:
as property of living organism, 10
regulation of, 129–130
Metamorphosis, 699–700
in insects, 881
ecdysone and, 305–306
Metaphase, of mitosis, 203f, 205–206
Metaphase I, of meiosis, 217f, 218–219
Metaphase II, of meiosis, 217f, 221
Metaphase plate, 217f, 219
Methane, 55f, 56
Methane-producing bacteria, 796, 798
Metric conversion chart, Appendix A
Microevolution, 731
Micronutrients, in plants, 388, 390t
Micropyle, 408
Microscopes, Appendix B
Microspores, 407, 408, 409f
Microvilli, 574, 575f
Midbrain, 481f
Milk, 892
intolerance of, 574
Millipedes, 876, 882
Mimicry, 971
Mineralocorticoids, 528
Minerals:
absorption in plants, 387–391
in human diet, 580
Minimum, Liebig's law of the, 938–939
"Missing link," 743
Mites, 876
Mitochondria, 102–104
and ATP synthesis, 172–173, 184
and origin of eukaryote cell, 110, 765, 962
Mitosis:
defined, 199
phases, 202–206
versus meiosis, 200f, 216, 222
Mitosis promoting factor (MPF), 195, 202
Mitotic chromosome, 273, 274f, 275f
Molds, 821
Molecular biology, 268, 750
Molecular defenses
nonspecific, 645
specific (antibodies), 647
Molecules, 55, 57
chemical structure, 69
Mollusca, 864t, 874–875
Molting, 878
Monera kingdom, 786–800. *See also* Bacteria; Prokaryotes
Monkeys, 893, 894f
Monoclonal antibodies, 654
Monocotyledons (monocots), 361, 411, 853
seedling development, 417, 418f
versus dicots, 361t, 372
Monocytes, 602
Monoecious plants, 407, 851
Monohybrid cross, 233, 234f
Monomer, 69, 70f
Monosaccharides, 70
Monotremes, 892–893
Mood-altering drugs, 487
Morgan, Thomas Hunt, 249–250
"Morning after" pill, 679
Morphine, 487
Morphogenesis, 696–698
Morphogenic induction, 695
Morphological adaptation, 728
Mortality, and population density, 988, 989
Mosaic, genetic, 256f, 257
Mother cell, 196
Motor cortex, 485
Motor neurons, 470
Mouth, in digestion, 568–569
MPF (mitosis promoting factor), 195, 202
mRNA, *see* Messenger RNA
Mucosa of digestive tract, 568
Mud flats, 912, 913f
Multicellular organisms, origin of, 765–766
Multichannel food chains, 942, 944f
Multiple fruits, 412, 413f
Multiple sclerosis, 473
Muscle fiber, 186–187, 553, 554f
Muscles, 454, 551–558
cardiac, 558, 559f
contraction of, 555–557
in nerve impulse transmission, 474f, 475t, 476f, 479–480, 556
skeletal, *see* Skeletal muscle
smooth, 558
Muscular dystrophy, 345–346
Muscular systems, 456f, 457
Mushrooms, 824–825
Mussels, 874
Mutagens, 252, 286–287
Mutant alleles, 241
Mutations, 241–242, 719
and genetic diversity, 242
molecular basis, 286–287
point mutations, 241
Mutualism, 978
Mycorrhizae, 386–387, 388
Myelin sheath of axon, 469f, 470, 473
Myofibrils, 553, 554f
Myosin, 555
Myriapods, 876, 881–882

N

NAD⁺, NADH (nicotinamide adenine dinucleotide):
energy yield from, 185
in fermentation versus respiration, 178
in glycolysis, 176f, 177
in Krebs cycle, 181, 182f
NADP⁺, NADPH (nicotinamide adenine dinucleotide phosphate):
as electron donor, 129
in photosynthesis, 159f, 161
as reducing power, 189
Names, scientific, 13
Nasal cavity, 624
Natal homing, 312–313
Natality, 988
Natural killer (NK) cells, 645
Natural selection, 23–25, 719, 721–727
patterns of, 724–727
pesticides and, 715
principles of, 24–25
Neanderthals, 742, 755f
Nectaries, 405
Nematoda, 864t, 871–873
Neodarwinism, 717
Nephron, 607–609
Nerve cells, *see* Neurons
Nerve gas, 478
Nerve growth factor (NGF), 466–467
Nerve impulse transmission, 470–473, 556
Nerve net, 492
Nerves, 479
Nerve tissue, 454f, 455
Nervous systems, 457, 466–496, 498–516
central, 479, 481–489
evolution, 491–493
of human, 456f, 481–491
origin of, 694–695
peripheral, 479, 489–491
role in digestion, 572
of vertebrates, 479–481
Neural circuits, 479–481
Neural plate, 694
Neural tube, 694, 884
Neuroendocrine system, 523–524
Neuroglial cells, 470
Neurons, 468–470
form and function, 468, 469f
target cells, 468
types, 470
Neurosecretory cells, 523, 525f
in evolution, 534, 536
Neurotoxins, 478, 818
Neurotransmission (synaptic transmission), 474–479
Neurotransmitters, 474–477, 475t
Neurulation, 694–695
Neutrons, 51
Neutrophils, 601–602
Niche overlap, 941, 963
Niches, 939–941
Nicotinamide adenine dinucleotide, *see* NAD⁺, NADH
Nicotinamide adenine dinucleotide phosphate, *see* NADP⁺, NADPH
Nicotine, 630
Nicotinic acid (niacin), 579t
Nitrifying bacteria, 950
Nitrogen, excretion and, 607, 613
Nitrogen cycle, 949–951
Nitrogen fixation, 390–391
by bacteria, 390, 799, 949–951
by cyanobacteria, 796
genetic engineering and, 316
Nitrogenous bases:
in DNA, 269–271
in RNA, 82, 83f, 278
Nodes of plant stem, 377
Nodes of Ranvier, 469f, 470, 473
Nomenclature, binomial system, 13
Noncovalent bonds, 57–58
Noncyclic photophosphorylation, 160f, 161, 162t
Nondisjunction, 336

Nongonococcal urethritis, 795
Nonpolar molecules, 56–57
Nonrandom mating, 719, 727–728, 730–731
Nonsense (stop) codons, 281, 285
Nonvascular plants, 839t
Norepinephrine (noradrenaline), 475t, 522t, 529
Norplant, 678
Nostrils, 624
Notochord, 694, 884
Nucleases, 573
Nucleic acids, 71t, 81–83. *See also* DNA; RNA
  sequencing of, 325
  structure, 82–83, 269–271
  in viruses, 800
Nucleoid of prokaryotic cell, 92
Nucleoli, 97
Nucleoplasm, 97
Nucleosome, 273, 274f
Nucleotides, 82–83
  in DNA, 269–271
  and genetic code, 278
  and genetic information, 272
  and protein assembly, 277
  in RNA, 278
  structure, 269–271
Nucleus, 92, 95–97
  envelope (nuclear membrane), 97
  matrix, 97
  transplantation of, 297
Nutrient cycles, 947–952
Nutrients, *see also* Food
  in human diet, 578–580
  in plants, 388, 390t
Nutritional deficiencies:
  in humans, 566–567
  in plants, 390t

◀ O ▶

Obesity, 76
Obligate anaerobes, 791
Ocean:
  ecosystems, 908
  floor, hydrothermal vent communities in, 167
Octopuses, 874–875
Oils, 75
Olfaction, 509, 510
Oligochaetes, 875–876
Oligotrophic lake, 916, 917f
Omnivore, 966
Oncogenes, 202
Oocyte, 673, 674
Oogenesis, in humans, 673–675
Oogonia, 674, 675f
Open growth of plants, 362
Operant conditioning, 1012
Operator region of operon, 299
Operon, 299–301
Opiates, 487
Opossum, 893
Optimality theory, 1016–1017
Oral contraceptives, 678–679
Orangutan, 894f
Order, in taxonomic hierarchy, 15
Organ donations, ethical issues, 489
Organelles, 91, 95–107
Organic chemicals, defined, 68
Organogenesis, 696, 697–698
Organs, defined, 455
Organ systems, 455–460
  energy expenditure, 457
  evolution of, 457–460
  in humans, *see* Humans, body systems
  types, 455–457
Organ transplants, 650
Orgasm, 672
Origin of Species, 24
Osmoregulation, 606–613
  adaptations of, 611–613
  defined, 607
Osmoregulatory (excretory, urinary) systems, 455, 456f
Osmosis, 140–141
Osteichthyes, 888f
Osteoblasts, 550
Osteoclasts, 550
Osteocytes, 546, 547f
Osteoporosis, 550
Ovarian cycle, 675
Ovaries of flower, 403
Ovaries of human, 522t, 532, 673–674
Overstory of forest, 917
Oviduct (fallopian tube), 673
Ovulation, 674, 676f
Ovules, 408, 852
Ovum, 669
  of human, 673–675
Oxidation, 128–129
Oxygen:
  appearance on earth, 174, 764
  as electron acceptor, 182, 184
  release into tissues, 594f, 627–629
  toxic effects, 174
  transport, 600–601
  uptake in lungs, 594f, 627
Oxyhemoglobin, 627
Oxytocin, 522t, 525
Oysters, 874
Ozone layer, 902–903

◀ P ▶

PABA (para-aminobenzoic acid), 126
Pacemaker of heart, 110, 598, 599
Pacinian corpuscle, 502
Pain receptor, 500, 501t
Paleozoic Era, 767–771
Palisade parenchyma, 378
Pancreas, 522t, 531–532, 573
Pancreatitis, 573
Pangaea, 769
Panting, 614
Pantothenic acid, 579t
Paper chromatography, 153
Parallel evolution, 734f, 735
Parapatric speciation, 732
Parasitism, 872, 972–974
Parasitoids, 972
Parasympathetic nervous system, 490–491
Parathyroid glands, 522t, 530f, 531
Parathyroid hormone (PTH), 522t, 531
Parenchyma cells, 363, 364f
Parthenocarpy, 418
Parthenogenesis, 664
Partial (incomplete) dominance, 239–240
Parturition (birth), 705
Passive immunity, 655
Pathogenic bacteria, 792
Pauling, Linus, 30–31, 39–40, 67
Pedicel, 403
Pedigree, 255, 256f
Peking Man, 743
Pelagic zone, 908
Penicillin, 35–37, 126
Penis, 665, 669, 670f
PEP (phosphoenolpyruvate) carboxylase, 166
Peppered moth, camouflage changes, 4–5, 722–723
Pepsin, 570
Pepsinogen, 570
Peptic ulcer, 570, 572
Peptide bonds, 78f, 80
Peptidoglycan, 787
Perennial plants, 361, 362–363, 366
Perfect flower, 407
Perforation plate in xylem, 370
Perforins, 647
Pericycle, 373, 376f, 377
Periderm, 369
Periosteum, 547f
Peripheral nervous system, 479, 489–491
Peristalsis, 569, 571f
Permafrost, 922
Pesticides, *see also* Insecticides
  bioconcentration of, 946
  fungi as, 823
  genetically engineered, 317
  natural selection and, 715
Petals, 403
Petiole of leaf, 377
PGA (phosphoglyceric acid), 163–164
PGAL (phosphoglyceraldehyde), 129, 164
pH, 61–62
  and blood gas exchange, 629
Phaeophyta, 838
Phage (bacteriophage), 268, 269f, 801
Phagocytic cells, 601–602, 644f, 645
Phagocytosis, 144–147
Pharyngeal gill slits, 706, 884
Pharynx, 569, 624, 636
Phenotype, 235
  environment and, 241
Phenylketonuria (PKU), 334–335, 337t
Pheromones, 1021
Philadelphia chromosome, 259, 260
Phimosis, 669
Phloem, 369–370, 372
  and food transport, 393–396
Phosphoenolpyruvate (PEP) carboxylase, 166
Phosphoglyceraldehyde (PGAL), 129, 164
Phosphoglyceric acid (PGA), 163–164
Phospholipids, 71t, 75, 77f
  in plasma membrane, 77f, 89, 138
Phosphorus, in diet, 580, 580t
Phosphorus cycle, 951–952
Phosphorylation, 175, 177–178. *See also* Photophosphorylation
"Photoexcited" electrons, 156, 159
Photolysis, 160
Photons, 156
Photoperiod, 434
  plant responses to, 400–401, 435–438
  and reproduction, 668
Photoperiodism, 435
Photophosphorylation, 161
  cyclic, 161–163, 162t
  noncyclic, 160f, 161, 162t
Photoreceptor, 500, 501t
Photorespiration, 166
Photosynthesis, 152–170
  in algae, 816
  ATP in, 160f, 161–162
  in bacteria, 791–792
  central role, 154
  early studies, 153–154
  and energy of electrons, 54f, 55
  light-dependent reactions, 155f, 156–163
  light-independent (dark) reactions, 155f, 156, 158f, 163–166
  overview, 155–156, 158f
  as two-stage process, 155f, 156
Photosynthetic autotrophs, 154
Photosynthetic pigments, 156, 157–159
Photosystems, 159
Phototropism, 438
Phyletic speciation, 732
Phylogenetic relationships, 14
Phylum, in taxonomic hierarchy, 15
Physical defenses, 970–971

Phytochrome, 435–436
and flowering, 437
and photoperiodism, 436
and seed germination, 438
and shoot development, 438
Phytoplankton, 816, 908
Pigmentation in humans, genetics of, 240, 241f
Pigments:
light-absorbing:
in eye, 506
in plants, 156, 157–159
photosynthetic, 156, 157–159
Piltdown Man, 743, 752–753
Pineal gland, 522t, 533
Pinocytosis, 144, 147–148
Pioneer community, 952
Pistil, 403
Pith of plants, 372
Pith rays, 372
Pituitary gland, 481f, 522t, 524–526
Placebo, 39
Placenta, 700
Placentals, 893
Planaria, 869–871
Planes of body symmetry, 862f
Plankton, 881, 908
Plant fracture properties, 375
Plant kingdom, 830–856
Plants, 358–382, 384–398, 400–422, 424–443, 830–856. *See also* Flowering plants
annual, biennial, perennial, 361
basic design, 360–361, 362f
carbohydrate synthesis in, 163–166
cell vacuoles, 102
cell wall, 108–109
circulatory system, 384–398
classification, 839t
communities, 932–933
competition between, 961
cytokinesis in, 206
desert adaptations, 164–166, 926
evolution, 833–834, 835f
fossil, 843, 844f
genetic engineering of, 316–319, 375
growth and development, 424–443
control of, 426–427
timing of, 434–438
growth signal in, 424–425
hormones, 427–434
discovery of, 424–425
as insecticides, 434, 975
longevity, 362–363, 366
nonphotosynthesizing, 834, 836
nutrients, 388, 390t
transport of, 393–396
organs, 372–378
flower, 378
leaf, 377–378
root, 373–377
stem, 372–373
osmosis in, 140–141, 142f
polyploidy in, 260–261
sex chromosomes in, 254
sexual reproduction in, 402–417
species differences, 830
transport in, 384–398
Plant tissues, 362–365
cambium, 362, 372–373
collenchyma, 364
parenchyma, 363
primary, 362
of root, 373–374, 376f, 377
of stem, 372
sclerenchyma, 364–365
secondary, 362
of root, 377
of stem, 372
Plant tissue systems, 365–372, 367f
dermal, 365, 366–369
ground, 365, 372
vascular, 365, 369–372
Plaques, in arteries, 75, 603
Plasma, 600
Plasma cells, 647, 653
Plasma membrane (cell membrane):
carrier proteins of, 141, 143f
components, 93–94
formation, 99
functions, 93–94
ion channels, 139, 140f, 471
permeability, 138, 139
receptors on, 94, 146, 147–148, 533–534, 535f
structure, 88–89, 93–94, 138
transport across, 136–144
active transport, 142–144
passive diffusion, 138–139
Plasmid, 321, 322f
Plasmodesmata, 387
Plasmodial slime molds, 819
Plasmodium, 813, 815f
Plasmolysis, 141, 142f
Platelets, 602
Plate tectonics, 767–769
Platyhelminthes, 864t, 869–871
Platypus, 892f, 893
Play, 1013
Pleiotropy, 240
Pleura, 626
Pleurisy, 626
Pneumococcus, role in molecular genetics, 266–267
Pneumocystis infection, 813
Point mutation, 286
Poison ivy, 656
Polar body, 675
Polarity in neurons, 471–473
Polar molecules, 56–57
Pollen, 403
protective properties, 833
Pollen:
sacs, 403f
tube, 409f, 410
Pollination, 402–407, 410f
Pollution
consequences, 918, 992, 993
genetically engineered degradation, 316–317
Polychaetes, 875–876
Polygenic inheritance, 240–241
Polymer, 69, 70f
Polymerase chain reaction, 326–327, 329
Polymorphic genes, 338
Polymorphism, 727
Polypeptide chain, 78f, 80
assembly, *see* Proteins, synthesis
Polyploidy, 260–261
and speciation, 733
Polyps, 868–869
Polysaccharides, 70–74
Polysome, 285
Polytene (giant) chromosomes, 252–253, 305
Polyunsaturated fats, 75
Pond habitats, 915–916
Pons, 481f
Population density, 986, 987, 996
Population ecology, 984–1004
Population growth, 988–1001
and earth's carrying capacity, 999–1001
factors affecting, 988–997
fertility and, 999
in humans, 997–1001
doubling time, 998
percent annual increase, 998
rate, 986
Populations, 11, 717
age–sex structure, 988, 999, 1000f
defined, 986
distribution patterns, 987–988
structure, 986–988
Porifera, 864–867, 864t
Porpoises, and echolocation (sonar), 499
Positional information, 696
Postembryonic development, 699–700
Potassium:
in diet, 580, 580t
transport across cell membrane, 142, 144f
Potential (stored) energy, 118
Pragmatism, 41
Prairies, 923
Predation, 966–972
Predator, 966
adaptations, 967–972
group behavior, 1018
Predator-prey dynamics, 966–967
Pregnancy, *see also* Embryo; Fetus
prevention of, 677–681
stages of, 700–705
Prepuce, 669
Pressure-flow transport of plant nutrients, 395–396
Pressure gradient, and respiration, 626–627
Prey, 966
defenses, 966t, 967–972
group behavior, 1018
Primary endosperm cell, 410, 411f
Primary growth of plants, 362
Primary plant tissues, 362
Primary organizer, 695
Primary producers, 937
Primary root tissues, 373–374, 376f, 377
Primary stem tissues, 372
Primary succession, 952, 953–955
Primary transcript, 306–307
Primates, 893–894
evolution, 894f
Primitive homologies, 746
Prion, 803
Probability, in genetics, 236
Progesterone, 522t, 676f, 677
Prokaryotes, 787–800. *See also* Bacteria
activities of, 798–800
fossilized, 763
gene regulation in, 298–301
as universal ancestor, 789f
versus eukaryotes, 91–92, 787
Prokaryotic cell, 91–92, 793f
cell division in, 197
chromosomes in, 273
and origin of eukaryote cell, 110, 765, 787–788
Prokaryotic fission, 197
Prolactin, 308, 522t, 525
Promoter, 299
Prophase, of mitosis, 203–205
Prophase I, of meiosis, 216–218, 217f
Prophase II, of meiosis, 217f, 219
Prosimians, 894f
Prostaglandins, 532, 705
Prostate gland, 672
Prostate specific antigen (PSA), 654
Protein kinases, 129, 195, 202
Proteins, 71t, 79–81
conformational changes in, 81
denaturing of, 127
as energy source, 188
in human diet, 578
of plasma membrane, 94, 138
structure, 66–67, 78f, 79–81
primary, 80
quaternary, 80f, 81
secondary, 80
tertiary, 80f, 81

synthesis, 277–285, 286f
chain elongation, 285
chain termination, 285
initiation, 284–285
site of, 98
transcription in, 277, 278, 279f, 286f
translation in, 277, 279, 286f
Proteolytic enzymes, 573
Proterozoic Era, 764–766
Prothallus, 847, 848f
Protist kingdom, 810–819
taxonomic problems, 810
Protists:
evolution and, 811–812
reproduction in, 222, 811
Protochordates, 884–886
Protocooperation, 977–978
Protoctista, 809n
Proton gradient, and ATP formation, 160f, 161, 184
Protons, 51, 61
Protoplast manipulation, 418–419
Protostomes, 873
Protozoa, 812–816
classification, 815t
motility, 814–816
pathogenic, 813
polymorphic, 813–814
Proximal convoluted tubule, 608–609, 610
Pseudocoelom, 861, 871–872
Pseudopodia, 814–816
Psilophyta, 839t, 844f, 845
Psychoactive drugs, 487
Pterophyta, 839t, 847
Pterosaurs, 771
Puberty, in females, 675
Puffballs, 824
Puffing in chromosomes, 305
Pulmonary circulation, 594–595
Punctuated equilibrium, 737
Punnett square, 235–236
Pupa, 881
Purines, 269, 270f
Pus, 645
Pyloric sphincter, 572
Pyramid of biomass, 944
Pyramid of energy, 943–944, 946
Pyramid of numbers, 944
Pyridoxine (vitamin B_6), 579t
Pyrimidines, 269, 270f
Pyrrophyta, 818
Pyruvic acid:
energy yield from, 185
in glycolysis, 176f, 178
in Krebs cycle, 181, 182f

◄ R ►

r (rate of increase), 990
r-selected reproductive strategies, 995–996
R group, of molecule, 71t, 79–80
and enzyme-substrate interaction, 125
Radiation, effect on DNA, 280–281, 287
Radicle, 411
Radioactivity, 52
discovery of, 48–49
Radioisotopes, 54, 746
Rainfall, 906
Rain forests, 428–429, 917, 919, 920f
Rainshadow, 926
Random distribution of population, 987–988
Rate of increase, intrinsic (r_o), 990
Rays, 887, 888f
Reactant, 123
Reaction center, 159
Reaction-center pigments, 158f, 159
Reactions, *see* Chemical reactions
Reading frame, 284
frameshift mutations, 286
Receptacle of flower, 403
Receptor-mediated endocytosis, 147
Receptors:
in cytoplasm, 534, 535f
for hormones, 533–534, 535f
for LDLs, 137, 146, 147f
on plasma membrane, 94, 147–148, 533–534, 535f
and medical treatment, 146
sensory, 479, 500–502
Recessive allele, characteristic, trait, 234–235, 237
and genetic disorders, 343, 345–346
Recessiveness, 234
Reciprocal altruism, 1023
Recombinant DNA, *see also* Biotechnology
amplification by cloning, 324
defined, 314
formation and use, 321–325
medicines from, 314, 315t, 321
Recruitment, 1019
Rectum, 574, 576f
Recycling nutrients in ecosystems, 947–952
Red algae, 838f-839f, 841–842
Red blood cells (erythrocytes), 600–601, 627
Red tide, 818
Reducing power of cell, 129
Reduction, 128–129
Reduction division, in meiosis, 216
Reefs, 911–912
Reflex, 479
conditioned, 1011
Reflex arc, 479–480
Region of elongation, 372, 375f
Region of maturation, 372, 375f
Regulation of biological activity, 15
Regulatory gene, 299
Releaser of fixed action pattern, 1009
Releasing factors of hypothalamus, 524
Renal tubule, 607–609
Renin, 522t, 591
Replication fork, 276, 277f
Replication of DNA, 275–277
Repressible operon, 300f, 301
Repressor protein, 299, 300f, 301
Reproduction:
asexual, 199
asexual versus sexual, 664–667
communication and, 1018
prevention of, 677–681
as property of living organism, 10
Reproductive adaptation, 729
Reproductive isolation, 732
Reproductive (biotic) potential, 986, 989–990
Reproductive strategies, 995
Reproductive systems, 457, 662–684
human, 456f, 667–677
female, 673–677
male, 669–673
Reproductive timing, 668
Reptiles, 889, 890f
Age of, 771
Research, scientific, 32–35
Resource partitioning, 964–965
Respiration, 174–192
aerobic, 180–184
versus fermentation, 175, 178
and surface area, 622–623, 626
versus breathing, 175n
Respiratory center, 632
Respiratory surfaces, 622–623
Respiratory systems, 455, 620–640
adaptation of, 632–635
evolution of, 636
in human, 456f, 623–627
Respiratory tract, 623–624
Response to stimuli, as property of living organism, 10
Resting potential, 471
Restriction enzymes, and recombinant DNA, 321–323
Restriction fragment length polymorphisms (RFLPs), 338, 339f
use in gene mapping, 341, 342f
Restriction fragments, 325
Reticular formation, 486
Retina, 503, 505f, 506
Retinoblastoma, 260
Retinol (vitamin A), 579t
Retrovirus, 643, 803
Rheumatic fever, 657
Rhizoids, 823, 842
Rhodophyta, 841
Rhyniophytes, 843
Rhythm method of contraception, 681
Riboflavin (vitamin B_2), 579t
Ribonucleic acid, *see* RNA
Ribose, 71f, 278
Ribosomal RNA (rRNA), 281–282
Ribosomes, 97, 98
assembly, 283f
in protein synthesis, 281–284, 286f
structure, 282–284
Ribozymes, 306, 307
Ribulose biphoshpate (RuBP), 164
Rigor mortis, 556
River habitats, 914–915
RNA, *see also* Nucleic acids; Transcription; Translation
enzymatic action, 306, 307
processing, 306–307
and control of gene expression, 304, 307, 308f
self-processing, 307
splicing, 307
structure, 82, 278
RNA polymerase, 278, 286f
and operon, 299
Rods of retina, 503, 505f, 506
Root cap, 373, 375f
Root hairs, 373, 374f
Root nodules, 389–390
Root pressure, 391
Root system of plants, 361, 362f, 373–377
Rough endoplasmic reticulum (RER), 98, 100
Roundworms, 864t, 871–873
rRNA (ribosomal RNA), 281–282
RU486, 678–679
Rubella virus, risk to embryo, 704
RuBP (ribulose biphoshpate), 164
Rumen, 581
Ruminants, 581, 813
"Runner's high," 487

◄ S ►

Saguaro cactus decline, 984–985
Salamanders, 888–889
Saliva, 568–569
Salivary glands, 568
Saltatory conduction, 473
Sand dollars, 882
Sandy beaches, 912, 913f
Sanitation, origins of, 784–785
SA (sinoatrial) node, 598
Saprobes, 822
Saprophytic decomposers, 798
Sarcodina, 815t
Sarcolemma, 554f
Sarcomeres, 553–556
Sarcoplasmic reticulum, 556–557
Saturated fats, 75
Savannas, 923–924
Scales of fish, 543

Scaling effects, 460
Scallops, 874
Schistosomes, 870f, 871
Schizophrenia, 230–231
Schwann cells, 469f, 470
SCID (severe combined immunodeficiency), 337t, 348–349
Science and certainty, 40–41
Scientific method, 32–35, 39f
caveats, 40–41
facts versus values, 909
practical applications, 35–40
Scientific names, 13
Scientific theory, 35
versus truth, 40–41
Sclereids, 365
Sclerenchyma cells, fibers, 364–365
Scorpions, 876
Scrotum, 669
Scurvy, 540–541
Sea anemones, 867–869
Sea cucumbers, 882
Sea stars, 864t, 882, 883f, 884
Sea turtles, natal homing, 312–313
Sea urchins, 864t, 882
Seasonal affective disorder syndrome (SADS), 533
Seaweeds, 838, 841–842
Secondary growth of plants, 362
Secondary plant tissues, 362
Secondary root tissues, 377
Secondary sex characteristics
female, 675
male, 672–673
Secondary stem tissues, 372
Secondary succession, 952–953, 955
"Second messenger," 533, 535f
Secretin, 522t, 573
Secretory pathway (chain), 99–100, 101f
Sedimentary nutrient cycles, 947, 951–952
Seedless vascular plants, 845–847
Seedling development, 416–417
Seed plants, 847–853
Seeds, 411, 412f
development, 411
dispersal, 412, 415f, 416
dormancy, 416
germination, 438
protective properties, 833
viability, 416
Segmentation of body, 863, 875–876
Segmented worms, 864t, 875–876, 877f
Segregation, Mendel's law of, 235
Selective gene expression, 294, 295f
Self-recognition, 654
Semen, 670, 672
Semicircular canals, 507f, 508
Semiconservative replication, 275
Seminal vesicle, 672
Seminiferous tubule, 669, 671f
Semmelweis, Ignaz, 784–785
Senescence in plants, 432
Sense organs, 498–516
diversity of, 511–513
in fishes, 512
how they work, 500–515
Sense strand of DNA, 278
Sensory cortex, 502, 503f, 511
Sensory neurons, 470
Sensory receptors, 479, 500–502
types, 500, 501t
Sepals, 403
Sequencing:
of amino acids, 66–67, 84, 750
of DNA, 325–326
and evolution, 329, 750
and frameshift mutations, 286
legal aspects, 315
and turtle migration, 313
of nucleic acid, 325
Sere, 952
Serosa of digestive tract, 568
Serotonin, 475t
Severe combined immunodeficiency (SCID), 337t, 348–349
Sex:
age–sex structure, 988, 999, 1000f
determination of, 253–254, 255
Sex changes in fish, 254
Sex chromosomes, 253
abnormal number, 336–337
and gender determination, 255
and gene dosage, 256–257
Sex linkage, 254–255, 345–346
Sexual dimorphism, 727–728
Sexual imprinting, 1014
Sexually transmitted diseases, 794–797
Sexual orientation, 663
Sexual reproduction
advantages, 664–665
in flowering plants, 402–417
and meiosis, 214, 215f
strategies, 665–667
Sexual rituals, 668
Sexual selection, 727–728
Sharing food, 1024–1025
Sharks, 887
Shell, 545
Shellfish, 874
poisoning by, 818
Shivering, 614
Shoot system of plants, 361, 362f
Short-day plants, 400, 437
Shrew, 893
Shrimp, 876
Shrublands, 924, 925f
Sickle cell anemia, 81, 82f, 337t, 346f, 347
and allele frequency, 717
codon change in, 286
and malaria, 723f, 724, 812
and natural selection, 723–724
Sieve plates, 370
Sieve-tube members, 370
Sight, 502–506, 512
Sigmoid (S-shaped) growth curve, 993–994
Sign stimulus, 1009
Silent Spring, 715
Silicosis, 101
Silk, 881
Simple fruits, 412, 413f
Simple leaf, 377
Singer-Nicolson model of membrane, 89
Sinks, and plant nutrient transport, 394
Sinoatrial (SA) node, 598
Skates, 887
Skeletal muscle, 552–557
cells in, 553–555
evolution, 561
fiber types in, 186–187
metabolism in, 186–187
Skeletal systems, 457, 544–551
evolution of, 560–561
in human, 456f, 548–551
Skin, 542–544
color in humans, 240, 241f
sensory receptors in, 502
Skull (cranium), 482
Slime molds, 819
Slugs, 819, 874
Small intestine, 572–574
Smell, 509, 510
Smoking, 630–631, 704
Smooth endoplasmic reticulum (SER), 98
Smooth muscle, 558
Snails, 864t, 874, 875f
Snakes, 889
Social behavior, 1018–1022
Social bonds, 1021–1022
Social learning, 1013
Social parasitism, 974
Sodium:
in diet, 580, 580t
transport across cell membrane, 142–144
Sodium—potassium pump, 142, 144f
Soft palate, 569f
Soil, formation of, 953
Solar energy, 154, 156f
dangers of, 280–281
Solutions, molecular properties, 59–60
Solvent, 59
Somatic cells versus germ cells, 214
Somatic division of peripheral nervous system, 490
Somatic sensation, 500–502
Somatic sensory cortex, 502, 503f
Somatotropin, *see* Growth hormone
Sori, 847
Speciation, 731–734
Species:
attributes, 731–732, 830
competition, 960–961
diversity and biomes, 917
origin of, 731–734
in taxonomic hierarchy, 14
Specific epithet, 13
Sperm, 669, 689, 670
Spermatid, 669
Spermatocytes, 669
Spermatogenesis, 669–670, 671f
Spermatogonia, 669
Spermatozoa, 669
Spermicides, 679
Sphenophyta, 839t, 846–847
Spiders, 864t, 876, 881
territorial behavior of, 1006–1007
Spina bifida, 694
Spinal cord, 488–489
Spinal nerve roots, 488f
Spinal nerves, 489–490
Spindle apparatus, 204–205
Spindle fibers, 204–205
Spine, *see* Vertebral column
Spiracles, 633, 635f
Split brain, 504
Split genes, 302f, 303
Sponges, 864–867, 864t
Spongy bone, 546, 547f
Spongy parenchyma, 378
Spontaneous generation, 32, 33
Sporangiospores, 823
Sporangium, 823, 847, 848f
Spores of bacteria, 792
Spores of fungi, 822
Spores of plants, 214, 407, 833, 843, 845
Sporophyll, 846
Sporophyte, 833
Sporozoa, 815t
Squids, 864t, 874
Stabilizing selection, 724–725
Stamens, 403
Starch, 73
Starfish (sea star), 882, 883f, 884
Start (initiation) codon, 281, 284
Stem of plants, 372–373
Sterilization, sexual, 681
Steroids, 75, 77f
and athletes, 528
and gene activation, 304–305
receptors for, 534, 535f
"Sticky ends" in recombinant DNA, 322f, 323

Stigma of flower, 403
Stimulus:
perception of, 511
receptors for, 500
sign stimulus, 1009
Stomach, 570–572
secretory cells of, 571f, 572
Stomates (stomatal pores), 165f, 166, 368, 392–393, 833
Stop (nonsense) codons, 281, 285
Stratigraphy, law of, 746
Stream habitats, 914–915
Streptococcus, in pregnancy, 704
Stretch reflex, 479, 480f
Strobilus, 846
Stroke, 590
Stromatolites, 763
Structural genes, 299, 300f
Style, of flower, 403
Subatomic particles, 51
Substrate, 124f, 125
Substrate-level phosphorylation, 176f, 177–178, 185
Succession in ecosystems, 952–955
Succulents, 926
Sucrose, 70
Sugars, 70–74. *See also* Glucose
Sulfa drugs, 126
Sun, formation of, 762
Sunlight:
and climate, 905–906
dangers of, 280–281
energy from, 154, 156f
undersea, 908
Superoxide radical, 56
Suppressor T cells, 649
Surface area, and respiration, 622–623, 626
Surface area–volume ratio, 460–462
Surface tension, 60, 61f
Surfactant, 60
Surrogate motherhood, 679
Survivorship curve, 989
Suspensor, of plant embryo, 411
Sutures of skull, 550
Swallowing, 569
Sweating, 614, 927
"Swollen glands," 606
Symbiosis, 962–963
and origin of eukaryote cell, 110, 765
Symmetry of body, 861, 862f
Sympathetic nervous system, 490–491
Sympatric speciation, 732
Synapses, 474–478
nerve poisons and, 478
Synapsis, 216, 217f, 218
Synaptic cleft, 474
Synaptic knob, 468, 469f, 474f
Synaptic transmission (neurotransmission), 474–479
Synaptic vesicles, 474
Synaptonemal complex, 218, 219f
Synovial cavities, 551
Syphilis, 795–796
Systemic circulation, 594f, 595
Systemic lupus erythematosus, 657

T

Tail, in chordates, 884
Tapeworms, 869, 870f, 871, 872
Tap root system, 373, 374f
Tarsier, 894f
Taste, 509–510
Taung Child (hominid), 743, 752
Taxonomy, 11–15, 808–809
Tay-Sachs syndrome, 337t
T cell receptors, 647
T cells, 647, 648, 649–650
Tectonic plates, 767–769
Teeth, 570
Telophase, of mitosis, 203f, 206
Telophase I, of meiosis, 217f, 219
Telophase II, of meiosis, 217f, 221
Temperature
of body, 449, 450, 451f, 588, 614
and chemical reactions, 127
and plant growth, 434–435
Temperature conversion chart, Appendix A
Template in DNA replication, 275
Tendon, 552
Teratogens, 704
Territorial behavior, 1006–1007, 1016–1017
Test cross, 237
Testes (testicles), 255, 522t, 532, 669
Testosterone, 75, 77f, 522t, 532, 672–673
dangers of, 528
Testosterone receptor, 304
Testosterone-receptor protein, 305
Test population, in experiments, 37
"Test-tube babies," 679
Tetanus, 478
Tetrad (chromosomal), 216, 218f, 219
Thalamus, 481f, 486
Thalassemia, 286
Thallus, 835
Themes recurring in this book, 15
Theory, scientific, 35
versus truth, 40–41
Theory of tolerance, 938
Therapsids, 774
Thermodynamically unfavorable reactions, 122
Thermodynamics, laws of, 119–120
Thermophilic bacteria, 167, 796
Thermoreceptor, 500, 501t
Thermoregulation, 449, 450, 451f, 588, 614
Thiamine (vitamin B_1), 567, 579t
Thigmotropism, 439
Thoracic cavity, 626
Thornwood, 924
Threatened species, 995–996
Throat (pharynx), 569, 624, 636
Thrombin, 602
Thrombus, 603
Thylakoids, 104, 156
and electron transport system, 160
photosynthesis in, 158f, 159, 160–162
Thymine, 269, 270f
base pairing, 271f, 272
Thymine dimer, 280
Thymus gland, 646f, 649
Thyroid gland, 522t, 529–531
Thyroid hormone, 529, 531
Thyroiditis, 657
Thyroid-stimulating hormone (TSH), 522t, 526
Thyroxine (T-4), 522t, 529
Ticks, 876
Tide pools, 912, 913f
Timberline, 920
Tissue plasminogen activator (TPA), 131, 603
Tissues:
in animals, 451, 452
gas exchange in, 627–629
in plants, 362–365
Tocopherols (vitamin E), 579t
Tolerance range, 938
Tomato, genetically engineered, 318
Top carnivore, 942
Totipotency, 297
Touching, 1021
Toxic chemicals, 946, 992, 993
Toxicity, testing for, 37–39
Toxic shock syndrome, 681
TPA (issue plasminogen activator), 131, 603
Trace elements in diet, 580, 580t
Trachea, in arthropods, 633, 635f, 878
Trachea, in humans, 625
Tracheids, 370
Tracheoles, 633, 635f
Tracheophytes, 835, 839t, 843–853
Transcription, 277, 278, 279f, 286f
and control of gene expression, 304–306
primary transcript processing, 306–307
Transfer RNA (tRNA), 279–280, 282f, 283f
role in translation, 284–285, 286f
Transgenic animals, 318f, 319, 321
Translation of RNA, 277, 279–285, 286f
chain elongation, 285
chain termination, 285
initiation, 284–285
and regulation of gene expression, 304, 307–308
Transpiration, 386, 391–393
Transpiration pull, 391–392
Transplanted organs, 489, 650
Transport:
across plasma membrane, 136–144
of electrons, 160–161, 182–185
of oxygen, 600–601, 627
in plants, 386–396
Transposition of genes, 293
Transverse tubules of muscle fiber, 556
Trematodes, 869, 870f, 871
Trichinosis, 873
Triglycerides, 71t
Triiodothyronine (T-3), 522t, 529
Trilobites, 878f
Trimesters of pregnancy, 700
Trisomy 21 (Down syndrome), 213, 221, 336, 337t
tRNA, *see* Transfer RNA
Trophic levels, 942
Trophoblast, 700, 701f, 703
Trophozoites, 813–814
Tropical rain forests, 428–429, 917, 919, 920f
Tropical thornwood, 924
Tropic hormones, 525, 526f
Tropisms in plants, 438–439
Trp (tryptophan) operon, 300f, 301
True bacteria, 791–796
Truth, and scientific theory, 40–41
Trypsin, 573
Tuatara, 889
Tubal ligation, 681
Tube cell, 408, 409f
Tubeworms, 875–876
Tubular reabsorption, 610
Tubular secretion, 610
Tubules of flatworm, 612f
Tubules of kidney, 607–609
Tumor-infiltrating lymphocytes (TILs), 648
Tumor-specific antigens, 648
Tumor-suppressor genes, 260
Tundra, 920, 922
Tunicates, 864t, 884, 885f
Turbellaria, 869–871
Turgor pressure, 141, 142f
Turner syndrome, 337
Turtles, 889
Tympanic membrane (eardrum), 507f, 508

U

Ultimate carnivore, 942
Ultraviolet radiation, *see also* Sunlight
effect on DNA, 280–281
Umbilical cord, 700
Uncertainty and scientific theories, 40–41
Understory of forest, 917
Uniform distribution patterns, 987
Uniramiates, 878f
Unity of life, 18
Unsaturated (double) bonds, 75
Uracil, 278, 279f

Ureters, 607
Urethra, 607, 669, 673
Urinary bladder, 607
Urinary (excretory, osmoregulatory) systems, 455, 456f
Urine, 607
  formation of, 609–611
Urkaryote, 798
Uterine (menstrual) cycle, 675, 677
Uterus (womb), 673

V

Vaccines, 654–655. *See also* Immunization
  for AIDS, 657
  for contraception, 677
Vacuoles, 102
Vagina, 673
Vaginal sponge, 681
Variable, in scientific experiment, 32
Vascular cambium, 362
Vascular plants, 361, 839t, 843–853
  higher (seed plants), 839t, 847–853
  lower (seedless), 839t, 845–847
Vascular tissue system of plants, 365, 367f, 369–372
Vas deferens, 670
Vasectomy, 681
Vasopressin, *see* Antidiuretic hormone
Vector, in recombinant DNA, 321
Vegetables, 412
Vegetative (asexual) reproduction in flowering plants, 417–418
Veins, 593, 595
Venation of leaf, 378
Venereal diseases, 794–797
Ventricles:
  of brain, 481f, 482
  of heart, 593f, 594
Venules, 593
Vernalization, 435
Vernix caseosa, 703
Vertebral column (backbone), 886
  origin of, 694
Vertebrates, 863, 864t, 884, 886–894
Vesicles, 97
Vessel members of xylem, 370
Vestibular apparatus, 507f, 508
Vestigial structures, 706, 749–750
Villi, chorionic, 700
Villi, intestinal, 574, 575f
Viroid, 803
Viruses, 787, 800–804
  diseases from, 794, 796–797, 803, 804
  evolutionary significance, 803–804
  gene transfer by, 804
  how they work, 801–802
  life cycle, 802f
  properties, 800–801
  as renegade genes, 803
  structure, 801
Vision, 502–506
  in fishes, 512
Visual cortex, 506
Vitamins, 567, 579–580
  as coenzymes, 127
  list of, 579t
Vocal cords, 624–625
Vulva, 673

W

Water:
  absorption in plants, 386–387, 389f
  biogeologic cycle, 947–948
  osmosis and, 140–141
  properties, 58f, 59–61
    photosynthesis and, 155, 160
"Water breathers," 632–633
Water molds, 822
Watson, James, 269–272
Waxes, 71t, 75
Weeds, 940–941
  control of, 941
  germination, 438
Whisk ferns, 845
White blood cells (leukocytes), 601–602
White matter:
  of brain, 481f, 482
  of spinal cord, 488f, 489
Wild type alleles, 241
Winds, prevailing, 906–907
Womb (uterus), 673
Wood, 370
Woody plants, 362
Worms:
  flat, 864t, 869–871, 872
  round, 864t, 871–873
  segmented, 864t, 875–876, 877f

X

Xanthophyta, 817
X chromosome, 253
  and gene dosage, 256–257
  X-linked characteristics, 254
  X-linked disorders, 345–346
Xeroderma pigmentosum, 280, 337t
X-ray crystallography, 67
Xylem, 369, 370
  water and mineral transport, 386–393

Y

Y chromosome, 253
  Y-linked characteristics, 255
Yeasts, 820
  in fermentation, 179
  infection from, 826
  reproduction in, 822
Yolk, 688
Yolk sac, 703

Z

Zidovudine (AZT), 657
Zooplankton, 813, 908
Zygomycetes, 822–823, 822t
Zygospore, 823
Zygote, 214, 215f, 690